CW01065268

Optical
Semiconductor
Devices

Optical Semiconductor Devices

MITSUO FUKUDA
NTT Opto-electronics Laboratories
Kanagawa, Japan

A WILEY INTERSCIENCE PUBLICATION
JOHN WILEY & SONS, INC.
NEW YORK/CHICHESTER/WEINHEIM/BRISBANE/SINGAPORE/TORONTO

Copyright © 1999 by John Wiley & Sons, Inc. All rights reserved.

Published simultaneously in Canada.

Library of Congress Cataloging-in-Publication Data:

Fukuda, Mitsuo.
    Optical semiconductor devices / by Mitsuo Fukuda.
       p.   cm.—(Wiley series in microwave and optical engineering)
    Includes index.
    ISBN 0-471-14959-4 (cloth : alk. paper)
    1. Optoelectronic devices. 2. Semiconductors. I. Title.
  II. Series.
  TK8320.F85 1998
  621.381′045—dc21

                                   98-16423

Printed in the United States of America

10 9 8 7 6 5 4 3 2 1

to my family

Contents

Preface

Optical semiconductor devices are widely used, in fields ranging from optical fiber communication systems to consumer electronics, and have become indispensable devices in the equipment and systems making up the infrastructure of our society. Most optical semiconductor devices are optoelectronic *pn*-junction devices, such as laser diodes, light-emitting diodes (LEDs), and photodiodes.

Many kinds of optoelectronic *pn*-junction devices have been developed over the past thirty years, and the main interest in this field has shifted from device physics and operation principles to device applications. Engineers and scientists working in this area require a wide range of knowledge related to optoelectronic semiconductor devices, and this book provides an introduction to optoelectronic *pn*-junction devices from the viewpoints of semiconductor material properties, operation principles, fabrication, packaging, reliability, and equipment/systems applications. It is written as a primer on these devices and the descriptions and explanations are therefore simplified to make them easier to understand. Waveguide theory, for example, is not treated; only calculated results are described.

This book describes all aspects of optical semiconductor devices and is divided into five parts—basics of semiconductor optoelectronic devices, operational characteristics, fabrication, reliability, and applications.

The first part, Chapter 1, describes the basics of optical semiconductor devices, especially *pn*-junction devices. Electrical and optical properties and processes in semiconductors, *pn*-junction properties, and the properties of double-heterostructure and quantum well structures are explained.

The second part—consisting of Chapters 2 through 5—discusses LEDs, laser diodes, photodiodes, and modulators used under reverse bias from a practical point of view. The basic characteristics of these devices, based on actual device data, are examined in detail.

The third part, Chapter 6, focuses on fabrication processes, such as crystal growth, device processing, and packaging. Some epitaxial methods of crystal growth are discussed, and electrodes and dielectric films are examined. Some typical examples are used in discussing important problems in the packaging of devices.

The fourth part, Chapter 7, treats typical reliability problems related to devices and packagings. First the basics of reliability are discussed followed by a discussion of device failure mechanisms due to the degradation of crystals, electrodes and surfaces. This is followed by an examination of packaging problems.

The final part, Chapter 8, describes some applications of optical semiconductor devices. Analog and digital communication systems, local area networks, and data links are explained as typical fiber-optic applications. Optical disk systems, laser printers, holographic scanners, and so forth, are described as some popular examples of beam optics applications.

The symbols used throughout the book are consistent except for a few cases where every attempt has been made to identify them with each chapter. Following common worldwide usage the MKS and cgs unit systems are used.

This book is based on the author's twenty years of experience at NTT Electrical Communications Laboratories and NTT Opto-electronics Laboratories. Many people guided and assisted me in my research. I have benefited significantly from discussions with my colleagues: Dr. K. Takahei, Dr. H. Nagai, and Prof. H. Asahi for semiconductor crystals and defects; Prof. S. Adachi for material properties; Prof. H. Kawaguchi, Prof. K. Wakita, Dr. Y. Yoshikuni, and Dr. Y. Itaya for the operation principles of devices; Drs. G. Iwane, Y. Nakano, H. Sudo, H. Oohashi, H. Mawatari, and M. Okayasu for reliability aspects of devices and device processing; and Drs. T. Sugie, K. Hagimoto, and H. Toba for optical fiber communication systems. I also wish to express my appreciation to Drs. S. Yamakoshi and O. Ueda of Fujitsu Laboratories; Drs. I. Mito and M. Kitamura of NEC; Dr. P. Devolder of CNET, France Telecom; Dr. F. Magistrali of Pirelli; and Dr. P. Whitney of Lasertron, for their valuable discussions.

I am further indebted to Dr. T. Ikegami, NTT-AT; Dr. J. Yoshida, NTT Opto-electronics Laboratories; Prof. Y. Suematsu, President of Kochi University of Technology; Prof. K. Iga, Tokyo Institute of Technology; Prof. H. Mase, Ibaraki University; Dr. P. W. Shumate, Bell Communications Research; Prof. F. Fantini, Padova University, Italy; and Dr. C. Harder, IBM Zurich (currently Uniphase); for their valuable advice and encouragement in my research life.

M. FUKUDA

Oiso, Syonan
Winter 1998

Optical Semiconductor Devices

CHAPTER ONE

Basics of Optoelectronic
pn-Junction Devices

Most semiconductor optoelectronic devices are *pn*-junction diodes, and their performance depends on the properties of the *pn*-junction and of the semiconductor material. The semiconductor material properties important for optoelectronic devices are related to optical processes, electronic processes, and the interaction of both kinds of processes. The injection process is important in devices in which the *pn*-junction is under forward electrical bias, such as laser diodes and light-emitting diodes (LEDs), whereas the electrical field generation is important in devices in which the *pn*-junction is under reverse electrical bias, such as photodiodes and modulators (or optical switches). This chapter describes the semiconductor materials and the *pn*-junction in a way that will lead to a better understanding of the principles of operation of laser diodes, LEDs, photodiodes, and optical modulators.

1.1 BASICS OF SEMICONDUCTORS FOR OPTOELECTRONIC *pn*-JUNCTION DEVICES

Various kinds of semiconductor materials consisting of single-crystal, polycrystalline, or amorphous substances have been used in optoelectronic devices. This section describes the material properties of the single-crystal semiconductor, because most optoelectronic devices, except for a few devices such as solar cells made of amorphous Si, are made using single-crystal semiconductors. A single-crystal material here is one with a completely ordered arrangement of atoms, whereas the atoms in a polycrystalline material are only partly ordered. The arrangement of atoms in an amorphous material shows no periodicity. The lattice constant of a semiconductor crystal influences its band structure and determines the electrical and optical processes in the crystal. The band-gap energy determines the wavelength range over which the device operates (light absorption, emission, and modulation). Recently developed optoelectronic devices are usually made using III-V compound semiconductors. Elementary semiconductors (Si and Ge) are used for photodiodes, and II-VI compound semiconductors have been investigated with the intention of using them in laser diodes and LEDs emitting visible light.

1.1.1   Bonding and Band Structure in Semiconductors

A single-crystal semiconductor is characterized by complete ordering of atomic arrangement. Elementary semiconductors such as Si and Ge crystallize in the diamond structure, and most III-V compound semiconductors crystallize in the zincblende structure. In II-VI compound semiconductors, the wurtzite structure often appears in addition to the zincblende structure. The atomic configuration and bonding arrangement for a covalently bonded elementary semiconductor are shown in Fig. 1.1. Each atom is bonded with two electrons. This type of bonding is called covalent bonding. The covalent bonding arises from the property that atoms tend to form closed outer subshells and it is characterized by the sharing of electrons between neighboring atoms. Solitary atoms of Si and Ge have four valence electrons in the outer (valence) subshell because they belong to IV-group element in the periodic table and four electrons are deficient to

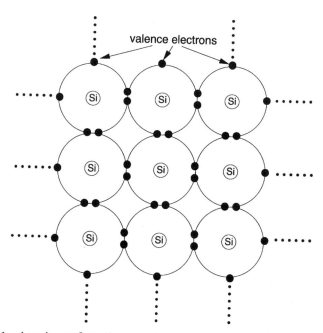

FIGURE 1.1 Atomic configuration and bonding arrangement in covalently bonded solid, Si.

completely fill it. Each covalently bonded lattice atom completes its outer sub-shell by sharing four of the electrons of its neighbors. Covalent bonding is found in nearly all semiconductors and results from the quantum mechanical interaction between the shared electrons. The bonding electrons are distributed uniformly among the lattice atoms of elementary semiconductors because the crystal consists of only one kind of atom. The bonding situation is similar in the zincblende compound semiconductors, such as GaAs (where Ga and As are lattice atoms) and InP (where In and P are lattice atoms). The bonding situation in the compound semiconductor, however, is different because the compound is composed of atoms from different groups in the periodic table. In a III-V compound semiconductor such as GaAs, the Ga lattice atoms bond with the As lattice atoms. The Ga atom is a group-III element and has three valence electrons in its outer subshell, whereas the As atom is a group-V element and has five valence electrons in the outer subshell. Although the bonding between Ga and As lattice atoms is created by two bonding electrons in a manner similar to that in the elementary semiconductors, Si and Ge, the bonding electrons are not shared equally but are shifted toward the As atoms. In addition to the covalent bond, this nonuniform distribution of the bonding electrons introduces an ionic character to the bond, which is chemical bonding by electrostatic force between positive and negative atoms. Although the bonding in III-V compound semiconductors is predominantly covalent and partly ionic, the bonding in II-VI compound semiconductors is mainly ionic and partly covalent because the difference between the numbers of valence electrons is larger. The covalent and the ionic bonds in compound semiconductors have been discussed in detail by Philips.[1] Schematic diagrams of diamond, zincblende, and wurtzite for primitive unit cell are shown in Fig. 1.2. A certain arrangement of atoms is repeated in crystal, and its minimum unit is called a primitive unit cell (or unit lattice). The primitive unit cell is defined by lattice constants a, b, and c and by the angles between them (α, β, and γ). In Si and Ge and most III-V compound semiconductors which have diamond and zincblende structures, $a = b = c$ and $\alpha = \beta = \gamma = 90°$. In the wurtzite structure on the other hand, $a = b \neq c$ and $\alpha = \beta = 90°$ and $\gamma = 120°$. The zincblende structure indicated in Fig. 1.2(b), which is the structure of most material used in optoelectronic devices, is generally represented with a family of planes by using ( ). The six faces of the primitive unit cell are the six equivalent (100) planes on which active layers of LED, laser diodes, and photodiodes are usually grown. The diagonal plane including both Ga and As atoms is one of the four (110) planes. These (110) planes are the natural cleavage planes commonly used to form the facets of the optical cavity in Fabry–Perot type laser diodes. The triangular cut through As atoms indicates one of the eight (111) planes. Slip dislocation along this (111) plane occurs under high mechanical stress and causes device degradation. The orientation in crystal is generally represented by using $\langle$ $\rangle$. The $\langle 100 \rangle$ means the direction perpendicular to the (100) planes.

The bonding structure in a semiconductor is characterized by covalent bonding and ionic bonding and is related to the band structure. Figure 1.3 shows an

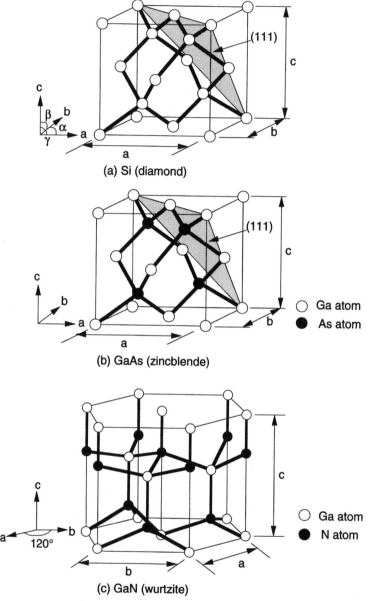

(a) Si (diamond)

(b) GaAs (zincblende)

○ Ga atom
● As atom

(c) GaN (wurtzite)

○ Ga atom
● N atom

FIGURE 1.2   Schematic diagrams of Si (diamond structure), GaAs (zincblende structure), and GaN (wurtzite structure).

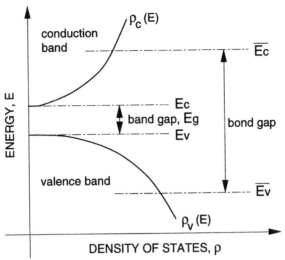

FIGURE 1.3 Allowed bands (conduction and valence bands) and forbidden band (band gap) in semiconductor crystals.

outline of the bond-gap and band-gap diagram. The notations E_c and E_v mark the edges of the conduction and valence bands, and $\overline{E_c}$ and $\overline{E_v}$ indicate the mean value of the conduction and the valence band. This energy band structure results from the periodic arrangement of atoms in a semiconductor crystal. In solitary atoms, the energy levels of electrons are discrete. When solitary atoms with discrete electron energy levels crystallize and are arranged as shown in Fig. 1.1, the valence electrons as well as conduction electrons interact with atoms (atomic nucleus). This interaction slightly shifts the discrete energy levels of the electrons, and many energy levels having very slightly different energies are created in the crystal. The number of the energy levels created corresponds to the number of atoms in the crystal. Pauli's exclusion principle allows two electrons having opposite spins to occupy the same energy level, and the total number of possible states for electrons is twice of the number of energy levels. Finally the allowed and forbidden energy bands are formed as shown in Fig. 1.3. The band-gap corresponds to a forbidden energy band for electrons and its energy E_g is given by

$$E_g = E_c - E_v, \tag{1.1}$$

where E_c and E_v are also called the conduction band minimum and valence band maximum. The horizontal axis indicates the number of states which can be occupied by electrons. This number of states is called the density of states. The band structure depends on the material and bonding structure described above, and thus also on the orientation (planes) in crystal (in atomic lattice) because the arrangement of atoms varies in different crystal orientations (see

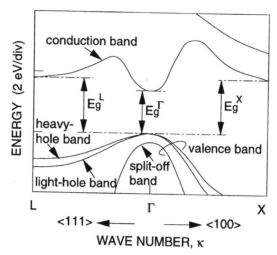

FIGURE 1.4 Band structure of zincblende crystal (a case of direct band-gap).

Fig. 1.2). The energy level (momentum) allowed to an electron depends on the orientation. In general, the momentum of an electron in a crystal is expressed by the crystal momentum **p** defined by

$$\mathbf{p} = (h/2\pi)\mathbf{k}, \tag{1.2}$$

where h is the Planck's constant; 4.136×10^{-5} eV sec, and **k** is the wavenumber vector. The wavenumber is given by $1/\lambda$ (λ: wavelength).

The crystal momentum depending on the orientation results in complicated band structure. Figure 1.4 shows common band structure of zincblende crystal along the $\langle 100 \rangle$ and $\langle 111 \rangle$ directions in **k**-space.[2] The allowed electron energy depends on the wavenumber (crystal momentum). In the energy-band, electrons are located near the minimum of the conduction band, whereas holes are located near the maximum of the valence band. Semiconductors can be roughly divided into two types: direct-gap and indirect-gap. Schematic diagrams of the simplified band structure for both types of semiconductors are shown in Fig. 1.5. In direct-gap semiconductors, the crystal momentum at the minimum of the conduction band coincides with that at the maximum of the valence band. In indirect-gap semiconductors the momentum of the minimum of the conduction band never coincides with that of the maximum of the valence band. Lattice constants, types of energy gaps, and the energy gap for Si, Ge, and the main binary compound semiconductors used in optoelectronic devices are summarized in Table 1.1. There is a strong correlation between lattice constant and energy gap in these materials: the energy gap of a semiconductor having a small lattice constant tends to be large.[1] In addition, the band-gap energy decreases with increasing temperature because thermal expansion increases the lattice constant.

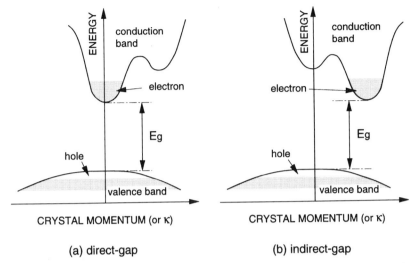

(a) direct-gap (b) indirect-gap

FIGURE 1.5 Simplified band structures of (a) direct-gap and (b) indirect-gap semiconductors.

If we mix two kinds of compound semiconductors, we can create a new compound semiconductor with properties intermediate between those of the original two. The physical parameters of the new compound semiconductor vary linearly in proportion to the alloy composition (mole fraction). This is called Vegard's law.[3] The mixed semiconductors most often used in optoelectronic devices are ternary compound semiconductors, such as AlGaAs and InGaAs, and quaternary compound semiconductors, such as InGaAsP and AlInGaP.[4,5] The band-gap energy of $Al_xGa_{1-x}As$ is shown in Fig. 1.6 as a function of AlAs mole fraction, x, where the mole fraction is set so that the sum of all III elements or of all V elements is equal to 1. The semiconductor materials, GaAs and AlAs, are the constituent binary compound semiconductors for AlGaAs. The direct (Γ valley) and indirect (X and L valleys) conduction band minima vary as composition varies between GaAs and AlAs. At around the composition of $x = 0.45$, $Al_xGa_{1-x}As$, the direct-gap changes to the indirect gap in corresponding to the band structures of both GaAs and AlAs. The correlation between band-gap energy and lattice constant is shown in Fig. 1.7 for binary, ternary, and quaternary compound semiconductors.[4,5] The vertical lines of AlGaAs and InGaAsP are very important for crystal growth with lattice matching to GaAs and InP.

1.1.2 Electrical Properties

1.1.2.1 Intrinsic and Extrinsic Semiconductors   The semiconductor crystals described in the previous section were assumed to be of ideal composition and not to contain impurities or lattice defects (or lattice imperfections). Pure semi-

TABLE 1.1 Lattice Constant, Type of Energy Gap, and Energy Gap for Si, Ge, and the Main Binary Compound Semiconductors Used in Optoelectronic Devices at Temperatures Near Room Temperature

Material	Lattice Constant (Å)	Crystal Structure	Type of Energy Gap	Energy Gap (eV)
		Elementary[a]		
Si	5.43	diamond	indirect	1.12 (X)
Ge	5.64	diamond	indirect	0.67 (L)
		III-V compound[a,b,c]		
AlP	5.45	zincblende	indirect	2.45 (X)
AlAs	5.66	zincblende	(indirect)	2.16 (X)
GaN	$a = 3.19, c = 5.19$	wurzite	direct	3.39
GaP	5.45	zincblende	indirect	2.26 (X)
GaAs	5.65	zincblende	direct	1.42
InN	$a = 3.54, c = 5.71$	wurzite	direct	2.4
InP	5.87	zincblende	direct	1.35
InAs	6.06	zincblende	direct	0.36
		II-VI compound[a,d,e]		
CdTe	6.48	zincblende	direct	1.44
HgS: β	5.84	zincblende	direct	2.5
HgSe	6.08	zincblende	direct	2.5
ZnS: α	$a = 3.82, c = 6.26$	wurzite	direct	3.8
ZnS: β	5.41	zincblende	direct	3.54
ZnSe	5.65	zincblende	direct	2.67

[a]S. M. Sze, *Physics of Semiconductor Devices*, John Wiley & Sons, New York, 1969.
[b]M. Neuberger, ed., *Handbook of Electronic Materials*, vol. 2, *III-V Semiconductor Compounds*, Plenum, 1973.
[c]R. K. Willardson and A. C. Beer, eds., *Semiconductors and Semimetals*, Academic Press, New York, vols. 1 and 2, 1966, vol. 4, 1968, and vol. 10, 1975.
[d]M. Aven and J. S. Prener, eds., *Physics and Chemistry of II-VI Compounds*, North-Holland, Dordrecht, 1967.
[e]K. Zanio, Cadmium Telluride, in *Semiconductors and Semimetals*, vol. 13, Academic Press, New York, 1978.

conductors are called intrinsic semiconductors and their electronic properties are not influenced by impurities. When an intrinsic semiconductor is excited by illumination, heating, or electrical biasing, the valence electrons are excited and away from their atoms are shown in Fig. 1.8(a). These free electrons carry negative electrical charge, so the creation of a free electron also results in a hole with a positive charge. Free electrons and holes move in opposite directions under an electrical field and can contribute to electrical conduction (two-carrier transport). These mobile electrons and holes are called carriers. In an intrinsic semiconductor the number of the free electrons is equal to the number

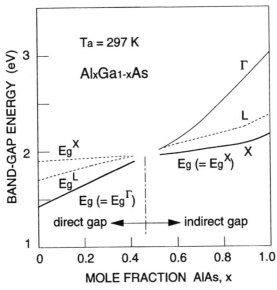

FIGURE 1.6 Compositional dependence of the direct energy gap, Γ, and the indirect gaps, X and L for $Al_xGa_{1-x}As$. From H. C. Casey, Jr. and M. B. Panish.[5] ©1978 Academic Press. Reprinted by permission of the author and Academic Press.

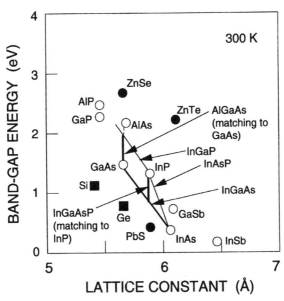

FIGURE 1.7 Energy band-gap versus lattice constant for compound semiconductors commonly used in optoelectronic devices. Values for Si and Ge are also for reference.

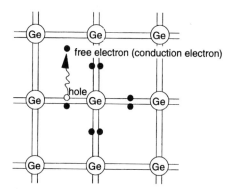

(a) intrinsic semiconductor

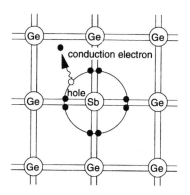

(b) n-type semiconductor

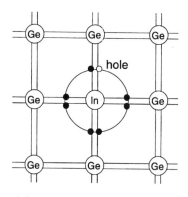

(c) p-type semiconductor

FIGURE 1.8 Atomic configuration and bonding arrangement of Ge: (a) intrinsic, (b) *n*-type, and (c) *p*-type.

of holes because excitation always results in the creation of an electron-hole pair. Therefore, at thermal equilibrium,

$$n_0 = p_0 = n_i, \qquad (1.3)$$

where n_0 and p_0 are electron density and hole density and n_i is called the intrinsic carrier concentration.

The electrical resistivity of intrinsic semiconductors for optoelectronic device use is very high at temperatures near room temperature because the carrier density (that is, n_i) is quite low. The electrical properties of the semiconductors can be changed, however, by adding impurities. Semiconductors whose electrical properties are determined by impurities are called extrinsic semiconductors, and these are the semiconductors usually used in optoelectronic devices. In semiconductor engineering, impurities generally means atoms other than those making up the host semiconductor, although defects, such as lattice imperfection, are also called impurities. Adding impurities to semiconductors is called doping and is usually done during crystal growth. Impurities can also be added by diffusion from the surface. Doping is necessary for creating carriers, electrons, and holes, and one type of impurity consists of atoms that have an additional electron after bonding, as shown in Fig. 1.8(b). In elements in column IV of the periodic table, such as Ge, four valence electrons form covalent bonds. When elements in column V, such as Sb with five valence electrons, are doped into Ge, four electrons form covalent bonds with neighboring Ge atoms, and one electron is weakly attracted to the nucleus of the Sb atom. This electron easily becomes a free electron at low temperatures, and then the electrical resistance of the semiconductor decreases tremendously. An electron-donating impurity is called the donor, and semiconductors with donors are *n*-type semiconductors. Donors are positively charged in semiconductor (ionization) after releasing electrons and create donor levels in the band-gap. The other type of impurity consists of the atoms that have a hole after bonding, as shown in Fig. 1.8(c). If an element in column III with three valence electrons, such as In, is doped into Ge, only three electrons form covalent bonds, and a hole is introduced. This type of impurity is called the acceptor, and semiconductors with acceptors are *p*-type semiconductors. Acceptors are negatively charged after capturing electrons and create acceptor levels in the band-gap. When both donors and acceptors exist in a semiconductor at the same time, the type of the semiconductor is determined by the type of dominant carrier. If both donors and acceptors are present in equal concentration, the characteristics of extrinsic semiconductors disappear and the semiconductor is similar to an intrinsic semiconductor. This is called the compensated intrinsic. For elements in column IV, such as Ge and Si, elements in column V (such as P, As, and Sb) are often used as donors and elements in column III (such as B, Ga, and In) are the typical acceptors. These impurities are of course selected in corresponding to the semiconductor material to be doped. In III-V compound semiconductors, elements in column VI (such as S, Se, and Te) act as donors and are substituted for lattice atoms in group-V. Elements in column II (such as Mg, Zn, and Cd) are acceptors and are substituted for lattice atoms in group-III.

Elements in column IV (such as Si and Ge) are amphoteric impurities (dopants) for III-V compound semiconductors. If they are substituted for the group-III lattice atoms, they act as donors. They are, however, acceptors when they are substituted for the group-V lattice atoms. The position at which they are incorporated depends on the crystal growth conditions: under relatively high temperatures, the elements in column IV tend to be substituted for group-III lattice atoms whereas under low temperatures they tend to be substituted for the group-V lattice atoms.

1.1.2.2   Carrier Effective Mass   Electrons in semiconductor crystals interact with forces originating from the periodic potential of lattice atoms. This interaction results in their characteristics differing from those of electrons in a vacuum. The kinetic energy of an electron in a crystal is given by

$$E - U_0 = (h/\pi)^2/2m^*k^2 (= P^2/2m^*), \qquad (1.4)$$

where E is the electron energy, U_0 is the potential energy, and P is the momentum. Differentiating Eq. (1.4) with respect to wavenumber k twice gives

$$d^2E/dk^2 = (h/\pi)^2/m^*, \qquad (1.5)$$

and the electron mass in a crystal, m^*, is given by

$$m^* = (h/\pi)^2/(d^2E/dk^2). \qquad (1.6)$$

The electron mass, m^*, in Eq. (1.6) is usually different from the rest mass and is called the effective mass. The velocity of the electron depends on the effective mass in a manner similar to the way it does in a vacuum. The electron velocity (strictly, the group velocity of the wavepacket for electron motion) is also given by the first derivative of Eq. (1.4) with respect to k:

$$v = [1/(h/\pi)](dE/dk)(= P/m^*). \qquad (1.7)$$

These relations are summarized in Fig. 1.9. The velocity is 0 at the band edge and the maximum near the center part of the energy band. The effective mass is positive on the lower-energy side of the energy band and negative on the higher-energy side. Quantum-mechanically, a particle having negative mass and negative charge is equivalently interpreted as a particle with positive mass and positive charge under an electric or magnetic field. This particle is the hole discussed in Section 1.1.2.1. In actual crystals, the correlation between the elec-

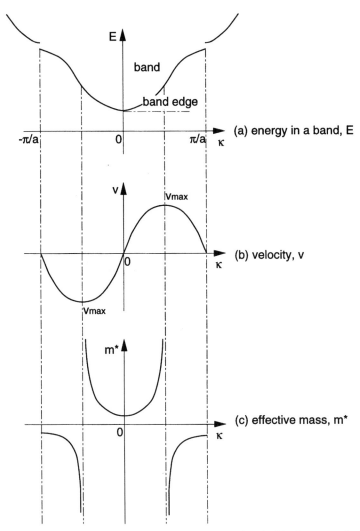

FIGURE 1.9 Schematic illustration of (a) energy in a band, (b) electron velocity, and (c) effective mass in crystals as a function of wavenumber. Here, $\kappa = \pi/a$ denotes the boundary of the first Brillouin zone.

tron energy, E, and the wavenumber, $\mathbf{k}$, depends on lattice orientations. Thus the effective mass also changes with the direction and is given by the mean value,

$$m_\mathrm{d} = (m_x^* m_y^* m_z^*)^{1/3}, \tag{1.8}$$

for the density of state effective mass at Γ in $\mathbf{k}$-space, and

TABLE 1.2   Effective Mass and Density of States at 300 K (m_0: Electron Mass in Vacuum)

Material	Electron Effective Mass	Hole Effective Mass	Effective Density of States in Conduction Band, N_c (cm^{-3})	Effective Density of States in Valence Band, N_v (cm^{-3})	Intrinsic Carrier Concentration, n_i (cm^{-3})
Si[a]	$0.33\,m_0$	$0.56\,m_0$	2.8×10^{19}	1.0×10^{19}	1.0×10^{10}
Ge[a]	$0.22\,m_0$	$0.31\,m_0$	1.0×10^{19}	6.1×10^{18}	1.5×10^{12}
GaAs[a,b]	$0.068\,m_0$	$0.5\,m_0$	4.7×10^{17}	7.0×10^{18}	2.1×10^{6}
InP[a,b]	$0.08\,m_0$	$0.4\,m_0$	5.7×10^{17}	6.3×10^{18}	8.7×10^{6}
Al$_{0.03}$Ga$_{0.97}$As[b]	$0.07\,m_0$	$0.5\,m_0$	4.6×10^{17}	9.1×10^{18}	1.1×10^{6}
In$_{0.76}$Ga$_{0.24}$As$_{0.55}$P$_{0.45}$[c]	$0.06\,m_0$	$0.43\,m_0$	3.6×10^{17}	7.1×10^{18}	1.7×10^{10}
In$_{0.47}$Ga$_{0.53}$As[c]	$0.04\,m_0$	$0.45\,m_0$	2.1×10^{17}	7.6×10^{18}	6.3×10^{11}

[a]S. M. Sze, *Physics and Technology of Semiconductor Devices*, John Wiley & Sons, New York, 1969.
[b]H. C. Casey Jr. and M. Panish, *Heterostructure Lasers*, Academic Press, New York, 1978.
[c]R. J. Nicholas et al., *Appl. Phys. Lett.*, **37**, 178, 1980.

$$m_c = (1/3)(1/m_x^* + 1/m_y^* + 1/m_z^*), \qquad (1.9)$$

for the conductivity or mobility effective mass, where m_x^*, m_y^*, and m_z^* are the effective masses along the principle axes. Hole effective mass is given by a combination of the light- and heavy-hole effective masses, m_{lh} and m_{hh}:

$$m_h = (m_{lh}^{3/2} + m_{hh}^{3/2})^{2/3}. \qquad (1.10)$$

The effective masses for some important semiconductors are summarized in Table 1.2.

1.2.3 Density of States   As described in Section 1.1.1, an electron is allowed to exist in a conduction band or a valence band. The density of electrons or holes per unit energy, however, is not infinite. Near the conduction and valence band edge, the number of states that can be occupied by an electron having an energy between E and $E + dE$ is given by the parabolic distribution as follows (see Fig. 1.10),[6]

$$\rho_c(E) = (4\pi/h^3)(2m_{de})^{3/2}(E - E_c)^{1/2} \qquad (1.11)$$

$$\rho_v(E) = (4\pi/h^3)(2m_{dh})^{3/2}(E_v - E)^{1/2} \qquad (1.12)$$

where m_{de} and m_{dh} are the density of state effective mass of electrons at the conduction band and of holes at the valence band, $\rho_c(E)$ is the density of states at energy E in the conduction band, and $\rho_v(E)$ is the density of states at energy E

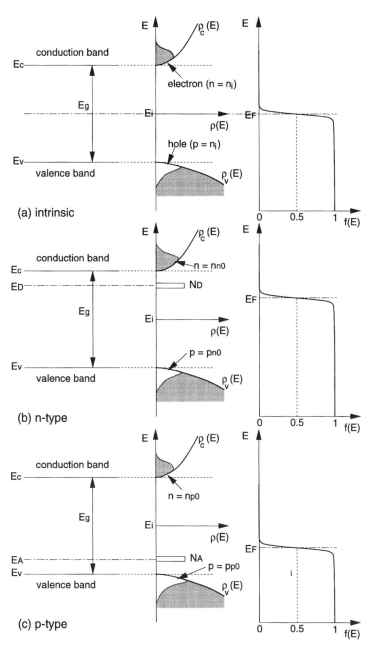

FIGURE 1.10   Energy band diagram, density of states, carrier concentration, and Fermi–Dirac distribution for intrinsic, *n*-type, and *p*-type semiconductors under thermal equilibrium. E_D and E_A indicate donor and acceptor level, and N_D and N_A indicate donor and acceptor concentration.

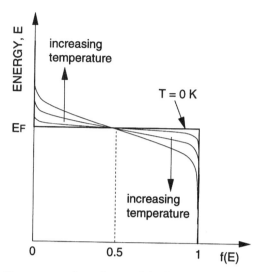

FIGURE 1.11 Temperature dependence of the Fermi–Dirac distribution function.

in the valence band. The effective mass influences the density of states with $\frac{3}{2}$ power as indicated in Eqs. (1.11) and (1.12). Therefore, the density of states for electrons in the conduction band, $\rho_c(E)$, is always smaller than that of electrons (holes) in the valence band, $\rho_v(E)$, because of the difference in the effective mass between electrons and holes. The density of states indicates the number of states that can be occupied by electrons (or holes). In the energy band, the occupation probability of an electron at energy E is given by the Fermi–Dirac distribution function, $f(E)$:

$$f(E) = 1/\{1 + \exp[(E - E_F)/k_B T]\}, \tag{1.13}$$

where E_F is the Fermi level—which, as indicated in Fig. 1.10, is the energy level at $f(E) = \frac{1}{2}$, k_B is Boltzmann's constant (8.617×10^{-5} eV K^{-1}), and T is absolute temperature. At $T = 0$K, $f(E)$ is equal to 1 in the energy range $E < E_F$ and is equal to 0 in the range $E > E_F$. This means that at 0 K all energy levels below the Fermi level are occupied by electrons and those above the Fermi level are perfectly empty. As temperature increases, the boundary gradually loses sharpness (see Fig. 1.11). The occupation probability of an electron at an energy higher than the Fermi level and of a hole at an energy lower than the Fermi level increases because the number of electrons having an energy higher than the Fermi level increases because of the thermal energy. If $(E - E_F) \gg k_B T$, Eq. (1.13) can be rewritten as

$$f(E) \approx \exp[-(E - E_F)/k_B T]. \tag{1.14}$$

Then

$$f(E) = A \exp\left(-E/k_B T\right), \tag{1.15}$$

where $A = \exp\left(E_F/k_B T\right)$. This equation is in the form of the Maxwell–Boltzmann (or simply Boltzmann) distribution function which is applied to gases. It indicates that Boltzmann statistics can be also applied to carriers having large energy and that the restrictions due to Pauli's exclusion principle can be relaxed. This is due to the difference between the numbers of available states and carriers. In higher-energy regions of the band, the number of available states may be much larger than that of carriers. The assumption, $(E - E_F) \gg k_B T$, is satisfied if $(E - E_F) > 3k_B T$, where $k_B T/q$ is nearly equal to 0.026 eV at 300 K. The condition expressed by Eq. (1.15) is called the Boltzmann approximation and is used to determine carrier statistics in nondegenerate semiconductors. The occupation probability of holes is correspondingly expressed by

$$1 - f(E) = 1/\{1 + \exp[(E_F - E)/k_B T]\}, \tag{1.16}$$

and if $(E_F - E) \gg k_B T$,

$$1 - f(E) \approx \exp[-(E_F - E)/k_B T]\} \tag{1.17}$$

holds.

1.1.2.4   *Carrier Concentration and the Fermi Level*   Carrier concentration generally indicates the density of the electrons in the conduction band of an *n*-type semiconductor and the density of the holes in the valence band of a *p*-type semiconductor. In extrinsic semiconductors, the concentration is controlled by the level of doping.

In intrinsic semiconductors the carrier density at energies between E and $E + dE$, dn, is given by the product of the density of states and occupation probability. Thus, the electron density at energies between E and $E + dE$ is expressed by

$$dn = f(E)\rho(E)dE. \tag{1.18}$$

The relation between energy and carrier density for intrinsic semiconductors is shown schematically in Fig. 1.12. The density of electrons and holes in the conduction and valence bands can be calculated by integrating Eq. (1.18) over the band. The occupation probability decreases rapidly as the energy increases. Consequently, the integration can be extended to the range from E_c to ∞ and the electron density in the conduction band, n, and the hole density in the valence band, p, are given by the following equations:

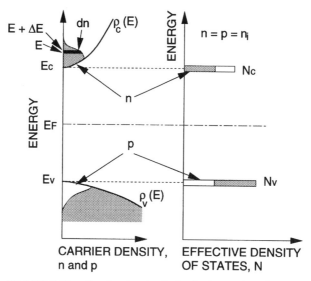

FIGURE 1.12   Carrier density in intrinsic semiconductors.

$$n = \int_{E_c}^{\infty} \rho_c(E)f(E)dE \qquad (1.19)$$

$$p = \int_{-\infty}^{E_v} \rho_v(E)[1 - f(E)]dE. \qquad (1.20)$$

Using Eqs. (1.11), (1.12), (1.15), and (1.17) and a formula for the Γ function, $\Gamma = \int_0^{\infty} x^{1/2} \exp(-x)dx = \pi^{1/2}/2$, we can rewrite Eqs. (1.19) and (1.20) as

$$n = N_c \exp[(E_F - E_c)/k_BT] = N_cf(E_c) \qquad (1.21)$$

and

$$p = N_v \exp[(E_v - E_F)/k_BT] = N_v[1 - f(E_v)], \qquad (1.22)$$

where N_c and N_v are called the effective density of states and are given by

$$N_c = 2(2\pi m_{de}k_BT/h^2)^{3/2} = 4.82 \times 10^{15}(m_{de}/m_0)^{3/2}T^{3/2} \qquad (1.23)$$

and

$$N_v = 2(2\pi m_{dh}k_BT/h^2)^{3/2} = 4.82 \times 10^{15}(m_{dh}/m_0)^{3/2}T^{3/2}. \qquad (1.24)$$

The effective density of states corresponds to the density of states under the assumption that all electrons (holes) in a conduction (valence) band occupy the conduction band minimum, E_c (valence band maximum, E_v) (see Fig. 1.12). The Fermi level can be easily treated by using carrier concentration and the effective density of states. It is calculated by using Eqs. (1.21) and (1.22), and it depends on the temperature if carrier concentration is given. As shown in Fig. 1.10, the Fermi level in an intrinsic semiconductor lies very close to the middle of the band gap. From the product of Eqs. (1.21) and (1.22),

$$np = N_c N_v \exp[-(E_c - E_v)/k_B T] = N_c N_v \exp(-E_g/k_B T) = n_i^2, \qquad (1.25)$$

where E_g is $E_c - E_v$ (band-gap energy) and n_i is the intrinsic carrier concentration. In a thermal equilibrium state, the np product is not associated with the Fermi level but is determined by the temperature, and then the product is equal to the square of the intrinsic carrier concentration. This relation, expressed in Eq. (1.25), is called the mass-action law. The effective density of states and intrinsic carrier concentrations for some important semiconductors are summarized in Table 1.2.

The relations described in Equations (1.21) to (1.25) also hold in extrinsic semiconductors. In thermal equilibrium, the product of electron density and hole density even in extrinsic semiconductors is scarcely influenced by the density of donors and acceptors but is governed by the temperature, and the np product is also equal to the square of intrinsic carrier density. An electron (hole) is released from the donor level (valence band) to the conduction band (acceptor level) by thermally getting the ionization energy. Actually, the ionization energies of shallow donors and acceptors in Ge, Si, and compound semiconductors such as GaAs and InP are less than 0.03 eV. At normal ambient temperatures, the thermal energy, $k_B T$, is nearly equal to 0.026 eV ($T = 300$ K), and most donors and acceptors are therefore ionized. The ionization energies measured for various impurities frequently used in Si and GaAs are summarized in Fig. 1.13.[6]

When donor or acceptor impurities are doped, the Fermi level adjusts itself to maintain charge neutrality. It must move from the middle point of the band-gap toward the conduction band edge if donors are added and toward the valence band edge if acceptors are added (see Fig. 1.10). A schematic diagram illustrating electron and hole density in n-type semiconductors with the impurity concentration N_D (cm^{-3}) is shown in Fig. 1.14. The total negative charge is equal to the total positive charge, so

$$n = n_D + n_e = N_D^+ + p, \qquad (1.26)$$

where n_D is the number of electrons thermally activated from the donor level (ionization of donor) and n_e is the number of electrons from the valence band,

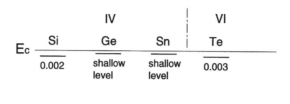

FIGURE 1.13   Measured ionization energies for various impurities in GaAs. The levels near the conduction (valence) band are measured from the bottom (top) of the conduction (valence) band. From S. M. Sze.[6] ©1981 John Wiley & Sons, Inc. Reprinted by permission of John Wiley & Sons, Inc.

N_D^+ is the number of ionized donors, and p is the number of holes in the valence band. The term n_D is given by the product of the number of donors and unoccupied probability of electrons as follows:

$$n_D = N_D^+ = N_D[1 - f(E_D)] = N_D/\{1 + \exp[-(E_D - E_F)/k_B T]\}. \qquad (1.27)$$

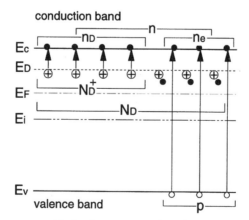

FIGURE 1.14   Electrons and holes in *n*-type semiconductors doped with impurity to a concentration, N_D.

By substituting Eqs. (1.21), (1.22), and (1.27) into Eq. (1.26), the Fermi level is approximately expressed at temperatures near 0 K:

$$E_F = (E_D + E_c)/2 - (k_B T/2)\ln(N_c/N_D). \tag{1.28}$$

The density of electrons can be obtained by substituting Eq. (1.28) into Eq. (1.21):

$$n = (N_c N_D)^{1/2} \exp[-(E_c - E_D)/2k_B T]. \tag{1.29}$$

As the temperature increases, donors become completely ionized and the electrons in the conduction band are almost activated from the donor level but not from the valence band. Consequently,

$$n = N_c \exp[-(E_c - E_F)/k_B T] = N_D, \tag{1.30}$$

and the Fermi level decreases according to the equation

$$E_F = E_c - k_B T \ln(N_c/N_D). \tag{1.31}$$

The temperature range in which Eqs. (1.30) and (1.31) hold is called the saturation range. At higher temperatures, most of the electrons in the conduction band are electrons excited from the valence band, and semiconductors behave as intrinsic semiconductors. In these temperatures carrier density and the Fermi level are given by the same equations applicable to the intrinsic semiconductors. These behaviors are summarized in Figs. 1.15 and 1.16. For *p*-type semiconductors, the Fermi level at temperatures near 0 K is given by

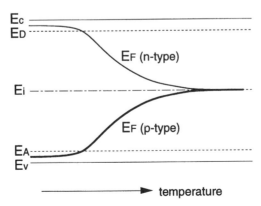

FIGURE 1.15   Temperature dependence of the Fermi level.

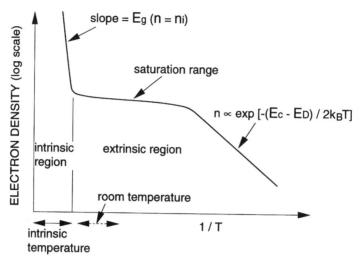

FIGURE 1.16   Temperature dependence of electron density in *n*-type semiconductors.

$$E_F = (E_A + E_v)/2 + (k_B T/2)\ln(N_v/N_A). \tag{1.32}$$

At relatively higher temperatures,

$$E_F = E_v + k_B T \ln(N_v/N_A) \tag{1.33}$$

holds. These behaviors are shown in Fig. 1.15. When the temperature is near room temperature (saturation range in Fig. 1.16), donors and acceptors are perfectly ionized and then $n = N_D^+ = N_D$ and $p = N_A^- = N_A$. Therefore, by using Eq. (1.25)

$$n = n_i^2/N_A \tag{1.34}$$

and

$$p = n_i^2/N_D. \tag{1.35}$$

These equations show that the density of the minority carrier is inversely proportional to the density of the doped impurity. The Fermi level in *n*-type semiconductors lies in the vicinity of the conduction band, whereas in *p*-type semiconductors it lies in the vicinity of the valence band (see Fig. 1.10).

1.1.2.5   Density of States and Band-Tailing in Heavily Doped Semiconductors
As the concentration of doped impurity in semiconductors increases, the Fermi

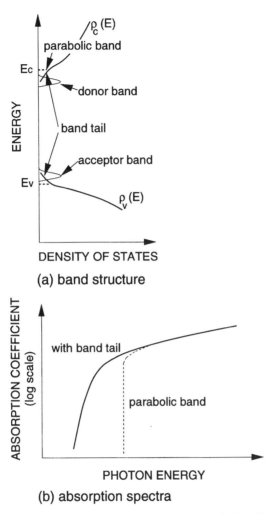

FIGURE 1.17   Band-gap narrowing and absorption spectra in heavily doped semiconductors.

level for *n*-type semiconductors gradually approaches the conduction band edge and the Fermi level for *p*-type semiconductors gradually approaches the valence band edge. Eventually the assumptions holding in lightly doped semiconductors, $(E_c - E_F) \gg 3k_B T$ or $(E_F - E_v) \gg 3k_B T$, no longer hold and the semiconductors degenerate. The band-gap narrowing occurs, as shown in Fig. 1.17, because the so-called band-tailing is generated in the density of states at around the band edge (band-tail states). As a result, the position of Fermi level moves within the conduction or valence band. The ionization energies of donors and acceptors goes to zero, and all donor and acceptor atoms are ionized. In such a situation, the carrier concentration becomes temperature independent. The free electron concentration is equal to the donor concentration, and the free hole concentration is also equal to the acceptor concentration. Thus the product of the free electron and hole concentration is no longer equal to the square of intrinsic carrier density. These free carriers screen the Coulomb interaction between the ionized impurity and the carriers, and the range of the interaction (screening length or Debye length) gradually decreases in inverse proportion to the square root of the free carrier concentration. Under these situations, the representation of localized impurity level separated from the band edges cannot be used. When the donor or acceptor introducing relatively deep levels are doped and the doping is not so heavy that the band-tailing is light, the impurity band is formed in the band-gap, and the density of states appears corresponding to the band [see Fig. 1.17(a)]. Light-emitting devices are often heavily doped with impurities to a concentration on the order of 10^{18} cm^{-3}. This heavy doping induces band-gap narrowing,[4,5,7,8] and this narrowing changes the band edge and thereby the light-emitting and light-absorbing processes as shown in Fig. 1.17(a). The theoretical analyses including the effect of band-tailing can explain experimental results better than analyses using the parabolic band calculation. This more exact treatment is especially important in explaining the operation of laser diodes, and their gain has been calculated taking the band-tail into consideration.[9] The light-emission processes of LEDs and laser diodes, however, depend on transitions between energy levels separated by more than the band-gap energy. At such high photon energies the discrepancy between the exact analyses and the parabolic calculation becomes small, as shown in Fig. 1.17(b), and the parabolic calculation can be used to roughly evaluate the performance of the light-emitting devices.

1.1.2.6 Conduction Process in Semiconductors: Scattering, Drift, and Diffusion

sion   Under the normal operating conditions, the semiconductors used in optoelectronic devices conduct an electric current. This conduction results from the movement of electrons and holes. If no carrier is excited, and the conduction and valence bands are respectively empty and filled, there is no electrical conduction. In extrinsic semiconductors a large number of electrons are thermally activated at room temperature and can contribute to the electrical conduction. Some of these carriers, however, disappear because of a recombination process in which electrons in the conduction band combine with holes in the valence band and again contribute to covalent bonding. In a thermal equilibrium, the rate of electron-hole pair generation is in equilibrium with the rate of recombination, and a carrier density sufficient for electrical conduction is maintained. This section discusses conduction processes due to electrons and holes: scattering, drift, and diffusion.

Scattering   Electrons in the semiconductors are thermally excited from the valence band to the conduction band, and holes are left in the valence band. These electrons and holes move and collide with other carriers, impurity centers, defects, and lattice imperfections. The direction of movement changes during collisions

within the semiconductor crystal, where the collision results from the electrostatic interaction between the carriers and the localized disordering of periodic
electrical field, which disordering is introduced by impurities, defects, and imperfections. This is called impurity scattering. The direction of carrier movement is
also changed by the lattice vibrations because at any finite temperature the atoms
in the lattice vibrate around their mean positions. This is called lattice scattering. The quanta of lattice vibrations are phonons, which are quantum-mechanical particles. The lattice scattering is also called phonon scattering. In elementary
semiconductors having monoatomic lattices, only acoustic phonons are induced,
whereas both acoustic and optical phonons (named according to their oscillation
frequencies) are induced in compound semiconductors such as GaAs, InP, and
InGaAs. In ternary- and quaternary-compound semiconductor crystals, such as
InGaAs and InGaAsP, not only these two kinds of scattering but also alloy scattering is important. This alloy scattering occurs because of the random atomic
arrangement in compound semiconductor crystals. Scattering results in the carriers having a scattering-limited thermal velocity, and the carrier movements are
random in all directions. Under these conditions, the semiconductors are noncon-
ducting because the carriers are moving randomly.

Drift, Mobility of Electrons and Holes, and Resistivity When an electric field,
F_{field}, is applied to the semiconductor crystal, the electrical force, qF_{field}, acts
on electrons and holes, where q is electron charge. The electrons and holes
in their respective bands are accelerated in opposite directions, and electrical
conduction becomes possible. The motion of the carriers under an electrical
field is called drift. The velocity of the carriers, however, is limited by collisions
as illustrated in Fig. 1.18. Part of the energy given to the carriers by the field is
lost during each collision and changes to Joule heat (lattice vibration). The mean
drift velocity, which is the mean velocity of the carriers under the influence of
the electric field, is expressed as

$$v_d = \mu F_{field}, \tag{1.36}$$

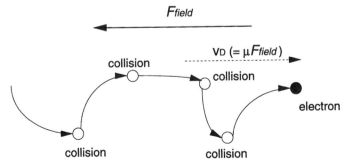

FIGURE 1.18 Schematic illustration of the motion of electrons in an electric field.

where the mobility, μ[cm^2/Vs or m^2/Vs], expresses the mean drift velocity per unit field. For electrons it is given by

$$\mu_e = -q\tau_{ce}/m_e^*,\tag{1.37}$$

and for holes it is given by

$$\mu_h = q\tau_{ch}/m_h^*,\tag{1.38}$$

where τ_{ce} and τ_{ch} are the mean free times between successive collisions for electrons and holes, and m_e^* and m_h^* are the effective masses of the electrons and the holes. The distance given by the product of the mean free time and the mean velocity, $\tau_c v_d$, is called the mean free path.

As the temperature of a crystal increases, the lattice vibration is intensified and the probability of phonon scattering increases. The mobility at high temperature is thus limited by phonon scattering. At low temperatures, mobility is high. When the number of impurity centers or other defects is large, the probability of carriers colliding with the impurity and the defects increases. At very low temperatures mobility is limited by the impurity scattering, is proportional to $T^{3/2}$, and is inversely proportional to the density of impurity centers. The mobility can therefore be expressed as

$$1/\mu = 1/\mu_I + 1/\mu_P,\tag{1.39}$$

where μ_I is the mobility limited by impurity scattering and μ_P is the mobility limited by phonon scattering. The contribution due to μ_I is dominant at low temperatures whereas that due to μ_P is dominant at high temperatures. At low temperatures, alloy scattering is also important in ternary and quaternary compound semiconductors. The mobility limited by the alloy scattering is combined with the other two mobilities by using Matthiessen's rule and is added as $1/\mu_A$ to the right-hand side of Eq. (1.39), where μ_A is the mobility limited by the alloy scattering. The electron mobilities, μ_e, of the quaternary alloy $In_{1-x}Ga_xAs_yP_{1-y}$ are shown in Fig. 1.19 as a function of the ambient temperature. The combined mobility depends on the composition of InGaAsP nonlinearly because of the alloy scattering.[11] The hole mobility is also influenced by the alloy scattering and is nonlinearly dependent on the alloy composition.[10] These mobilities are important device parameters, especially for high-speed devices.

As discussed above, the drift of electrons and holes under an electrical field results in a current. Therefore the current due to electrons in the conduction band, J_{dr-e}, and that due to holes in the valence band, J_{dr-h}, are given by the following equations:

$$J_{dr-e} = -qnv_{d-e}\tag{1.40}$$

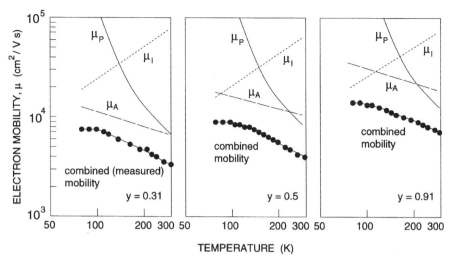

FIGURE 1.19 Electron mobilities of three different composition quaternary semiconductors, $In_{1-x}Ga_xAs_yP_{1y}$ as a function of ambient temperature. The measured data shown as points can be matched by a combination of the mobility limits imposed by (polar optical) phonon scattering, μ_P, ionized-impurity scattering, μ_I, and alloy scattering, μ_A. From J. R. Hayes et al.[10] ©1982 John Wiley & Sons, Inc. Reprinted by permission of John Wiley & Sons, Inc.

and

$$J_{\text{dr-h}} = qpv_{\text{d-h}}, \tag{1.41}$$

where n is the electron density, $v_{\text{d-e}}$ is the mean drift velocity for electrons, p is the hole density, and $v_{\text{d-h}}$ is the mean drift velocity for holes. Here $v_{\text{d-e}}$ and $v_{\text{d-h}}$ are calculated by using Eqs. (1.36), (1.37), and (1.38). The total current density in the steady state is, from Eqs. (1.40) and (1.41),

$$J_{\text{dr}} = q(\mu_e n + \mu_h p)F_{\text{field}}. \tag{1.42}$$

This equation expresses Ohm's law, and conductivity, σ_{cond}, is obtained from the Eq. (1.42) as

$$\sigma_{\text{cond}} = q(\mu_e n + \mu_h p) \qquad [(\Omega\,\text{cm})^{-1}]. \tag{1.43}$$

Resistivity, ρ_{resist}, is the reciprocal of the conductivity:

$$\rho_{\text{resist}} = 1/q(\mu_e n + \mu_h p) \qquad [\Omega\,\text{cm}]. \tag{1.44}$$

This equation indicates that the resistance of semiconductors decreases as carrier density and mobility increase. For extrinsic semiconductors, n and p in Eqs. (1.43) and (1.44) can respectively be replaced with the N_D and N_A values at temperatures near room temperature.

The resistivity (or conductivity) of a semiconductor crystal changes with temperature according to the variation of carrier density and mobility discussed above. In the intrinsic region shown in Fig. 1.16, in which extrinsic semiconductors behave like intrinsic semiconductors, the resistivity of intrinsic and extrinsic semiconductors decreases rapidly with the increase in the crystal temperature because the influence of electron-hole pair generation is more dominant than that of the decrease in mobility limited by phonon scattering. For the compensated intrinsic semiconductors, where donor and acceptor concentrations are equal, the resistivity is higher than that for intrinsic semiconductors because the mobility limited by the impurity scattering in compensated intrinsic semiconductors is higher than that in intrinsic semiconductors, even though the carrier density for both semiconductors is given by Eq. (1.25).

Diffusion of Electrons and Holes   The diffusion of electrons and holes in semiconductor crystals is a result of nonuniform density distribution. If the excess carriers are generated in a part of a semiconductor, they thermodynamically diffuse from the high-density region to the low-density region and are eventually distributed homogeneously. The excess carrier here means that the carrier is present at a concentration greater than that in thermal equilibrium state. This excess carrier concentration is produced by a method such as heating or illumination. The diffusion process is illustrated in Fig. 1.20 for one direction. The diffusion is a thermodynamic process, and the velocity of diffusing carriers is also limited by collision and scattering in a manner similar to that in which drift is limited. Therefore, the diffusion velocity of carriers, v_{diff}, is given by

$$v_{\text{diff}} = -(D/N)(dN/dx), \tag{1.45}$$

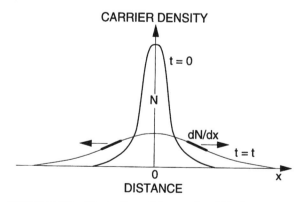

FIGURE 1.20   Illustration of the carrier diffusion process.

where

$$D = (\tau_c k_B T)/m^*, \tag{1.46}$$

and D is called the diffusion constant (or coefficient) for carriers. The terms τ_c and m^* are the mean free time and the effective mass of carriers, and N is the density of carriers at $x = 0$ and at $t = 0$. The velocity is proportional to the concentration gradient, dN/dx, and the negative sign means that the carriers move in a direction opposite to that in which the concentration gradient increases. The carrier density passing through a unit area per unit time is determined by the velocity of carriers given by Eq. (1.45), and thus the current density due to the diffusion process for electrons is expressed as

$$J_{\text{e-diff}} = qD_e(dn/dx), \tag{1.47}$$

where n is electron density. The corresponding equation for holes is

$$J_{\text{h-diff}} = -qD_h(dp/dx), \tag{1.48}$$

where p is hole density. Using Eqs. (1.37) and (1.38), the diffusion constants can be given as follows:

$$D_e = \tau_{ce} k_B T/m_e^* = -\mu_e k_B T/q \qquad [\text{cm}^2/\text{s or m}^2/\text{s}] \tag{1.49}$$

$$D_h = \tau_{ch} k_B T/m_h^* = -\mu_h k_B T/q \qquad [\text{cm}^2/\text{s or m}^2/\text{s}] \tag{1.50}$$

and from these equations, the next equations can be derived:

$$|D_e/\mu_e| = k_B T/q \qquad \text{and} \qquad |D_h/\mu_h| = k_B T/q, \tag{1.51}$$

which are each well known as the Einstein relation. The value of D/μ is about 26 meV near room temperature (about 300 K). From Eqs. (1.47) and (1.48), the total current due to the diffusion of electrons and holes can be calculated as

$$J_{\text{diff}}(x) = J_{\text{e-diff}} + J_{\text{h-diff}} = q[D_e(dn/dx) - D_h(dp/dx)]. \tag{1.52}$$

When an electric field is applied to a semiconductor with concentration gradient of carriers, the electron current density can be expressed by combining Eq. (1.37) with Eq. (1.47),

$$J_e(x) = q\mu_e n F_{\text{field}} + qD_e(dn/dx), \tag{1.53a}$$

and the hole current density can be expressed by combining Eq. (1.38) with Eq. (1.48),

$$J_h(x) = q\mu_h p F_{field} - qD_h(dp/dx). \tag{1.53b}$$

The total current density is given by

$$J(x) = J_e(x) + J_h(x). \tag{1.54}$$

The diffusion described above was considered to occur in only one direction, although carriers generally diffuse in three-dimensional space. The exact solutions is thus obtained by substituting grad n (or p) into dn (or $dp)/dx$ in the equations.

The excess (minority) carriers in semiconductors diffuse a certain distance and then vanish by recombination if the carriers are not taken out of the semiconductors. The mean time, which corresponds to the average time interval between generating and vanishing for the excess carriers, is called the lifetime. The distance which the excess carriers move during the average time interval is called the diffusion length and is given by

$$L = (D\tau)^{1/2}, \tag{1.55}$$

where L is the diffusion length, D is the diffusion constant, and τ is the carrier lifetime. The recombination processes will be discussed in the following section.

1.1.2.7 Nonradiative Recombination of Carriers (Excess Minority Carriers)

Excess carriers created by excitation (by means of illumination or by means of injection under electrical bias) vanish through recombination processes, and the systems return to the thermal equilibrium state. There are two kinds of recombination processes: direct recombination and recombination via recombination centers in the band-gap. The direct recombination includes radiative and nonradiative processes. The radiative recombination is direct recombination in which electrons in the conduction band and holes in the valence band recombine directly (band-to-band recombination) and is the basic process in light-emitting devices. These radiative recombination processes will be discussed in the next section. As discussed later in this section, the nonradiative direct recombination process is Auger recombination. Most recombination via recombination centers is nonradiative.

Nonradiative Recombination Processes in Semiconductor Crystals   The nonradiative recombination is related to crystalline defects, impurities, and the Auger effect. It influences device characteristics and is directly related to device degradation, or reliability. The energy of the nonradiative recombination is usu-

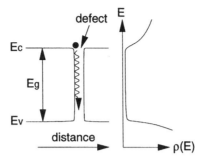

(a) defect in crystal (continuum of states)

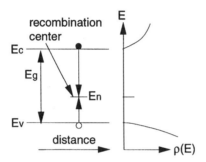

(b) defect in crystal (deep level)

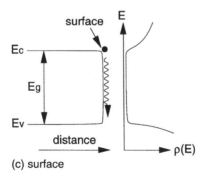

(c) surface

FIGURE 1.21   Nonradiative recombination processes in semiconductors.

ally emitted as phonons and contributes to device degradation. The crystalline defects include crystal imperfections, such as dislocation and stacking faults, and point defects, such as interstitial atoms and vacancies. These defects change the tight binding state between electrons and atoms, and as shown in Fig. 1.21(a), they may give rise to the continuum of states in the band-gap joining the conduction band to the valence band. Through the continuum of states, electrons and holes recombine nonradiatively. As shown in Fig. 1.21(b), the point defects often produce discrete deep levels in the band gap. Nonradiative recombination also occurs via the levels in the band-gap. Recombination through the deep level is often accompanied by the emission of a lower-energy photon (longer-wavelength light). This process is not nonradiative recombination, but device characteristics are nevertheless degraded by such a process. Such a recombination process is thus often considered, from the viewpoint of device performance, along with the nonradiative recombination. Electrons and holes within a diffusion length expressed by Eq. (1.55) from the position of the defect are trapped and recombine nonradiatively.

Bulk Recombination and Recombination Via Nonradiative Recombination Centers   The recombination via a single deep level in the band-gap, where only one trapping energy level is present in the band-gap, consists of four steps: electron capture, electron emission, hole capture, and hole emission (see Fig. 1.22). The rate of recombination via the deep level, R_d (cm^{-3}s^{-1}), can be expressed using the Schokley–Read recombination model under steady-state nonequilibrium conditions and is given by:[12,13]

$$R_{\mathrm{d}} = \sigma_{\mathrm{h}}\sigma_{\mathrm{n}}v_{\mathrm{th}}(np n_{\mathrm{i}}^2)N_{\mathrm{t}}/\{\sigma_{\mathrm{n}}[n + n_{\mathrm{i}}\exp((E_{\mathrm{t}} - E_{\mathrm{i}})/k_{\mathrm{B}}T)]$$
$$+ \sigma_{\mathrm{h}}[p + n_{\mathrm{i}}\exp(-(E_{\mathrm{t}} - E_{\mathrm{i}})/k_{\mathrm{B}}T)]\}, \tag{1.56}$$

where σ_{h} and σ_{n} are respectively the hole and electron capture cross sections,

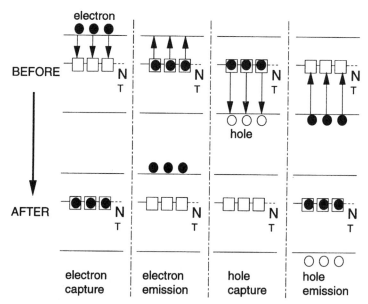

FIGURE 1.22   Recombination via a single deep level. From S. M. Sze.[6] Copyright ©1981 John Wiley & Sons, Inc. Reprinted by permission of John Wiley & Sons, Inc.

v_{th} is the carrier thermal velocity $[= (3k_B T/m^*)^{1/2}]$, N_t is the trap density, E_t is the trap energy level, E_i is the intrinsic Fermi level, and n_i is the intrinsic carrier density. The value of $(np - n_i^2)$ indicates the deviation from the thermal equilibrium conditions ($np = n_i^2$ and then $R_d = 0$ under the thermal equilibrium condition). The capture cross section is normally 10^{-13} to 10^{-15} cm^2, and $\sigma_n \gg \sigma_h$ for electron traps and $\sigma_h \gg \sigma_n$ for hole traps. When the traps act like nonradiative recombination centers, $\sigma_h = \sigma_n = \sigma_r$. Consequently, the recombination rate in Eq. (1.56) can be simplified as follows:

$$R_d = \sigma_r v_{th} N_t (np - n_i^2)/\{n + p + 2n_i \cosh[(E_t - E_i)/k_B T]\}. \tag{1.57}$$

This equation indicates that the nonradiative recombination rate increases as the energy level of the recombination center, E_t, approaches the mid-gap, E_i. The nonradiative ecombination centers located near the mid-band-gap (deep level) influence the device performance more than do those near the band edges (shallow level). Under low excited conditions, the density of the excited carriers (electrons and holes) is much less than that of majority carriers. Therefore, if the semiconductor is *n*-type, $n \gg p$ and $n \gg n_i \exp[(E_t - E_i)/k_B T]$ and the recombination rate is expressed as

$$R_{dn} = \sigma_h v_{th} N_t (np - n_i^2)/n. \tag{1.58}$$

Here $n_0 p_0 = n_i^2$ and n is nearly equal to the equilibrium value, n_0. The equation can thus be rewritten as

$$R_{dn} = \sigma_h v_{th} N_t (p - p_0). \tag{1.59}$$

Similarly, for *p*-type semiconductors

$$R_{dh} = \sigma_e v_{th} N_t (n - n_0). \tag{1.60}$$

The net rate of recombination is generally given by the minority carrier lifetime and the carrier density in the thermal equilibrium and nonequilibrium conditions. Thus if the semiconductor is *n*-type, the rate is given by

$$R_{dn} = (p - p_0)/\tau_{rh}, \tag{1.61}$$

where τ_{rh} is the hole lifetime in the *n*-type semiconductor. Comparing Eq. (1.58) with Eq. (1.61),

$$\tau_{rh} = 1/\sigma_h v_{th} N_t. \tag{1.62}$$

The electron lifetime in a *p*-type semiconductor is similarly given by

$$\tau_{re} = 1/\sigma_n v_{th} N_t. \tag{1.63}$$

Surface Recombination Crystals end at the surface of semiconductors and there are thus many dangling bonds at the surface, as well as impurities from the ambient atmosphere (such as oxygen). These dangling bonds and impurities result in the formation of surface states with energy levels in the band-gap and in band-bending at the surface. These surface states give rise to enhanced non-radiative recombination and introduce continuum states to the surface as shown in Fig. 1.21(c). The nonradiative recombination usually occurs at a higher rate at the surface than in the inner region. The carrier density within the diffusion length from the surface is always reduced because of the surface recombination. The basic concept is illustrated in Fig. 1.23. The recombination rate at the surface of *n*-type semiconductors, R_{sh}, can be expressed in a manner similar to that expressing the recombination via nonradiative recombination center in the inner region:[6,14,15]

$$\begin{aligned} R_{sh} &= \sigma_{sh} v_{th} N_{ts} A_s (p - p_0) \\ &= S_h A (p - p_0). \end{aligned} \tag{1.64}$$

Here

$$S_h = \sigma_{sh} v_{th} N_{ts} \tag{1.65}$$

is the surface recombination velocity in cm/s, where σ_{sh} is the capture cross section at the surface, N_{ts} is surface-state density, and A_s is the surface area. Similarly, for the surface of *p*-type semiconductors,

$$\begin{aligned} R_{se} &= \sigma_{se} v_{th} N_{ts} A_s (n - n_0) \\ &= S_e A (n - n_0), \end{aligned} \tag{1.66}$$

and the surface recombination velocity of *p*-type semiconductors, S_e, is given by

$$S_e = \sigma_{se} v_{th} N_{ts}. \tag{1.67}$$

The surface recombination velocity is, for example, $S_h \approx S_e \leq 10^3$ cm/s for Si, $S_h \approx S_e \approx 10^6$ cm/s for GaAs, and $S_h \leq 10^4$ cm/s and $S_e > 10^4$ cm/s for InP.[16,17] The surface recombination also influences device performance and reliability, especially degradation of the facets of laser diodes and degradation at the perimeters of *pn*-junctions in photodiodes.

Auger Recombination Although Auger recombination processes are nonradiative recombination, they are not related to the recombination processes via defects and deep levels in the band-gap.[14,18,19] During Auger recombination

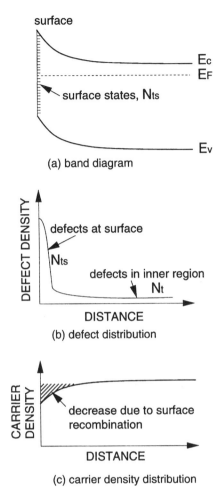

FIGURE 1.23 Illustration of (a) band diagram for an *n*-type semiconductor, (b) defect distribution, and (c) carrier distribution at the semiconductor surface.

the energy released by electron-hole recombination is immediately transferred to other electrons or holes. Thus photons are never created and the transferred energy finally dissipates by emitting phonons. This process can be understood as the inverse of impact ionization, in which electron-hole pairs are generated when accelerated electrons collide with electrons bound to the lattice atoms. Several types of Auger recombination processes occur in semiconductor crystals. Typical processes are band-to-band recombination (direct recombination): CHCC, CHSH, and CHLH. These are shown in Fig. 1.24, where C, H, L, and S respectively indicate the conduction band, heavy hole, light hole, and split-off valence band. These processes are governed by electrons at positions marked E_1 and E_2 and holes at E_1' and E_2' in the figure. In the CHCC process, an electron at

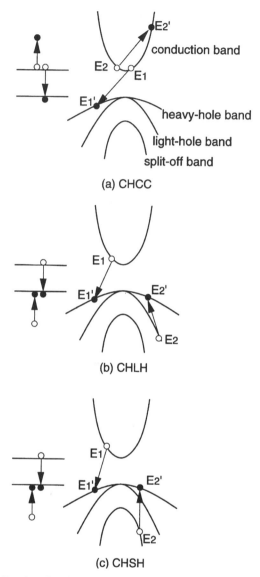

FIGURE 1.24   Band-to-band Auger processes in direct band-gap semiconductors.

E_1 recombines with a hole at E_1', and the excess energy generated is transferred to an electron at E_2. Then the electron is excited to E_2' and loses the excess energy by emitting phonons during lattice-electron interaction (phonon scattering). The electron eventually returns to the thermal equilibrium state. The Auger processes are three-carrier nonradiative recombination processes, and two electrons and one hole are associated with the CHCC process. The CHSH and the CHLH processes, on the other hand, are limited by one electron and two heavy holes. Therefore the Auguer recombination rate, R_A, is proportional to the cube of the carrier density and is given as

$$R_A = Cn^2p \tag{1.68}$$

for the transitions within the conduction band and as

$$R_A = Cnp^2 \tag{1.69}$$

for the transitions within the valence band or intravalence band. Here C is the Auger coefficient, and n and p are the carrier densities of electrons and holes, respectively. These carrier-carrier interactions will be intensified as the carrier concentration increases, and the carrier concentration is proportional to $(k_B T/E_g)^{3/2} \exp(-E_g/k_B T)$.[14] Consequently, the Auger recombination rate becomes high as the temperature increases and as the band-gap energy decreases. The Auger recombination is thus an important factor with regard to device performance at high temperatures and it often determines the temperature dependence of device characteristics for light-emitting devices, especially narrow band-gap (long-wavelength) devices such as 1550 nm-band InGaAsP/InP laser diodes.

In addition to the band-to-band Auger transitions, there are Auger transitions between donor level and valence band, conduction band and acceptor level, and donor level and acceptor level. Furthermore, phonon-assisted Auger transition occurs, although its probability is low and it can be neglected when one analyzes the device performances.

1.1.3 Optical Properties

The operation of semiconductor optoelectronic devices is based on the optical properties of semiconductors as well as on their electrical properties. The important optical properties here are light emission, light absorption, and the interaction between light and refractive index. They are discussed in this section.

1.1.3.1 Radiative Recombination and Light Absorption in Semiconductors
As shown in Table 1.1, there are two types of semiconductors: direct and indirect band-gap semiconductors. In direct band-gap semiconductors, the conduction band minimum coincides with the valence band maximum in **k**-space (at Γ valley). Direct band-to-band transition, in which the momentum must be kept constant, can occur between the conduction band minimum and the valence band maximum (see Fig. 1.25). In indirect band-gap semiconductors—such as Si, Ge, and GaP—the position of the conduction band minimum is different with respect to the position of the valence band maximum in **k**-space. Consequently, transition of electrons cannot occur without a change in the momentum, and

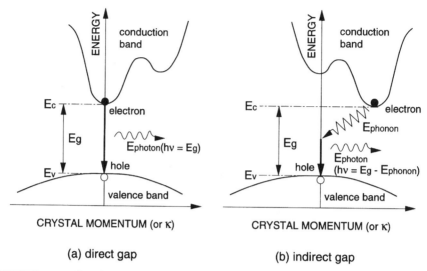

(a) direct gap   (b) indirect gap

FIGURE 1.25 Band-to-band recombination processes in (a) direct and (b) indirect band-gap semiconductors.

indirect transition occurs with the assistance of a phonon (lattice vibration), as shown in Fig. 1.25. Here, the energy is conserved during the transitions. These different types of transitions are due to the difference between the tight binding states of electrons and lattice atoms (see Section 1.1.1). The probability of indirect transition is much smaller than that of direct transition because of the requirement for phonon assistance. Consequently, direct band-gap semiconductors are usually used in light-emitting devices. The physical phenomena related to radiative recombination in semiconductors are spontaneous emission, stimulated emission, and absorption. These phenomena are closely related to one another and, for direct band-gap semiconductors, are summarized in Fig. 1.26.

Quasi-Fermi Level   In the thermal equilibrium state, the occupation probability of electrons and holes in each band is given by the Fermi level and is expressed with a Fermi–Dirac distribution function as discussed in Section 1.1.2.3. In an excited state, however, a large number of electrons and holes are generated in each band, and the occupation probability of electrons in the conduction band and holes in the valence band cannot be expressed by one Fermi level. Those electrons and holes nearly obey the Fermi–Dirac distribution, and the occupation probability can be expressed by using two Fermi levels in each band as shown in Fig. 1.27. The two Fermi levels are called the quasi-Fermi levels. For a nonequilibrium state, each occupation probability is given by substituting E_{F_c} and E_{F_v} into E_F in Eq. (1.13). For electrons this is

$$f_n(E) = 1/\{1 + \exp[(E - E_{F_c})/k_B T]\}, \tag{1.70}$$

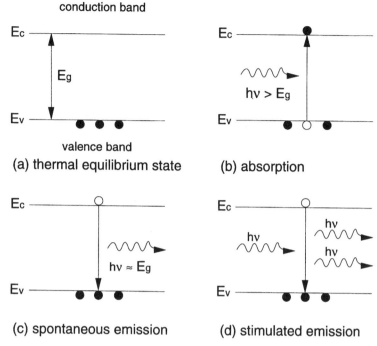

FIGURE 1.26   Transition processes in direct band-gap semiconductors.

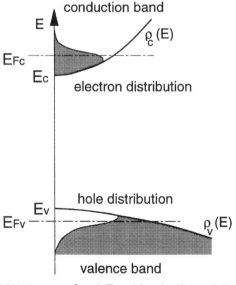

FIGURE 1.27   Quasi–Fermi levels, E_{F_c} and E_{F_v}.

and for holes it is

$$f_p(E) = 1/\{1 + \exp[(E - E_{F_v})/k_BT]\}. \tag{1.71}$$

Here the difference between the two quasi-Fermi levels $(E_{F_c} - E_{F_v})$, indicates the deviation from the equilibrium, and $E_{F_c} = E_{F_v} = E_F$ holds again when the excitation ends and the nonequilibrium state returns to the equilibrium state. For nondegenerate semiconductors the electron density in the conduction band and hole density in the valence band are calculated by using Eqs. (1.14) and (1.17) and by substituting Eq. (1.70) into Eq. (1.21) and Eq. (1.71) into (1.22). For electrons this is

$$n = N_c f(E_c) = N_c \exp[(E_{F_c} - E_c)/k_BT], \tag{1.72}$$

and for holes it is

$$p = N_v[1 - f(E_v)] = N_v \exp[(E_v - E_{F_v})/k_BT]. \tag{1.73}$$

By using the quasi-Fermi levels, the carrier distribution in the conduction band and the valence band can be described for LEDs and laser diodes under injection excitation.

Spontaneous Emission   When semiconductors in the thermal equilibrium state are illuminated with light having an energy larger than the band-gap energy, electrons are excited from the valence band to the conduction band (optical excitation or pumping) and the corresponding holes are left in the valence band. The excited electrons stay in the conduction band for a certain time (lifetime) and then return to the valence band through direct or indirect recombination processes. Then the semiconductors return to the thermal equilibrium state. As shown in Fig. 1.25, in the recombination processes, the energy corresponding to the direct transition is emitted as a photon. The wavelength, λ, of this photon is given by

$$\lambda = hc/E_g \approx 1.24/E_g \qquad [\mu\text{m}] \tag{1.74}$$

for direct band-gap semiconductors and by

$$\lambda = hc/(E_g - E_{\text{phonon}}) \approx 1.24/(E_g - E_{\text{phonon}}) \qquad [\mu\text{m}] \tag{1.75}$$

for indirect band-gap semiconductors. Here h and c are Planck's constant and the velocity of light in a vacuum, and E_g and E_{phonon} are the band-gap energy of the semiconductor and the phonon energy (in eV). This phenomenon related to the radiative recombination is spontaneous emission. Ordinary light-emission processes in semiconductors, such as photoluminescence, correspond to

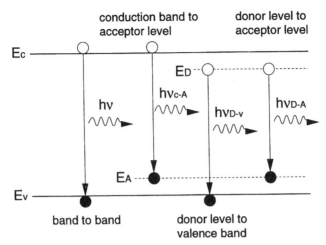

FIGURE 1.28 Simplified illustration of the spontaneous emission processes in direct band-gap semiconductors.

this spontaneous emission. In the strict sense, for the transition in direct band-gap semiconductors, the emitted energy is nearly equal to the band-gap energy, E_g, and the spectral widths of the emitted light increase in a range between 10 meV and 100 meV at room temperature because of the distribution of the electrons and the holes in each band (see Fig. 1.27). In the spontaneous emission, the excited electrons move randomly from the conduction band to the valence band. The phase of the emitted light is essentially random (incoherent radiation). Not only band-to-band transition but also band-to–impurity level transition, such as conduction band-to–acceptor level and donor level–to–valence band, and impurity level–to–impurity level transitions occur in extrinsic semiconductors. These are shown in Fig. 1.28 for direct band-gap semiconductors. In indirect band-gap semiconductors, the radiative transition is related mainly to impurity levels, such as excitonic transitions (for bound excitons) at isoelectronic traps. This type of transition in light-emitting diodes will be discussed in Section 2.2.2.

   Under excited conditions, the radiative recombination rate increases in proportion to the product of the electron and hole densities because the probability of the recombination increases as electron and hole densities increase. Consequently, the spontaneous emission rate, R_{sp}, in unit volume is given by

$$R_{sp} = B_{sp}np, \qquad (1.76)$$

where B_{sp} is the radiative recombination coefficient (transition probability of spontaneous emission), in cm^3 s^{-1}, and depends on the band-gap energy and on whether the semiconductor has a direct or an indirect band-gap.[4,5] The value of B_{sp} is usually between 10^{-9} and 10^{-11} cm^3 s^{-1} in the direct band-gap semiconductors and 10^{-13} and 10^{-15} cm^3 s^{-1} in the indirect band-gap semiconductors. The spontaneous emission rate is the sum of the radiative recombination rate in the thermal equilibrium state, R_{sp0}, and the radiative recombination rate for the excited excess carriers, R_r. If the excited carrier densities are expressed by Δn and Δp (where $\Delta n = \Delta p$), Eq. (1.76) can be rewritten as

$$
\begin{aligned}
R_{sp} &= R_{sp0} + R_r \\
&= B_{sp}(n_0 + \Delta n)(p_0 + \Delta p) \\
&= B_{sp}[n_0 p_0 + \Delta n(n_0 + p_0 + \Delta n)],
\end{aligned}
\tag{1.77}
$$

where

$$
R_{sp0} = B_{sp} n_0 p_0
\tag{1.78}
$$

and

$$
\begin{aligned}
R_r &= R_{sp} - R_{sp0} \\
&= B_{sp} \Delta n(p_0 + n_0 + \Delta n).
\end{aligned}
\tag{1.79}
$$

Under high-excitation conditions, $\Delta n \gg n_0$, p_0 holds and

$$
R_{sp} \approx R_r \approx B_{sp} \Delta n^2 \approx B_{sp} n^2.
\tag{1.80}
$$

The intensity of the spontaneous emission becomes high as the electron density in the conduction band and the hole density in the valence band are increased by the excitation, because the spontaneous emission rate is proportional to the square of the carrier density. Simultaneously, the emission peak energy (or wavelength) becomes higher (or shorter). As the carrier density increases, the electrons in the conduction band and the holes in the valence band are filled from the band edges in order. As the result, spontaneous emission with higher energy increases. This introduces the peak shift (band-filling effect).

Optical Absorption   There are several optical absorption processes: fundamental absorption, free carrier absorption, absorption via energy levels in the band-gap, and so forth. The amount of absorption is expressed by using the absorption coefficient described by Lambert's law, as

$$
dI = -\alpha_{ab}I \ dx
\tag{1.81a}
$$

or

$$
I = I_0 \exp(-\alpha_{ab}x),
\tag{1.81b}
$$

where I is the light intensity, I_0 is the initial intensity, x indicates the distance, and α_{ab} is the absorption coefficient (in cm^{-1}). This coefficient is a function of the wavelength of the light. The inverse of the absorption coefficient is called the absorption length (or the penetration depth). At the absorption length, the light intensity has decreased to $1/e$ [substituting $x = 1/\alpha_{ab}$ in Eq. (1.81b)].

FUNDAMENTAL ABSORPTION   The fundamental absorption corresponds to the excitation of electrons from the valence band to the conduction band. Consequently, when the semiconductor is illuminated, light having an energy greater than the band-gap energy of the semiconductor is absorbed and light having an energy less than the band-gap energy passes through the semiconductor without being absorbed (see Fig. 1.29). The energy (or wavelength) at which the fundamental absorption starts is called the absorption edge. At energies near the absorption edge, the absorption constant is expressed as follows:[20]

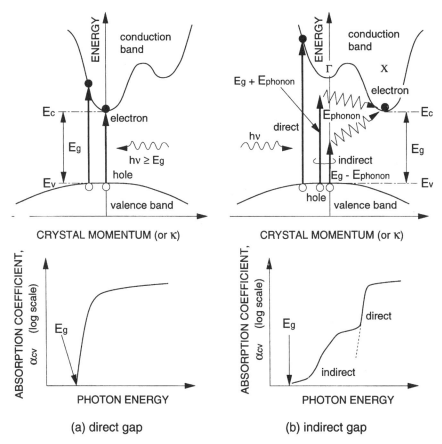

FIGURE 1.29 Simplified illustration of the fundamental absorption processes in (a) direct and (b) indirect band-gap semiconductors.

$$\alpha_{cv} \propto (h\nu - E_g)^{1/2} \tag{1.82}$$

for direct transition, and

$$\alpha_{cvin} \propto (h\nu - E_g \pm E_{phonon})^2 \tag{1.83}$$

for indirect transition. Here $h\nu$ is the energy of the illuminating photon, and E_{phonon} is the phonon energy related to the indirect transition. The band-gap energy, E_g, for the indirect transition usually corresponds to the energy difference between the valence band maximum at the Γ valley and the conduction band minimum at X. Energy conservation holds in direct and indirect transitions, and momentum scarcely changes in direct transition because the momentum of the photon is small. In indirect band-gap semiconductors the absorption process changes from indirect to direct as photon energy increases (see Fig. 1.29). As evident from Eqs. (1.82) and (1.83), the absorption coefficient at the near absorption edge increases more rapidly in direct band-gap semiconductors than in indirect band-gap semiconductors. The absorption coefficient is the function of the energy (wavelength) of the incident photon. At the absorption edge, the absorption coefficient is small because the density of states in the conduction band is low. As the photon energy increases, the electron can transit to a higher energy level in the conduction band, at which the density of states is larger. Thus the absorption coefficient increases exponentially and tends to saturate. These behaviors are essentially governed by the density of states in the conduction band, expressed by Eq. (1.11). The transition probability in the optical absorption processes is higher in direct band-gap semiconductors than in indirect band-gap semiconductors because, as in the case of spontaneous emission, indirect transition is impossible without a change in the momentum by means of phonon emission or absorption.

The processes in the fundamental absorption are influenced by the band-tailing at the absorption edge and by the other effects, and the absorption edge therefore becomes indistinct. For example, if the doping concentration in the *n*-type direct band-gap semiconductors increases, the Fermi level gradually increases and lies above the conduction band edge. In this degenerated situation, the states for transition of electrons near the conduction band-edge are occupied and the transition to higher-energy states occurs. Consequently, the absorption edge shifts to the higher-energy (or shorter-wavelength) region. This shift introduced by the doping-induced band-filling is called the Burstein–Moss shift. As discussed in Section 1.1.2.5, in highly doped semiconductors, band-tailing (shrinkage of the band-gap) also occurs. In addition, an impurity band is formed in the band-gap when the band-tailing is not severe (see Fig. 1.17). These result in the deviation of the absorption edge to the lower-energy region, and a tailing to longer wavelength range is observed in the absorption spectrum.

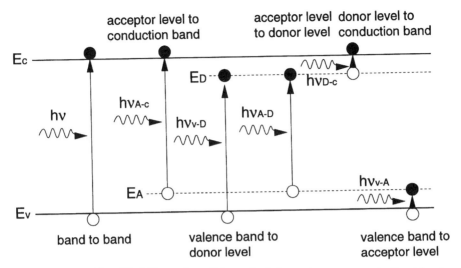

FIGURE 1.30 Simplified illustration of the optical absorption processes in direct band-gap semiconductors.

ABSORPTION RELATED TO IMPURITY LEVEL   Absorption in the extrinsic semi-conductors is due not only to the band-to-band transition but also to band-to–impurity level transitions such as conduction band-to–acceptor level, donor level–to–valence band, and impurity level–to–impurity level transitions. Optical absorption processes are shown in Fig. 1.30 for the direct band-gap semiconductors. For conduction band–to–acceptor level, donor level–to–valence band, and impurity level–to–impurity level transitions, the absorption is observed as shoulders on the low-energy side of the absorption edge. In addition, optical absorption also appears at wavelengths in the far-infrared region corresponding to the energies of donor level–to–conduction band and valence band–to–acceptor level transitions.

EXCITON ABSORPTION   Excitons are electron-hole pairs held together by their mutual Coulomb interaction and exciton absorption is observed in the very pure semiconductors at low temperature. There are two kinds of excitons: free excitons and bound excitons localized in the vicinity of a donor, acceptor, or neutral atom. Excitons are created only in very pure semiconductors because the Coulomb interaction is easily screened by free carriers. Because of the exciton energy, the transition in the optical absorption process occurs at a photon energy lower than that corresponding to the band-gap energy. The emission process related to the exciton is the reverse of the absorption process. The energy of an exciton is a few millielectron-volts, and thus in direct band-gap semiconductors at low temperature a sharp line transition is usually observed at an energy slightly below the band-gap energy.

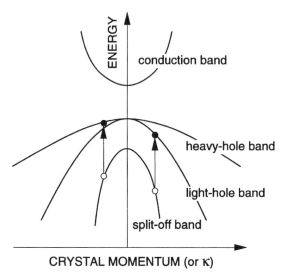

FIGURE 1.31   Simplified illustration of intervalence band absorption.

INTRABAND TRANSITION   Intraband transition occurs in both *n*- and *p*-type semiconductors but the most important transition for optoelectronic devices is that in the *p*-type semiconductors. The intraband absorption is a result of the valence band of most semiconductors being separated into the light-hole and the heavy-hole bands and the split-off-band. The separation of the three sub-bands is in turn a result of spin-orbit interaction. As shown in Fig. 1.31, electron transitions from the split-off-band to the light-hole band or the heavy-hole band and from the light-hole band to the heavy-hole band occur during the optical absorption process. The transitions depend on the doping and on temperature. The absorption coefficient of *p*-type III-V compound semiconductors doped to 10^{18} cm^{-3} increases by about 10 cm^{-1}. This intervalence-band absorption is thought to be an important factor limiting the temperature characteristics of InGaAsP/InP laser diodes operating at wavelengths between 1300 and 1550 nm. These absorption processes or transitions are normally not observed in *n*-type semiconductors.

FREE-CARRIER ABSORPTION   Free electrons in a vacuum never interact with light because they cannot satisfy the requirements for conservation of the momentum of the electrons and for conservation of the energy of the whole system. Free-carriers within the energy band, however, interact with light and induce the absorption of photon. Within the same valley, the carriers transit to higher energy levels under momentum conservation.[21] This process requires a change in the momentum of the carrier during the transition. The momentum change is provided by phonon and impurity scattering, satisfying the requirement for momentum conservation. The absorption coefficient introduced by this type of free-carrier absorption, α_{fc}, is proportional to the square of wavelength, λ, and is expressed as follows:

$$\alpha_{fc} \propto N\lambda^2, \tag{1.84}$$

where N is the free carrier concentration. Free-carrier absorption is also induced by the interband transitions, such as the indirect transition processes between Γ, X, and L valleys.

The total absorption coefficient is the sum of each of the coefficients determined by the different absorption processes:

$$\alpha_{ab} = \alpha_{cv} + \alpha_{il} + \alpha_{ex} + \alpha_{ib} + \alpha_{fc}, \tag{1.85}$$

where α_{il}, α_{ex}, and α_{ib} are respectively the coefficients of absorption related to impurity level absorption, exciton absorption, and intraband absorption. The schematic illustration shown in Fig. 1.32 is for undoped pure semiconductors, so it ignores the absorption related to the impurity levels.

Relationships Between Stimulated Emission, Spontaneous Emission, and Fundamental Absorption

STIMULATED EMISSION   Stimulated emission is, like spontaneous emission, a radiative recombination process. The stimulated emission leads to the coherent emission from a semiconductor laser diode. When light (photons) having an energy of E_g is incident to the excited state, the excited electrons are stimulated

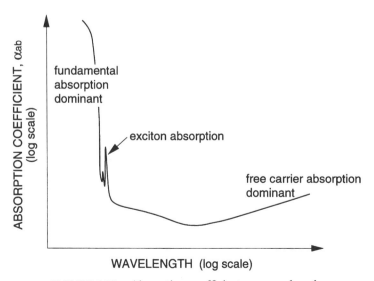

FIGURE 1.32   Absorption coefficient vs. wavelength.

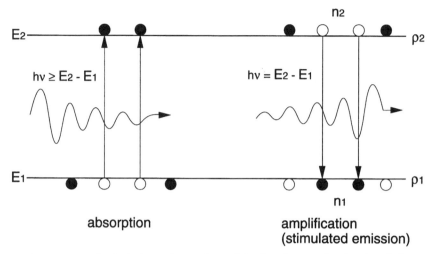

FIGURE 1.33 Optical absorption and amplification due to stimulated emission.

and move in the same phase as the incident light. This phenomenon results from the vibration of the dipole, which is formed by polarization of atoms under light incidence, being synchronized to the frequency of the incident light. The energy (frequency) of the emitted light also coincides with that of the incident light, E_g. According to Eqs. (1.21) and (1.70) the electron densities n_1 at energy level E_1 and n_2 at level E_2 ($E_2 > E_1$) follow Boltzmann statistics, and the ratio is given as

$$(n_2/\rho_2)/(n_1/\rho_1) = \exp[-(E_2 - E_1)/k_B T], \tag{1.86}$$

where ρ_1 and ρ_2 are densities of the states of electrons at E_1 and E_2 (see Fig. 1.33). Equation (1.86) indicates that in a thermal equilibrium state the electron density at a higher energy level is less than that at a lower energy level. Therefore, only the absorption of the incident light from outside the semiconductors is observed in the thermal equilibrium state. If $n_2/\rho_2 > n_1/\rho_1$ as a result of excitation, however, the incident light is amplified and stimulated emission is observed (see Fig. 1.33). From Eq. (1.86), the relation $n_2/\rho_2 > n_1/\rho_1$ is satisfied if equivalently $T < 0$, and this situation is called negative temperature or population inversion because the electron density at a higher energy level is larger than that at a lower energy level.

BASIC RELATIONSHIPS   As shown in Fig. 1.34, if photons having an energy of $E_{21} = E_2 - E_1$ and the density, $n_{ph}(E_{21})$, are incident to an excited semiconductor, the rate of stimulated emission from E_2 to E_1, $r_{stim}(E_{21})$, is proportional to the product of the electron density at E_2, $\rho_c(E_2 - E_c)f_2$, hole density at E_1, $\rho_v(E_v - E_1)(1 - f_1)$, and the incident photon density:[22]

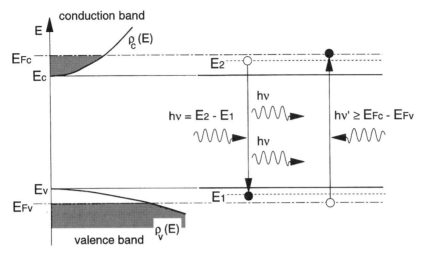

FIGURE 1.34 Simplified illustration of stimulated emission and optical absorption resulting from transitions between E_1 and E_2 at absolute zero.

$$r_{\text{stim}}(E_{21}) = B_{21}n_{\text{ph}}(E_{21})\rho_c(E_2 - E_c)f_2\rho_v(E_v - E_1)(1 - f_1), \qquad (1.87)$$

where B_{21} is the transition probability of the stimulated emission from E_2 to E_1 and

$$f_1 = 1/\{1 + \exp[(E_1 - E_{F_v})/k_B T]\} \qquad (1.88a)$$

and

$$f_2 = 1/\{1 + \exp[(E_2 - E_{F_c})/k_B T]\}. \qquad (1.88b)$$

The stimulated emission is closely related to the fundamental absorption, and the transition rate for the absorption from E_1 to E_2, $r_{\text{abs}}(E_{12})$, is thus proportional to the product of the electron density at E_1, $\rho_v(E_v - E_1)f_1$, hole density at E_2, $\rho_c(E_2 - E_c)(1 - f_2)$, and the photon density:

$$r_{\text{abs}}(E_{12}) = B_{12}n_{\text{ph}}(E_{21})\rho_c(E_2 - E_c)(1 - f_2)\rho_v(E_v - E_1)f_1, \qquad (1.89)$$

where B_{12} is the transition probability due to the optical absorption from E_1 to E_2. The transition probability under conditions in which stimulated emission and absorption occur is different from that under the condition in which spontaneous emission occurs because the spontaneous emission has no interaction with light (photons) through the recombination processes. Consequently, the transition probability of spontaneous emission, r_{sp}, has no relation to the photon density and is proportional to the product of the electron density at E_2, $\rho_c(E_2 - E_c)f_2$ and hole density at E_1, $\rho_v(E_v - E_1)(1 - f_1)$. Thus Eq. (1.76) is exactly rewritten as

$$r_{sp}(E_{21}) = A_{21}\rho_c(E_2 - E_c)f_2\rho_v(E_v - E_1)(1 - f_1), \qquad (1.90)$$

where A_{21} is the transition probability of the spontaneous emission from E_2 to E_1.

THE RELATIONSHIPS AT THERMAL EQUILIBRIUM   At thermal equilibrium, the rates of the radiative transitions for spontaneous and stimulated emission and the rate of the reverse transition for absorption are balanced, and thus

$$r_{stim}(E_{21}) + r_{sp}(E_{21}) = r_{abs}(E_{12}). \qquad (1.91)$$

As discussed in Section 1.1.3.1, $E_{F_c} = E_{F_v} = E_F$ holds here. From Eq. (1.91) and the theory for blackbody radiation,[5]

$$A_{21} = (8\pi n_r^3 E_{21}^2/h^3 c^3)B_{21} \qquad (1.92)$$

and

$$B_{12} = B_{21}, \qquad (1.93)$$

where n_r is the refractive index and h and c are Planck's constant and the velocity of light in a vacuum. Equations (1.92) and (1.93) are called the Einstein relations. The stimulated emission is closely related to the spontaneous emission and the absorption, and the transition probability of stimulated emission is equal to that of absorption.

CONDITION REQUIRED FOR STIMULATED EMISSION   Stimulated emission will be observed when the rate of transition from the conduction band to the valence band is larger than that of the transition from the valence band to the conduction band. Consequently, the condition for stimulated emission is given, by comparing Eq. (1.87) with Eq. (1.89), as follows:[4,5,23]

$$r_{stim}(E_{21}) > r_{abs}(E_{12}), \qquad (1.94)$$

and thus, by using $B_{12} = B_{21}$ [see Eq. (1.93)],

$$f_2(1 - f_1) > (1 - f_2)f_1. \qquad (1.95)$$

By substituting Eq. (1.88) into Eq. (1.95),

$$\exp[(E_{F_c} - E_{F_v})/k_B T > \exp[(E_2 - E_1)/k_B T] \tag{1.96a}$$

or

$$E_{F_c} - E_{F_v} > E_2 - E_1. \tag{1.96b}$$

These equations indicate that the emitted photon energy due to the stimulated emission is always lower than the energy difference between the quasi-Fermi levels (see Fig. 1.34). In addition, for stimulated emission to occur the quasi-Fermi levels have to be set at the position above the conduction band minimum and below the valence band maximum. This means that a high carrier density is required in the conduction and the valence band; this condition occurs in laser diodes.

NET RATES OF STIMULATED EMISSION, SPONTANEOUS EMISSION, AND ABSORPTION   The net rate for stimulated emission, R_{stim}, is given by taking the difference between the transition rates of stimulated emission and of absorption,

$$R_{stim}(E_{21}) = r_{stim}(E_{21}) - r_{abs}(E_{12}), \tag{1.97}$$

and using $B_{12} = B_{21}$ [see Eq. (1.93)]:

$$R_{stim}(E_{21}) = B_{21} n_{ph}(E_{21})\rho_c(E_2 - E_c)\rho_v(E_v - E_1)(f_2 - f_1). \tag{1.98}$$

In the similar manner, the net rate of absorption is given from the difference between the transition rates of absorption and of stimulated emission as

$$R_{abs}(E_{12}) = r_{abs}(E_{12}) - r_{stim}(E_{21}), \tag{1.99}$$

and using Eqs. (1.87) and (1.89):

$$R_{abs}(E_{12}) = B_{21} n_{ph}(E_{21})\rho_c(E_2 - E_c)\rho_v(E_v - E_1)(f_1 - f_2). \tag{1.100}$$

The net absorption rate here will be given by the product of the absorption coefficient, $\alpha_{cv}(E_{12})$, and photon flux (number of incident photons per unit time). The velocity of light in a medium with refractive index, n_r, is c/n_r, where c is the velocity of light in a vacuum. Consequently, the absorption coefficient $\alpha_{cv}(E_{12})$ is given by

$$\alpha_{cv}(E_{12}) = B_{21}\rho_c(E_2 - E_c)\rho_v(E_v - E_1)(f_1 - f_2)/(c/n_r). \tag{1.101}$$

This equation shows that the absorption coefficient is related by several constant terms to the stimulated emission rate given by Eq. (1.98). This relation will be important when the lasing condition in laser diodes is discussed. The net rate of spontaneous emission is already given by Eq. (1.90) because the spontaneous emission never interacts with optical absorption processes.

The basic net rates of stimulated emission, absorption, and spontaneous emission are given by Eqs. (1.98), (1.100), and (1.90). These are relations between two levels—E_2 and E_1, with the energy difference, $E_{21} = E_2 - E_1$—in semiconductors. In actual semiconductors there is a continuum of states in the conduction and valence bands, and the actual net rates of emission and absorption—and thus the absorption coefficient—must be the sum of the basic net rates between the two levels for all of the energy levels separated by energy E_{21}. Integrating Eqs. (1.98), (1.101), and (1.90) over the whole energy in the conduction and valence band therefore gives the values observed experimentally. The absorption coefficient is usually derived experimentally, and then stimulated and spontaneous emission rates are calculated. The integration of the equations has been discussed in detail in text books.[5] When the stimulated emission rate is high, the absorption coefficient becomes negative. This negative absorption coefficient means the amplification of light (i.e., it means gain) and is a very important parameter in laser diodes. The rate of spontaneous emission must be integrated again over the whole energy range of emitted photons. This integration gives the following equation:[5,24]

$$R_{sp} = A_{21}np = B_{sp}np,$$ (1.102)

which corresponds to Eq. (1.76).

1.1.3.2 Radiative and Nonradiative Recombination Rates and Carrier Lifetime
As discussed in Section 1.1.2.7 for nonradiative recombination and in Section 1.1.3.1 for spontaneous emission, the rates of radiative and nonradiative recombination are governed by several kinds of processes. The rate of carrier recombination in *p*-type semiconductors, $R(n)$, will be given roughly by the sum of Eqs. (1.60), (1.66), (1.69), and (1.79):

$$R(n) = \sigma_e v_{th} N_t(n - n_0) + S_e A_s(n - n_0) + Cnp^2 + B_{sp}\Delta n(n_0 + p_0 + \Delta n),$$ (1.103)

and here the minority carriers are electrons. Under relatively high excitation conditions, $n \approx \Delta n \gg n_0$, $p \approx \Delta p \gg p_0$, and $\Delta n = \Delta p$ (requirement of electrical neutrality), and Eq. (1.103) and thus be simplified to

$$R(n) = \sigma_e v_{th} N_t n + S_e A_s n + Cn^3 + B_{sp}n^2.$$ (1.104)

The third and the fourth terms on the right-hand side (Auger and radiative recombination) are largely determined by the material and its band structure. The first and second terms (bulk and surface recombination) depend on the crystal quality, which is in turn influenced by crystal growth techniques and device process techniques. Equation (1.104) also holds for *n*-type semiconductors, in which the minority carriers are holes. In semiconductors the excited carrier recombines at the rate given by Eq. (1.104). The lifetime of the carriers is inversely proportional to the recombination rate. If the excited carriers in a unit volume recombine at the rate $R(n)$, the mean residence time of the excited electrons in the conduction band is given as

$$\tau_n = n/R(n)$$
$$= 1/[\sigma_e v_{th} N_t + S_e A_s + C n^2 + B_{sp} n], \tag{1.105}$$

where τ_n is the carrier lifetime and corresponds to the time when the excited carrier density is reduced by recombination to $1/e$ of the initial value.

The carrier lifetime in Eq. (1.105) combines the lifetimes due to nonradiative and radiative recombinations and can be divided into nonradiative and radiative terms, τ_{nr} and τ_r, as follows:

$$1/\tau_n = 1/\tau_{nr} + 1/\tau_r, \tag{1.106}$$

where τ_r is determined by the spotaneous emission process and by using Eq. (1.79) to be

$$\tau_r = \Delta n/R_r = 1/B_{sp}(p_0 + n_0 + \Delta n). \tag{1.107}$$

If the excitation in lightly doped p-type semiconductors is weak and most of the excited electrons combine with holes introduced by doping, $p_0 \gg n_0$ and $p_0 > \Delta n$, and Eq. (1.107) can thus be simplified to

$$\tau_r \approx 1/B_{sp} p_0. \tag{1.108}$$

Under high excitation, $p_0 \gg n_0$ and $\Delta n > p_0$, and

$$\tau_r \approx 1/B_{sp} \Delta n. \tag{1.109}$$

This means that the excited electrons recombine with the excited holes, a process called bimolecular recombination that is observed in undoped or lightly doped semiconductors.

The optical output power of spontaneous emission from light-emitting devices is basically governed by the factors indicated in Eqs. (1.104) to (1.106) if the absorption in the semiconductor layers and the reflection at the surface of the devices are ignored. The efficiency of radiative recombination (internal quantum efficiency), η_i, in the semiconductor crystals can be expressed as the ratio of radiative carrier lifetime to total carrier lifetime; that is, the ratio of radiative recombination to total recombination of the excited electrons per unit time:

$$\eta_i = (1/\tau_r)/(1/\tau_n)$$
$$= 1/(1 + \tau_r/\tau_{nr}). \tag{1.110}$$

By combining Eqs. (1.107) and (1.110),

$$\eta_i = B_{sp}\tau_n(p_0 + n_0 + \Delta n). \tag{1.111}$$

In photodiodes, part of the photoinduced current disappears by means of the recombination indicated in Eqs. (1.104) to (1.106), and device efficiency is also influenced by the nonradiative recombination. The actual cases in photodiodes as well as in LEDs will be discussed again in the descriptions of each device. The characteristics of laser diodes are similar to those of LEDs before lasing. After lasing, the injected carriers recombine within a time on the order of 10^{-12} s or less because of the stimulated emission. The lifetime limited by nonradiative recombination, given by Eqs. (1.105) and (1.106), in contrast is of the order of 10^{-9} s or less. Consequently, the nonradiative recombinations influence the lasing threshold but scarcely change the characteristics after lasing.

***1.1.3.3 Refractive Index, Dielectric Constant, Absorption Coefficient, and the Kramers–Kronig Relations**[25]* The main optical processes in semiconductors have been discussed in the previous sections. This section describes other important optical constants, those related to the refractive index of semiconductors. The refractive index, which is closely related to the dielectric constant and absorption coefficient, is a very important parameter for laser diodes and other waveguide devices, such as semiconductor optical modulators.

The complex refractive index in a semiconductor is defined by

$$n^* = n_r + ik_{extinc}, \tag{1.112}$$

where n_r and k_{extinc} are called the refractive index and the extinction coefficient. The dielectric constant is expressed with

$$\epsilon_s = \epsilon_\infty + \epsilon(f), \tag{1.113}$$

where ϵ_s is the static dielectric constant, ϵ_∞ is the high-frequency dielectric constant, and $\epsilon(f)$ is a term determined by optical phonon frequency, crystal mass, and so forth. Under an alternating electrical field, the dielectric constant abruptly decreases at some frequencies due to the term $\epsilon(f)$. This frequency dependence of the dielectric constant is governed by the frequency response for inversion of a dipole under an alternating electrical field. In the range of infrared or visible-light frequencies, the dielectric constant is determined by ionic polarization and electronic polarization.[26] At lower frequencies, other effects such as space charge polarization are added. The complex dielectric constant of a material is used when the dielectric loss is considered and it is defined by

$$\epsilon^* = \epsilon_1 + i\epsilon_2, \tag{1.114}$$

where

$$\epsilon_1 = n_r^2 - k_{extinc}^2 \tag{1.115}$$

and

$$\epsilon_2 = 2n_r k_{extinc}. \tag{1.116}$$

Here the absorption coefficient discussed in Section 1.1.3.1 is expressed by

$$\alpha_{ab} = 2\omega k_{extinc}/c = 4\pi\nu k_{extinc}/c$$
$$= 2\pi\nu\epsilon_2/cn_r, \tag{1.117}$$

where ω and ν are the angular frequency ($= 2\pi\nu$) and the frequency of light and c is the velocity of light in a vacuum. When the electrical conductivity of the material is nearly zero, from Eqs. (1.115) and (1.116),

$$n_r \approx (\epsilon_1)^{1/2} \tag{1.118}$$

and

$$k_{extinc} \approx \epsilon_2 \approx 0. \tag{1.119}$$

Equation (1.119) indicates that the material is transparent. From Eqs. (1.114), (1.118), and (1.119), a relation $n_r \approx (\epsilon^*)^{1/2}$ can be derived, where ϵ^* corresponds to the static dielectric constant ϵ_s. The real and imaginary parts of the complex dielectric constant, ϵ_1 and ϵ_2, are interdependent according to the Kramers–Kronig relations.[15,25] From the Kramers–Kronig relations and Eq. (1.117), the following equation can be deduced as

$$n_r(E) - 1 = (ch/2\pi^2)P\int_0^\infty \alpha_{ab}(E')dE'/[(E')^2 - E^2], \tag{1.120}$$

where P is the principal value of the Cauchy integrals expressed with $P\int_0^\infty = \lim_{a \to \infty} [\int_0^{w-a} + \int_{w-a}^\infty]$. Equation (1.120) indicates that the refractive index can be obtained from the absorption spectrum.

1.1.4 Electrical and Optical Properties and Electric Field

The electrical and optical properties of a semiconductor change when it is under an electrical field, and some of the phenomena due to these changes are very important in the optoelectronic devices used under reverse bias, such as photodiodes and semiconductor optical modulators. Typical phenomena are the avalanche multiplication related to the process of impact ionization, the electrooptic effect (Pockels effect and Kerr effect) inducing a change in dielectric constant and thus in refractive index, the Franz–Keldysh and Stark effects resulting in changes in refractive index and absorption coefficient, and the quantum-confined Stark effect related to exciton absorption. These phenomena will be discussed in detail in Sections 1.2.3.2, 4.4.1, and 5.1.

1.2 *pn*-JUNCTIONS AND HETEROSTRUCTURES

Most optical semiconductor devices include a *pn*-junction and their operation is based on the properties of the *pn*-junction. Carrier injection through the *pn*-junction into the conduction and valence bands is required in order to generate the radiative recombination in LEDs and laser diodes and to change the absorption coefficient and the refractive index of the optical waveguide in optical modulators used under forward bias. In photodiodes, electron-hole pairs produced by illumination are separated in opposite directions at the *pn*-junction and the resultant photocurrent is utilized as an electrical output. The electrical field generated at the *pn*-junction under a reverse bias changes the absorption coefficient and the refractive index of the optical waveguide in optical modulators used under a reverse biasing. For these devices, the *pn*-junction is indispensable. The heterostructure is also indispensable in these devices for improving their efficiency. Most optoelectronic *pn*-junction devices consist of heterostructures, and most of those heterostructures are actually double heterostructures.

The junction in semiconductors generally means the transition region between the two regions having different electrical properties. From the material point of view, there are two types of junctions: homojunctions consist of one kind of material and heterojunctions consist of various materials. From the electrical point of view, there are two kinds of junctions, *pn*-junctions and junctions formed of the same material with different carrier concentrations (homojunction) or of different materials with the same conduction type (heterojunction) such as p^+p- and n^-n-junctions. The *pn*-junction is formed when a *p*-type semiconductor is in contact with an *n*-type semiconductor. This contact is primarily produced by impurity diffusion from the surface of the semiconductors after crystal growth or by impurity doping during crystal growth. The *pn*-junction formed by impurity diffusion is called a diffused junction and that formed by impurity doping is called a grown junction. There is another type of junction, the alloy junction, in which the impurities are in contact with semiconductors as an alloy, such as a combination of *n*-type Ge with In. The *pn*-junction is usually, however, a diffused or grown junction. As shown in Fig. 1.35, the junction profile can be abrupt or graded—abrupt (or step) junction and the graded junction.

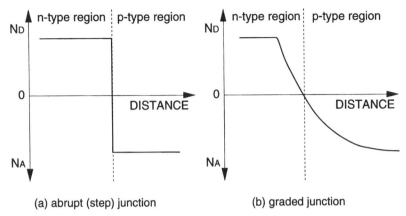

(a) abrupt (step) junction   (b) graded junction

FIGURE 1.35 Impurity distribution at *pn*-junctions: (a) abrupt junction and (b) graded junction.

1.2.1 Formation of *pn*-Homojunctions

The *pn*-junction in the thermal equilibrium state is shown schematically in Fig. 1.36. When *p*- and *n*-type semiconductors are in contact with each other at room temperature, electrons in the *n*-type semiconductor diffuse to the *p*-type semiconductor, while holes in the *p*-type semiconductor diffuse to the *n*-type semiconductor because of their different electron and hole concentrations. According to Eqs. (1.47) and (1.48), the diffused electron and hole densities are $-D_e(dn/dx)$ and $-D_h(dp/dx)$, respectively [see Fig. 1.36(a)]. The diffused electrons and holes leave donor and acceptor ions bound to the crystal lattice, in numbers equal to those of the diffused carriers. These donor and acceptor ions become the space charge and form an electric double layer at about the *pn*-boundary [see Fig. 1.36(b)]. This region is called the depletion layer or space-charge region (layer). As shown in Fig. 1.36(c), the space charge induces an electrical field limiting the electron and hole diffusion, and the built-in electrical field is the maximum at the *pn*-boundary and goes to zero at the edges of the depletion region. The potential of the *n*-side region is higher than that of the *p*-side region by an amount V_D [see Fig. 1.36(d)], which is called the built-in potential (or the diffusion potential). This is the result of the band-bending V_{Dn} in the *n*-type region and $-V_{Dp}$ in the *p*-type region. Here the built-in potential is given by

$$V_D = -\int F_{\text{D-field}} \, dx, \tag{1.121}$$

where $F_{\text{D-field}}$ is the built-in field. The Fermi level of the *n*-type region coincides with that of the *p*-type region [see Fig. 1.36(e)]. If the two Fermi levels do not coincide, the electrons or holes diffuse until the levels coincide. In the ther-

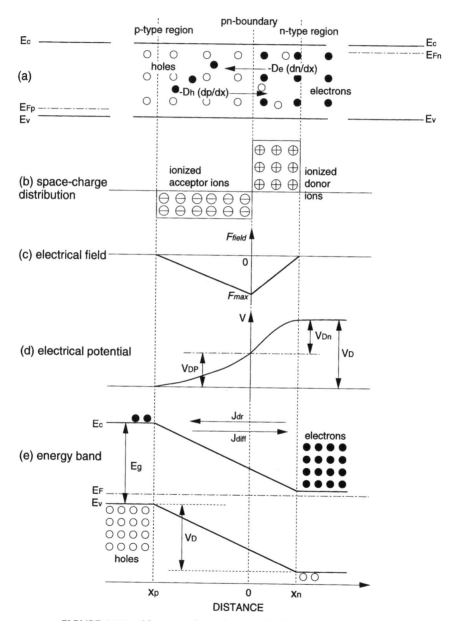

FIGURE 1.36 Abrupt *pn*-homojunction in thermal equilibrium.

mal equilibrium state, the diffusion current, J_{diff}, expressed by Eq. (1.52) and resulting from the differences of carrier concentrations, and the drift current, J_{dr}, expressed by Eq. (1.42) and induced by the electrical field, are balanced and no current flows outside of the diode.

1.2.2 Built-in Potential and Depletion Layer

The built-in potential described in the previous section is given by the difference between the energies of the two Fermi levels of the n- and p-side regions or between the conduction (valence) band edges in the n- and p-side regions. Therefore the following equation holds:

$$-qV_{\text{D}} = E_{\text{F}n} - E_{\text{F}p} = E_{cn} - E_{cp} = E_{vn} - E_{vp}, \qquad (1.122)$$

where $E_{\text{F}n}$ and $E_{\text{F}p}$ are the Fermi levels of the n- and p-side regions, E_{cn} and E_{cp} are the conduction band edges in the n- and p-side regions, and E_{vn} and E_{vp} are the valence band edges in the n- and p-side regions. The carrier density can be given by using Eqs. (1.21) and (1.22). For electrons in the n-side region,

$$n_n = N_c \exp[(E_{\text{F}n} - E_{cn})/k_{\text{B}}T]. \qquad (1.123)$$

For electrons in the p-side region,

$$n_p = N_c \exp[(E_{\text{F}p} - E_{cp})/k_{\text{B}}T]. \qquad (1.124)$$

At thermal equilibrium and room temperature, $E_{\text{F}n} = E_{\text{F}p} = E_{\text{F}}$, and thus from Eqs. (1.123) and (1.124),

$$n_p/n_n = \exp[(E_{cn} - E_{cp})/k_{\text{B}}T]. \qquad (1.125)$$

From Eqs. (1.122) and (1.125),

$$V_{\text{D}} = (k_{\text{B}}T/q)\ln(n_n/n_p); \qquad (1.126)$$

similarly,

$$V_{\text{D}} = (k_{\text{B}}T/q)\ln(p_p/p_n), \qquad (1.127)$$

where p_n and p_p are the hole densities in the n- and p-side regions. As pointed out in Section 1.1.2.4, the donors and acceptors are fully ionized at room temperature. Consequently, the electron density in the n-side region (majority carrier), n_n, is equal to the density of donors, N_{D}, and the electron density in the p-side region (minority carrier), n_p, is given by n_i^2/N_{A} from Eq. (1.34). Thus, Eq. (1.127) can be rewritten as

$$V_D = (k_B T/q) \ln(N_D N_A / n_i^2). \tag{1.128}$$

The same result is obtained when treating holes instead of electrons. The built-in potential is always smaller than the band-gap energy and increases as the donor and acceptor concentration increases.

Within the depletion layer, carriers are swept away by the built-in field and the space charge distributes as shown in Fig. 1.36(b). In the abrupt junction, the impurity type changes in a steplike fashion. The following equation therefore holds:

$$q N_D x_n = q N_A x_p = \epsilon_s \epsilon_0 F_{max}, \tag{1.129}$$

where ϵ_s is the static dielectric constant and F_{max} is the maximum electrical field [see Fig. 1.36(c)]. The space charge density is equal to $q N_D$ in the *n*-side region and $q N_A$ in the *p*-side region. In addition, if $N_D \gg N_A$, $x_n \ll x_p$, the depletion layer expands into the region with the lower doping concentration. Outside of the depletion layer the space charge density is zero because it is easily screened by the carriers, and the region is electrically neutral. The built-in voltage, V_D, is calculated as the area bounded by the solid line indicating the built-in electrical field in Fig. 1.36(c) [see Eq. (1.121)] as

$$|V_D| = F_{max}(x_n + x_p)/2. \tag{1.130}$$

Using Eqs. (1.129) and (1.130), the width of the depletion layer, W, is given as

$$\begin{aligned} W &= x_n + x_p \\ &= [(2\epsilon_s \epsilon_0/q)(1/N_D + 1/N_A)|V_D|]^{1/2}. \end{aligned} \tag{1.131}$$

When the voltage, V, is applied between the *n*- and *p*-side regions (positive to the *n*-side region and negative to the *p*-side region, called reverse bias), Eq. (1.131) can be rewritten as

$$W = [(2\epsilon_s \epsilon_0/q)(1/N_D + 1/N_A)|V_D + V|]^{1/2}. \tag{1.132}$$

The width of the depletion layer in an abrupt junction changes in proportion to the square root of the potential difference, $(V_D + V)^{1/2}$.

By using Eqs. (1.129), (1.130), and (1.132), the positive charge in the *n*-type region, Q, per unit area of the *pn*-junction is expressed as

$$Q = qN_Dx_n$$
$$= \epsilon_s\epsilon_0 F_{max}$$
$$= [(2\epsilon_s\epsilon_0 q)N_D N_A(V_D + V)/(N_D + N_A)]^{1/2}. \quad (1.133)$$

In the abrupt *pn*-junction, positive charge, $+Q$, and negative charge, $-Q$, are facing each other at the depletion layer [see Fig. 1.36(b)], and the charge changes with the bias voltage, V. Consequently, a parallel plate capacitor is formed at the depletion layer, and the capacitance per unit area (F/cm^2) is given by the first derivative of the charge with respect to V:

$$C = dQ/dV$$
$$= [\epsilon_s\epsilon_0 q N_D N_A/2(N_D + N_A)(V_D + V)]^{1/2}. \quad (1.134)$$

By substituting Eq. (1.132) into Eq. (1.134),

$$C = \epsilon_s\epsilon_0/W, \quad (1.135)$$

and this equation gives the capacitance of the parallel plate capacitor having a unit area and a distance, W. The capacitance expressed by Eq. (1.135) is called the junction capacitance, the depletion layer capacitance, or the barrier capacitance, and it changes with V. If $N_A \gg N_D$, Eqs. (1.132) and (1.134) are simplified as follows:

$$W = [2\epsilon_s\epsilon_0(V_D + V)/qN_D]^{1/2} \quad (1.136)$$

$$C = [\epsilon_s\epsilon_0 q N_D/2(V_D + V)]^{1/2}. \quad (1.137)$$

From Eq. (1.137),

$$1/C^2 = 2(V_D + V)/A^2 q\epsilon_s\epsilon_0 N_D, \quad (1.138)$$

where A is the area of the *pn*-junction. This relation can be drawn as shown in Fig. 1.37. The voltage at $1/C^2 = 0$ corresponds to the built-in potential, V_D, and the point is estimated by extrapolation of the straight line. The built-in potential can be calculated from an experimentally obtained relation like that shown in Fig. 1.37. The donor density can also be calculated from the slope of the line. A similar procedure can be used when $N_A \ll N_D$.

   The abrupt junction appears in the grown junction, whereas the diffused junction which is formed by the impurity diffusion is often a graded junction (see Fig. 1.35). The basic concepts for the analyses of graded junction are similar to those for analyses of the abrupt junction. The behaviors of the graded junction can be examined by solving a Poisson's equation for a space-charge distribution. The space charge (or impurity) distribution can be expressed as

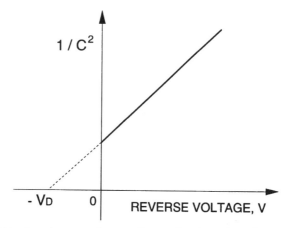

FIGURE 1.37   Junction capacitance, C, as a function of applied voltage, V_{appl}.

$\rho(x) = q a_{\text{grad}} x$ for a linearly-graded junction, where a_{grad} (cm^{-4}) is the impurity concentration gradient. The results obtained are as follows:

$$V_D = q a_{\text{grad}} W^3 / 12 \epsilon_s \epsilon_0, \tag{1.139}$$

$$W = [12 \epsilon_s \epsilon_0 (V_D + V) / q a_{\text{grad}}]^{1/3}, \tag{1.140}$$

and for a unit area of the *pn*-junction,

$$C = [\epsilon_s^2 \epsilon_0^2 q a_{\text{grad}} / 12 (V_D + V)]^{1/3} = \epsilon_s \epsilon_0 / W. \tag{1.141}$$

As indicated in Eqs. (1.139), (1.140), and (1.141), the built-in potential, the width of the depletion layer, and the junction capacitance are governed by the impurity concentration gradient. As the impurity concentration gradient decreases, the built-in potential and capacitance decrease and the depletion layer broadens. The width of the depletion layer is proportional to $(V_D + V)^{1/3}$ and the capacitance is inversely proportional to $(V_D + V)^{1/3}$.

1.2.3 Current-Voltage Characteristics

1.2.3.1 Forward Bias   The energy band diagram and carrier density distribution for a *pn*-homojunction at thermal equilibrium state and under the forward and reverse biases are shown in Fig. 1.38, where E_c, E_v, and E_F indicate the conduction band edge, valence band edge, and Fermi level. In the thermal equilibrium the state, the diffusion current and the drift current balance each other, and the total current is zero. The density of electrons having energies higher than the potential barrier, E_{cp}, in the *n*-side region, n_{n0}, is equal to the density of electrons in the *p*-side region, n_p. By using Eq. (1.126), this situation can be expressed with

$$n_{n0} = n_p = n_n \exp(-q V_D/k_B T), \qquad (1.142a)$$

where n_p is the minority carrier density in the *p*-side region and n_n is the majority carrier density in the *n*-side region. Similarly for holes,

$$p_{p0} = p_n = p_p \exp(-q V_D/k_B T), \qquad (1.142b)$$

where p_n is the minority carrier density in the *n*-side region and p_p is the majority carrier density in the *p*-side region. By positively biasing the *p*-side region and negatively biasing the *n*-side region, the built-in potential can be reduced and current starts to flow. This is the forward bias condition. The potential of the *n*-side region is decreased by the applied voltage, V, and the potential barrier decreases to $q(V_D - V)$ as shown in Fig. 1.38. This results in an increase in electron energy by qV, and the density of electrons having energies higher than the potential barrier, E_{cp}, increases to $n_p \exp(q V/k_B T)$ in the *n*-side region. This increase is obtained by replacing V_D with $(V_D - V)$ in Eq. (1.142). Consequently, the difference in the electron density between the *p*- and *n*-side regions, Δn, can be written

$$\Delta n = n_p \exp(q V/k_B T) - n_p = n_p[\exp(q V/k_B T) - 1], \qquad (1.143a)$$

and at room temperature $(k_B T/q \approx 26 \text{ mV})$ and at biases larger than 0.1 V,

$$\Delta n = n_p \exp(q V/k_B T). \qquad (1.143b)$$

The electron density gradients expressed by Eq. (1.143) result in electrons diffusing from the *n*-side region to the *p*-side region and thus in diffusion current flowing through the *pn*-junction. The holes correspondingly diffuse from the *p*-side region to the *n*-side region because of a difference in hole density, Δp, given by

$$\Delta p = p_n \exp(q V/k_B T) - p_n = p_n[\exp(q V/k_B T) - 1], \qquad (1.144a)$$

and at room temperature and biases larger than 0.1 V,

$$\Delta p = p_n \exp(q V/k_B T) \qquad (1.144b)$$

The density of minority carriers increases exponentially with increasing applied forward bias. Increasing the minority carrier density through the *pn*-junction to a level above that at thermal equilibrium is called injection. Injection is the principal process on which the function of optoelectronic *pn*-junction devices

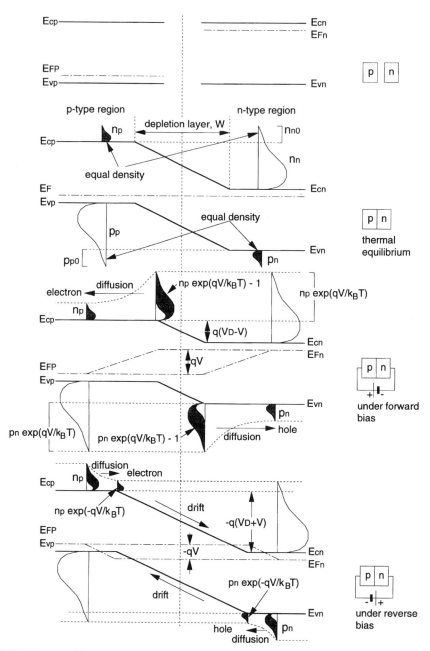

FIGURE 1.38 Energy band diagrams for and carrier density distributions of the *pn*-junction at thermal equilibrium and under forward and reverse biases.

is based. The diffused minority carriers recombine with the majority carriers and the density of the diffused minority carriers exponentially approaches the value it has in the thermal equilibrium state. The distance at which the carrier density decreases by $1/e$ from the density at the edge of the depletion layer corresponds to the carrier diffusion length. Using Eq. (1.55), the diffusion lengths of electrons, L_{ep}, and of holes, L_{hn}, are rewritten as follows:

$$L_{ep} = (D_{ep}\tau_{ep})^{1/2} \qquad (1.145a)$$

$$L_{hn} = (D_{hn}\tau_{hn})^{1/2}. \qquad (1.145b)$$

Here D_{ep} and D_{hn} are the diffusion constants of electrons in the p-type semiconductor and of holes in the n-type semiconductor, and τ_{ep} and τ_{hn} are the injected electron lifetime in the p-type semiconductor and the injected hole lifetime in the n-type semiconductor. Under the forward bias, the width of the depletion layer, W, is less than it is at thermal equilibrium [see Eq. (1.132)] and $W \ll L_{ep}$ and L_{hn}. Consequently, each diffusion length is nearly equal to the distance from the pn-boundary under the forward bias and the injected carrier density changes within the region corresponding to the diffusion length, $(L_{ep} + L_{hn})$. In the light-emitting devices, the radiative recombination occurs within the region of $(L_{ep} + L_{hn})$, in which the large number of excess minority carriers are injected and persist. This region is called the injection excitation region, and the diffusion current is called the injected current. As discussed in Section 1.1.2.6, the mobility of electrons is larger than that of holes, and thus the diffusion length of electrons is usually larger than that of holes. Consequently, the injection excitation region is wider in the p-side region than in the n-side region. The Fermi level is constant in the depletion layer because no carriers are generated or disappear there. In the same manner as Eqs. (1.72) and (1.73) and using Eq. (1.25), the product of electron density and hole density is given as

$$np = n_i^2 \exp[(E_{Fn} - E_{Fp})/k_B T] = n_i^2 \exp(qV/k_B T). \qquad (1.146)$$

If $E_{Fn} = E_{Fp}$, the relationship $np = n_i^2$ holds. As shown in Fig. 1.38, this relation holds in the region far from the pn-junction. Within the injection excitation region, however, $(E_{Fn} - E_{Fp}) = qV > 0$ and $np > n_i^2$. This results from the increase in the excess minority carriers due to injection under the forward bias.

The diffusion electron and hole currents are obtained by solving Eq. (1.47) at the p-side edge of the depletion layer and Eq. (1.48) at the n-side edge of the depletion layer (see Fig. 1.38). The change in the injected minority carrier density per unit time is determined by the rates of injection and recombination, and thus the change in electron density in the p-side region is given by

$$d\Delta n(x)/dt = D_{ep}(d^2\Delta n(x)/dx^2) - (\Delta n(x) - n_p)/\tau_{ep}. \qquad (1.147)$$

Under steady-state conditions, $d\Delta n(x)/dt = 0$, and $\Delta n(x = \infty)$ and n_p and $\Delta n(x = x_p)$ is given by Eq. (1.143b). Thus

$$\Delta n(x) = n_p[\exp(qV/k_BT) - 1]\exp[(x - x_p)/L_{ep}] + n_p. \qquad (1.148)$$

Similarly, for holes,

$$\Delta p(x) = p_n[\exp(qV/k_BT) - 1]\exp[-(x - x_n)/L_{hn}] + p_n. \qquad (1.149)$$

The injected electron current at the p-side edge of the depletion layer ($x = x_p$) and the injected hole current at the n-side edge of the depletion layer ($x = x_n$) are given by substituting Eq. (1.148) into Eq. (1.47) and (1.149) into Eq. (1.48). Then

$$\begin{aligned} J_{\text{e-diff}} &= qD_{ep}[d\Delta n(x)/dx]|_{x=x_p} \\ &= -(qD_{ep}n_p/L_{ep})[\exp(qV/k_BT) - 1] \end{aligned} \qquad (1.150a)$$

and

$$\begin{aligned} J_{\text{h-diff}} &= qD_{hn}[d\Delta p(x)/dx]|_{x=x_n} \\ &= -(qD_{hn}p_n/L_{hn})[\exp(qV/k_BT) - 1]. \end{aligned} \qquad (1.150b)$$

As indicated in Eq. (1.52), the total current density is given by the sum of the injected electron and hole currents:

$$\begin{aligned} J_{\text{diff}} &= (|J_{\text{e-diff}}| + |J_{\text{h-diff}}|) \\ &= J_{s0}[\exp(qV/k_BT) - 1]. \end{aligned} \qquad (1.151)$$

Here, J_{s0} is called the saturation current and is a constant because

$$J_{s0} = q(D_{ep}n_p/L_{ep} + D_{hn}p_n/L_{hn}). \qquad (1.152a)$$

The saturation current can be rewritten, using Eq. (1.25), as

$$J_{s0} = q(D_{ep}/L_{ep}p_p + D_{hn}/L_{hn}n_n)N_cN_v \exp(-E_g/k_BT). \qquad (1.152b)$$

This equation indicates that the saturation current is determined by the band-gap energy. The injected (diffusion) current is therefore proportional to $\exp[-(E_g - qV)/k_BT]$. A schematic illustration for each injected (diffusion) current density and total current density is sketched in Fig. 1.39.

Under the high-injection excitation, the density of the injected minority carriers is comparable to the density of the majority carriers and is proportional to $\exp(qV/2k_BT)$.[6] Consequently, Eq. (1.151) is slightly changed as follows:

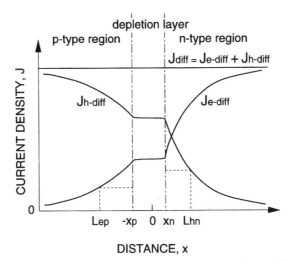

FIGURE 1.39 Change in current density under forward bias.

$$J_{\text{diff}} = J_{s0}[\exp(qV/2k_BT) - 1], \tag{1.153}$$

and

$$
\begin{aligned}
J_{s0} &= q(D_{ep}n_p/L_{ep} + D_{hn}p_n/L_{hn}) \\
&= qn_i(D_{ep}/L_{ep} + D_{hn}/L_{hn}). \tag{1.154}
\end{aligned}
$$

This high injection condition often appears in the operation of LEDs and laser diodes.

Most injected carriers contribute to the injected current. As discussed in Section 1.1.2.7, however, part of the injected carrier recombines at defects and at the surface within the depletion layer. For both recombinations, the current is proportional to $\exp(qV/2k_BT)$.[6,16,27] Given the description in Section 1.1.2.7, the recombination or generation current caused by deep levels in the depletion layer, J_{NR}, is expressed as

$$J_{NR} = (qW/2)n_i\sigma_r v_{th}N_t \exp(qV/2k_BT), \tag{1.155}$$

where W is the width of the depletion layer ($\sigma_r = \sigma_n = \sigma_h$ in Section 1.1.2.7). The term J_{NR} is called the generation-recombination current and is dominant at low bias. Given the description in Section 1.1.2.7, the surface recombination current, J_{SUR}, is expressed as

$$J_{SUR} = A_{\text{const}} \exp(qV/2k_BT), \tag{1.156}$$

where A_{const} is a constant. This surface recombination current governs the current-voltage characteristics in the low bias range for diodes composed of III-V compound semiconductors. The effect of this surface recombination current on the current-voltage characteristics of light-emitting diodes will be discussed in Chapters 2 and 3. The total forward current is, therefore, the sum of the currents expressed by Eqs. (1.151), (1.155), and (1.156):

$$I_F = A_{pn}(J_{diff} + J_{NR}) + A_{sur}J_{SUR}$$
$$\propto \exp(q V_F / a_f k_B T), \qquad\qquad (1.157)$$

where a_f is ideally between 1 and 2, A_{pn} is the area of the *pn*-junction, and A_{sur} is the area of the *pn*-junction surface (roughly determined by the *pn*-junction perimeter and the width of the injection excitation region). The Auger recombination is negligible here because it becomes dominant in the high-excitation range. Linear and semilogarithmic plots of ideal current-voltage characteristics are sketched in Fig. 1.40. The dominant current changes with the forward bias voltage. In the low-bias range, where the forward current is less than 1 mA in most cases, the generation-recombination current is dominant for Si (and Ge) devices, whereas the surface recombination current is usually dominant for III-V compound devices.[27,28] As the bias increases, the injected (diffusion) current appears and gradually surpasses the generation-recombination current and the surface recombination current. When the injected current is relatively large, there is a saturation effect in the forward current. This saturation is caused by the ohmic losses (voltage drops) in the neutral *n*- and *p*-type regions. In addition, Joule's heating occurs under such high-injection excitation (high current), and the current tends to be large even at the constant bias voltage as the temperature increases because of the temperature dependence of the band-gap energy [see Eq. (1.152b)]. If there is no saturation effect, the current expressed by Eq. (1.153) becomes dominant at high biases. The operating conditions of LEDs and laser diodes are in the higher excitation range, and the resistance of a *pn*-junction in such a range is given by the first derivative of Eq. (1.157) with respect to the forward current, I_F:

$$R_j = (k_B T / a_f q)(1/I_F). \qquad\qquad (1.158)$$

The junction resistance decreases with increases in the injected current and gradually approaches the value determined by the resistance of the outside of the *pn*-junction.

1.2.3.2 *Reverse Bias*   When the *p*-side region is negatively biased and the *n*-side region is positively biased, the potential of the *n*-side region is increased by the applied voltage, V, and the potential barrier increases to $q(V_D + V)$ as shown in Fig. 1.38. The density of electrons having energies higher than the potential barrier, E_{cp}, is given by replacing V_D with $(V_D + V)$ in Eq. (1.142) as

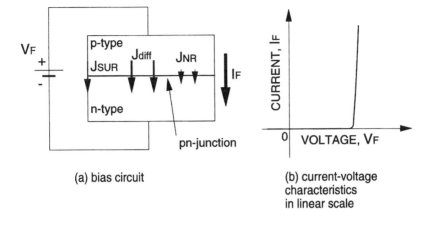

(a) bias circuit

(b) current-voltage
characteristics
in linear scale

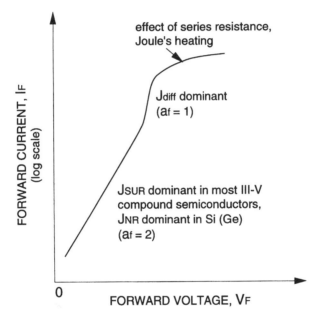

(c) current-voltage characteristics in semi-log scale

FIGURE 1.40   Ideal forward current-voltage characteristics.

$$\Delta n' = n_p \exp(-qV/k_\mathrm{B}T) < n_p. \qquad (1.159)$$

The density of electrons having energies higher than the potential barrier, E_{cp}, in the *n*-side region is lower than the density of electrons in the *p*-side region (minority carriers). The electrons in the vicinity of the *p*-side edge of the depletion layer drift to the *n*-side region because of the electric field generated in the layer, and their density becomes lower than that in the thermal equilibrium state. This leads to Eq. (1.159). Decreasing the minority carrier density below that at thermal equilibrium through the *pn*-junction is called extraction. The density difference results in the diffusion of electrons from the *p*-side region, and the diffused electrons drift further to the *n*-side region after reaching the *p*-side edge of the depletion layer. The difference in the electron density between the *p*- and *n*-side regions (more exactly, between the *p*-side region and the *p*-side edge of the depletion layer), Δn, can be written as

$$\Delta n = n_p - n_p \exp(-qV/k_BT) = n_p[1 - \exp(-qV/k_BT)]. \qquad (1.160)$$

This equation indicates that the extracted carrier density is a constant, n_p, when the reverse bias is high. The behavior of holes can be treated in a manner similar to that used in treating electron behavior above, and $np < n_i^2$ (extraction) is found to hold within the depletion layer and in the region within the diffusion lengths of electrons and holes because the relation $(E_{Fn} - E_{Fp}) < 0$ [see Eq. (1.146)]. A schematic illustration for each diffusion current density and total current density is sketched in Fig. 1.41.

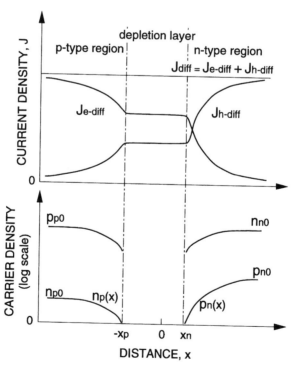

FIGURE 1.41 Changes in current density and carrier distribution under reverse bias.

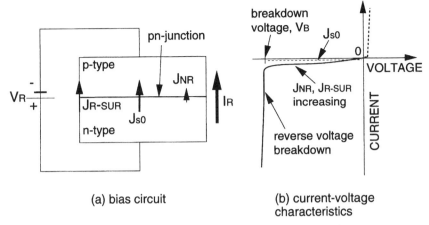

(a) bias circuit

(b) current-voltage
characteristics

FIGURE 1.42   Reverse current-voltage characteristics.

As reverse bias becomes large, $\exp(-qV/a_{\mathrm{f}}k_{\mathrm{B}}T) \ll 1$, and thus the right-hand side of Eq. (1.151) becomes nearly equal to J_{s0}. This saturation current (diffusion current) is nearly constant over the wide range of applied voltage. Consequently, Eq. (1.157) is modified for reverse current-voltage characteristics to

$$I_{\mathrm{R}} = A_{pn}(J_{s0} + J_{\mathrm{NR}}) + A_{\text{r-sur}}J_{\text{R-SUR}}, \qquad (1.161)$$

where I_{R} is the reverse current, J_{NR} and $J_{\text{R-SUR}}$ are the generation-recombination current and the surface current, and $A_{\text{r-sur}}$ is the surface area of the *pn*-junction. For the generation-recombination current, a recombination current is dominant under forward bias and a generation current is dominant under reverse bias. The surface current is caused by some complicated processes and is different from the surface recombination current given by Eq. (1.156). Reverse current-voltage characteristics are illustrated schematically in Fig. 1.42. Although the reverse current is ideally determined by the saturation current, it actually increases with the reverse bias because the generation-recombination current and the surface current increase.

At a high reverse bias, there is a point at which the reverse current abruptly increases. This is called the reverse voltage breakdown, and the voltage at that point is called the breakdown voltage (see Fig. 1.42). The range in which the reverse current increases gradually just before the breakdown is called the multiplication range. The breakdown is caused by avalanche breakdown or Zener breakdown (see Fig. 1.43). When a high electrical field within the depletion layer is generated under reverse bias, electrons diffusing from the outside of the depletion layer are accelerated within the depletion layer by the high field and they collide with lattice atoms. These collisions create new electrons and new holes by ionization. This impact ionization process is repeated within the deple-

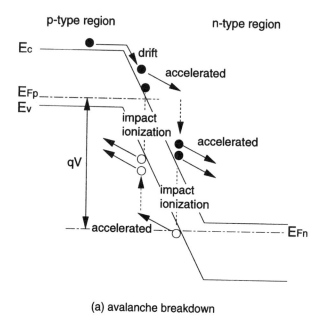

(a) avalanche breakdown

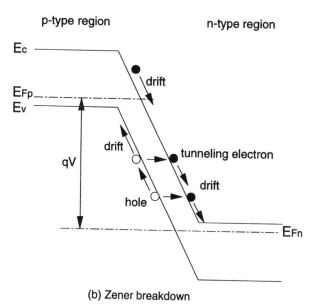

(b) Zener breakdown

FIGURE 1.43 Simplified illustrations of (a) avalanche breakdown and (b) Zener breakdown.

tion layer, and the number of electrons increases quickly (electron avalanche). The avalanche breakdown due to this electron avalanche can be considered to be the reverse of the Auger recombination discussed in Section 1.1.2.7. This avalanche breakdown will be discussed again when avalanche photodiodes are treated in detail in Chapter 4. If the *pn*-junction is formed with semiconductors having high purity concentration, the depletion layer generated is thin as indicated in Eq. (1.132). Under high reverse bias, the thin layer is subjected to a strong electrical field, and quantum-mechanical tunneling occurs (see Fig. 1.43). The tunneling electrons increase the reverse current quickly. This phenomenon is called the Zener effect, and the breakdown due to the Zener effect is called the Zener breakdown. The dominant breakdown in a *pn*-junction is determined by the breakdown voltages of the two mechanisms. In ordinary circumstances, the avalanche breakdown is dominant in *pn*-junctions with low impurity concentrations and the Zener breakdown is dominant in thin *pn*-junctions with high impurity concentrations.

1.2.4 Heterojunctions

1.2.4.1 pn-Heterojunctions A heterojunction is a junction formed by connecting two different materials, and there are two kinds of heterojunctions related to semiconductors: junctions consisting of two kinds of semiconductors, and junctions consisting of a semiconductor and a metal (contact). The contact of semiconductors with metals is discussed in Section 6.2.1. The semiconductor heterojunctions can also be classified into two types from the electrical point of view: *pn*-junctions and isojunctions. The *pn*-heterojunction consists of two different kinds of semiconductors having different conduction types, whereas the isojunction is formed by two different kinds of semiconductors having the same conduction type (such as *n-n* or *p-p* junctions). The properties of heterojunctions are basically similar to those of homojunctions, although some important differences appear.[4,5] One of the most widely used and important heterojunctions for semiconductor optoelectronic devices is formed by combining an *n*-type semiconductor having a wide band-gap and a *p*-type semiconductor having a narrow band-gap. The band diagram for this kind of heterojunction is shown in Fig. 1.44, where E_c, E_v, and E_F indicate the conduction band edge, valence band edge, and Fermi level, and E_g is the band-gap energy. The electron affinity and the work function are respectively denoted χ and ϕ. The electron affinity corresponds to the energy required to take an electron from the conduction band edge to the vacuum level. The work function is the energy required to take an electron from the Fermi level to the vacuum level, and it depends on the carrier concentration and the conduction type of the semiconductor. When different semiconductors are in contact with each other, the energy band is connected in the electron affinity relative to the vacuum level. This introduces discontinuities in the conduction band, ΔE_c, and in the valence band, ΔE_v, at the heterojunction. In Fig. 1.44, the band offsets are given by

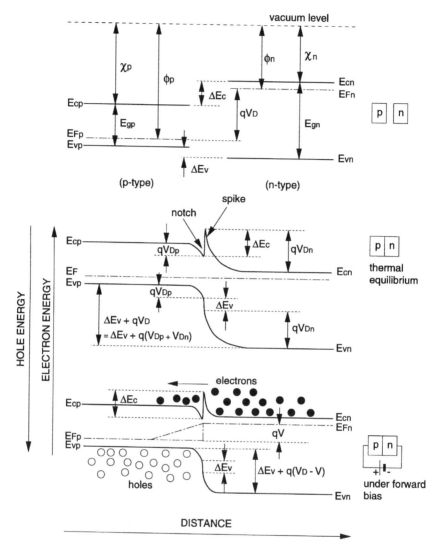

FIGURE 1.44   Energy band diagram for a *pn*-heterojunction.

$$\Delta E_{\mathrm{c}} = \chi_p - \chi_n \tag{1.162}$$

and

$$\Delta E_{\mathrm{v}} = (\chi_n + E_{gn}) - (\chi_p + E_{gp}) = \Delta E_{\mathrm{g}} - \Delta E_{\mathrm{c}}. \tag{1.163}$$

When *n*- and *p*-type semiconductors are connected, electrons and holes diffuse to the *p*- and *n*-side regions until the Fermi levels in both regions coincide just as they do in homojunctions. This carrier diffusion results in the formation of a depletion layer and in upward band-bending in the *n*-side region and downward band-bending in the *p*-side region diffusion. Spikes and notches thus appear at the heterointerface. As shown in Fig. 1.44, the built-in potentials are given as qV_{Dn} in the *n*-side region and qV_{Dp} in the *p*-side region. These built-in potentials are essentially similar to those of homojunctions, and the total built-in potential is given by the sum of the built-in potentials:

$$-qV_D = E_{Fn} - E_{Fp} = -q(V_{Dn} + V_{Dp}). \tag{1.164}$$

For an abrupt junction under a bias, V, the width of the depletion layers in the *n*- and *p*-side regions is obtained by solving Poisson's equation in a manner similar to that in the case of homojunctions. The results are given as follows:

$$x_n = [2\epsilon_{sn}\epsilon_{sp}\epsilon_0 N_A(V_D - V)/qN_D(\epsilon_{sn}N_D + \epsilon_{sp}N_A)]^{1/2} \tag{1.165a}$$

and

$$x_p = [2\epsilon_{sn}\epsilon_{sp}\epsilon_0 N_D(V_D - V)/qN_A(\epsilon_{sn}N_D + \epsilon_{sp}N_A)]^{1/2}. \tag{1.165b}$$

The total width of the depletion layer is thus

$$\begin{aligned} W &= x_n + x_p \\ &= [2\epsilon_{sn}\epsilon_{sp}\epsilon_0(N_D + N_A)^2(V_D - V)/qN_DN_A(\epsilon_{sn}N_D + \epsilon_{sp}N_A)]^{1/2}. \end{aligned} \tag{1.166}$$

The potentials in the *n*- and *p*-side regions are related as described by the following equation:[29]

$$N_D\epsilon_{sn}(V_{Dn} - V_n) = N_A\epsilon_{sp}(V_{Dp} - V_p), \tag{1.167}$$

where V_n and V_p are the biases applied to the *n*- and *p*-side regions, and $V = V_n + V_p$. If the static refractive indexes of the two regions are nearly equal, the applied voltage is mainly biased to the region having a low impurity concentration. The junction capacitance can be estimated in a manner similar to that used in the case of homojunctions and is expressed (per unit area) by the following equation:

$$C = [qN_DN_A\epsilon_{sn}\epsilon_{sp}\epsilon_0/2(\epsilon_{sn}N_D + \epsilon_{sp}N_A)(V_D - V)]^{1/2}. \tag{1.168}$$

1.2.4.2   pn-Heterojunctions Under Bias   Under a forward bias, the built-in potential is reduced by the bias voltage as shown in Fig. 1.44. This built-in potential reduction leads to the diffusion of electrons from the *n*-side region to the *p*-side region, while the diffusion of most holes is prevented by the large potential barrier, $\Delta E_v + q(V_D - V)$. These electron and hole diffusion currents (or injected currents) can be expressed by the same formulas that apply to the homojunction [see Eqs. (1.150a) and (1.150b)]. Therefore, the ratio between them is given by

$$
\begin{aligned}
|J_{\text{e-diff}}/J_{\text{h-diff}}| &= (D_{ep}n_p/L_{ep})/(D_{hn}p_n/L_{hn}) \\
&= (D_{ep}L_{hn}n_n n_{ip}^2/(D_{hn}L_{ep}p_p n_{in}^2) \\
&= (D_{ep}L_{hn}n_n/D_{hn}L_{ep}p_p)(m_{dep}m_{dhp}/m_{den}m_{dhn})^{3/2} \\
&\quad \cdot \exp[(E_{gn} - E_{gp})/k_B T]
\end{aligned}
\tag{1.169}
$$

where n_{in} and n_{ip} are the intrinsic carrier densities in the *n*-type and *p*-type semiconductor. Here the following relations deduced from Eqs. (1.23), (1.24), and (1.25) are used for *n*-type and *p*-type semiconductors:

$$
\begin{aligned}
n_n p_n = n_{in}^2 &= N_{cn}N_{vn}\exp(-E_{gn}/k_B T) \\
&= 4(2\pi k_B T/h^2)^3 m_{den}^{3/2} m_{dhn}^{3/2}\exp(-E_{gn}/k_B T)
\end{aligned}
\tag{1.170a}
$$

and

$$
\begin{aligned}
n_p p_p = n_{ip}^2 &= N_{cp}N_{vp}\exp(-E_{gp}/k_B T) \\
&= 4(2\pi k_B T/h^2)^3 m_{dep}^{3/2} m_{dhp}^{3/2}\exp(-E_{gp}/k_B T),
\end{aligned}
\tag{1.170b}
$$

where N_{cn} and N_{vn} are the effective densities of states in the *n*-type semiconductor and N_{cp} and N_{vp} are the effective densities of states in the *p*-type semiconductor. The terms m_{den} and m_{dhn} are the density of states effective mass for electrons and holes in the *n*-type semiconductor and m_{dep} and m_{dhp} are the density of states effective mass for electrons and holes in the *p*-type semiconductor. As shown in Fig. 1.44, $(E_{gn} - E_{gp}) > 0$ in Eq. (1.169) and thus the exponential term dominates and $|J_{\text{e-diff}}| \gg |J_{\text{h-diff}}|$ at temperatures near room temperature. This indicates that the electron diffusion current is predominant in a *pn*-heterojunction composed of a wide band-gap *n*-type semiconductor and a narrow band-gap *p*-type semiconductor. The characteristics mentioned above also hold for the combination of narrow band-gap *n*-type and wide band-gap *p*-type semiconductors, in which holes diffuse from the *p*-side region to the *n*-side region and electron diffusion is prevented by the potential barrier. In general, at the *pn*-heterojunction only the majority carrier in the wide band-gap semiconductor is injected into the narrow band-gap semiconductor.

As in the *pn*-homojunction, the forward current in a *pn*-heterojunction is composed mainly of the diffusion current, the surface recombination current, and the generation-recombination current. Consequently, Eq. (1.157) can be rewritten for the *pn*-heterojunction as

$$I_F = A_{pn}(J_{\text{diff w} \to \text{n}} + J_{NR}) + A_{sur}J_{SUR}$$
$$\propto \exp(q V_F/a_f k_B T), \tag{1.171}$$

where the injected current is different from that in a homojunction and $J_{\text{diff w} \to \text{n}}$ indicates the injected (diffusion) current from the wide band-gap semiconductor to the narrow band-gap semiconductor. In the case shown in Fig. 1.44, $J_{\text{diff w} \to \text{n}} = J_{e\text{-diff}}$, and the diffusion current for the *pn*-heterojunction is expressed as

$$J_{\text{diff w} \to \text{n}} = J_{e\text{-diff}}$$
$$= J_{sh0}[\exp(q V/k_B T) - 1], \tag{1.172a}$$

where J_{sh0} is the saturation current for the *pn*-heterojunction and is given by

$$J_{sh0} = q D_{ep} n_p / L_{ep}$$
$$= q D_{ep}/L_{ep} p_p N_{cp} N_{vp} \exp(-E_{gp}/k_B T). \tag{1.72b}$$

Therefore, electron diffusion current increases quickly after the applied forward voltage becomes larger than the narrow band-gap energy, E_{gp}/q. Under high-injection excitation, in which the density of the injected electrons (minority carrier) is comparable to the density of holes (majority carrier), Eqs. (1.172a) and (1.172b) can be rewritten in a manner similar to that used in the homojunction case as

$$J_{\text{diff w} \to \text{n}} = J_{sh0}[\exp(q V/2k_B T) - 1] \tag{1.173a}$$

and

$$J_{sh0} = q D_{ep} n_{ip} / L_{ep}. \tag{1.173b}$$

The relation in Eq. (1.173a), however, becomes imprecise because of the influence from the resistance of the semiconductors and Joule's heating. When narrow band-gap *n*-type and wide band-gap *p*-type semiconductors are combined, a similar formula can be obtained for hole diffusion.

In addition to the components of the current in Eq. (1.171), there is a current via interfacial states at heterojunctions that deteriorates the current-voltage characteristics when the bias is relatively low. One example of the current via interfacial states is sketched in Fig. 1.45. These interfacial states originate primarily from the crystal defects—such as stacking faults and the dangling bonds

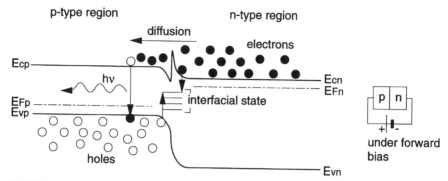

FIGURE 1.45 A recombination process via interfacial states at a heterojunction under forward bias.

of lattice atoms—which are mainly caused by the lattice mismatch occurring during crystal growth because the heterojunction is formed by semiconductor materials having different lattice constants. The additional current can be taken into account, roughly, by being included in the generation-recombination current or the surface recombination current. Such additional currents are in any case proportional to the bias voltage by the factor $\exp(qV/2k_{B}T)$.

The current-voltage characteristics under reverse bias can be approximated in a manner similar to that used to describe the characteristics of a *pn*-homojunction, and avalanche breakdown and Zener breakdown also occur when *pn*-junctions are under a high reverse bias. The analyses of actual characteristics, however, need to take into account the spike and the notch at the heterointerface and are more complicated. The actual cases for the reverse current-voltage characteristics will be discussed in detail in Chapter 4.

1.2.4.3 Isotype Heterojunction In the same manner as for the *pn*-heterojunctions, when two different *n*-type semiconductors or two different *p*-type semiconductors are connected, electrons or holes diffuse from the semiconductor with the wider band-gap to the other semiconductor until the Fermi levels in both regions coincide. For *p-p* isotype heterojunctions, therefore, the downward band-bending in the wider band-gap side and the upward band-bending in the other side are generated by the hole diffusion, causing spikes and notches to appear at the heterointerface as shown in Fig. 1.46. The depletion layer is formed in the wider band-gap side, and holes accumulate at the notch in the narrower band-gap side. The notch is usually less than 20 nm wide and forms a triangular potential well; the accumulating holes form a two-dimensional gas. The energy band diagrams for the isotype heterojunctions can be also obtained by taking into consideration the electron affinity (see Fig. 1.46). The band offset is expressed with the next equations as

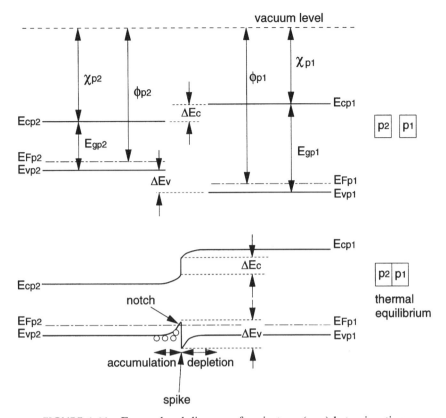

FIGURE 1.46   Energy band diagram of an isotype (p-p) heterojunction.

$$\Delta E_{\rm c} = \chi_{p2} - \chi_{p1}, \tag{1.174}$$

$$\Delta E_{\rm v} = (\chi_{p1} + E_{{\rm g}p1}) - (\chi_{p2} + E_{{\rm g}p2}) = \Delta E_{\rm g} - \Delta E_{\rm c}, \tag{1.175}$$

and

$$-qV_{{\rm D}pp} = E_{{\rm F}p1} - E_{{\rm F}p2} = \phi_{p2} - \phi_{p1}, \tag{1.176}$$

where $V_{{\rm D}pp}$ is the built-in potential at the p-p isotype heterojunction.

At n-n isotype heterojunctions, the upward band-bending in the wider band-gap side and the downward band-bending in the other side are generated by the electron diffusion. The band diagram is similarly sketched by using electron affinity. The accumulated carriers at the notch are electrons and they form a two-dimensional quantized electron gas in the well. This well structure is very important for high-speed electron devices.

1.2.5 Double Heterostructures

A structure with a heterojunction is called a single heterostructure, and a structure with two heterojunctions (which normally consist of a *pn*-junction and an isotype junction) is called a double heterostructure. This double heterostructure is widely applied in optoelectronic *pn*-junction devices, from devices used in optical fiber communication systems to devices used in consumer electronics products such as compact discs, printers, and autofocusing cameras. Most of the double heterostructures used in optoelectronic devices have consisted of a wide band-gap semiconductor/a narrow band-gap semiconductor/a wide band-gap semiconductor with respective thicknesses of more than 500 nm/100 nm to 1500 nm/more than 500 nm. The thickness of the narrow band-gap layer depends on the device, for example, 100–200 nm for laser diodes and 500–1500 nm for LEDs.

The band diagram of the double heterostructure made up of *p*-type/*p*-type/*n*-type materials is shown in Fig. 1.47, where the narrow band-gap semiconductor is shown to be sandwiched between the wide band-gap semiconductors. The phenomena generated at each heterojunction have already been described in Sections 1.2.1 to 1.2.4. Just after the semiconductors are connected, their Fermi levels are unified by electron or hole diffusion, and then the built-in potentials appear at the two interfaces. The built-in potential, V_D, at the *pn*-junction and the built-in potential, V_{Dpp}, at the *p-p* isotype junction are given by Eqs. (1.164) and (1.176). Under forward bias, V, those built-in potentials are reduced by V_{pn} and V_{pp}, respectively, where $V = V_{pn} + V_{pp}$. The V_{pp} is usually small, and thus $V \approx V_{pn}$. In this situation electrons are injected through the *pn*-heterojunction by the forward bias, $V \approx V_{pn}$, and most of them are confined within the *p*-type narrow band-gap region because of the heterobarrier in the conduction band, $\Delta E_c + q(V_{Dpp} - V_{pp})$, at the isotype heterojunction. If the injected electron density is greater than the hole concentration in the *p*-type narrow band-gap region, holes are injected from the *p*-type wide band-gap region because of the requirement for charge neutrality. This simultaneous injection of electrons and holes is called a double injection. Similar results can be obtained for double heterostructures formed by *p*-type wide band-gap/*n*-type narrow band-gap/*n*-type wide band-gap layers, where hole injection begins under the forward bias. For both the double heterostructures, high concentrations of electrons and holes can be achieved within the narrow band-gap region as a result of the double injection. This is a big advantage for light-emitting devices that have to have radiative recombination take place at a high rate.

Under forward bias conditions the current-voltage characteristics are essentially governed by the behaviors of the *pn*-heterojunction and are obtained by modifying Eq. (1.171) to take into account the additional current via interfacial states. For application to actual devices, Eq. (1.171) must also be slightly modified to account for the series resistance, such as the contact resistance at the electrode and the resistance of the material itself. If this resistance is expressed with R_s, the voltage drop it causes is included in Eq. (1.171) as follows:

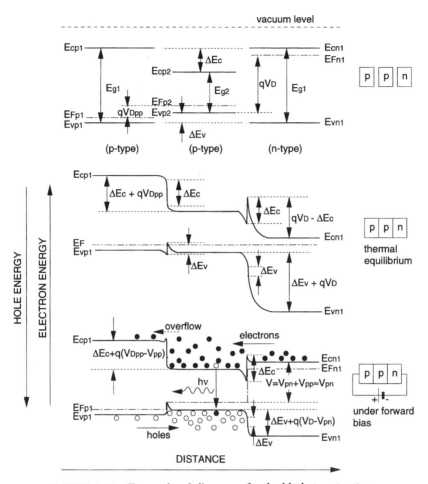

FIGURE 1.47 Energy band diagram of a double heterostructure.

$$I_F \propto \exp[q(V_F - R_s I_F)/a_f k_B T]. \qquad (1.177)$$

The current in the operating range of a light-emitting device is dominated by the injected (diffusion) current and is experimentally expressed by using Eq. (1.172) as

$$I_F \approx J_{\text{diff} w \to n} = J_{sh0}\{\exp[q(V_F - R_s I_F)/a_f k_B T] - 1\}, \qquad (1.178)$$

where a_f for diffusion current is ideally 1 but is actually between 1 and 2. In a high injection range, the value of a_f is 2, although the relation specified in Eq. (1.178) is also imprecise because of the electrical resistance and Joule's heating.

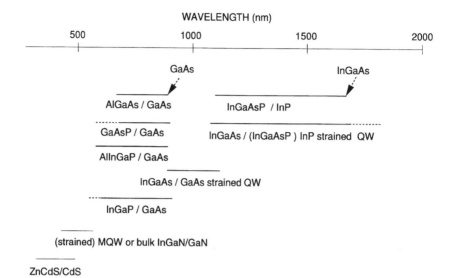

FIGURE 1.48 Typical double heterostructures and the wavelength corresponding to the band-gap energy.

The reverse current-voltage characteristics of double heterostructures are also determined by the *pn*-junctions. They are, however, too complicated. The actual characteristics will be discussed with regard to the characteristics of photodiodes (see Section 4.1.2).

The combinations of semiconductors used to form heterojunctions and the double heterostructures cannot be selected randomly. Their lattice constants under crystal growth conditions need to coincide with each other (lattice matching). If the lattice constants do not coincide, a semiconductor layer of one kind cannot be grown on another kind of semiconductor. Even if the crystal growth can be carried out, a large number of interfacial states is introduced by misfit dislocations that are generated at the heterojunction because dangling bonds of the lattice atoms are left along the dislocations. Consequently, combinations of semiconductor materials for forming the heterojunctions have been carefully selected. As shown in Fig. 1.7, several combinations have been established. Some of them are summarized in Fig. 1.48.

1.3 LOW-DIMENSIONAL STRUCTURES—QUANTUM WELLS, STRAINED QUANTUM WELLS, AND OTHER QUANTUM STRUCTURES

1.3.1 Quantum Wells

The quantum well structures used in optoelectronic *pn*-junction devices are, except for their thicknesses, basically similar to the double heterostructures in that they consist of a wide band-gap layer/a narrow band-gap layer/a wide band-gap layer. The characteristics of optoelectronic *pn*-junction devices gradually deteriorate with decreasing thickness of the sandwiched narrow band-gap layer. For example, the threshold current of laser diodes increases because of the reduction of the confinement for injected carrier and optical field, and the photocurrent in photodiodes decreases because of the decreased thickness of the absorption layer. When the thickness of the narrow band-gap layer becomes comparable to the de Broglie wavelength of an electron or hole, however, some new (quantum size) effects become apparent. The de Broglie wavelength is given by dividing Planck's constant by the momentum of the electron or hole and is a few tens of nanometers. In the thin layer, carrier motion perpendicular to the layer is restricted, and the kinetic energy is quantized into discrete energy levels. Structures consisting of such extremely thin layers having quantized energy levels are called quantum well (QW) structures, the thin layer is called a potential well (or simply a well), and one of the neighboring layers with wide band-gap energies is called a potential barrier (or simply a barrier). The well thickness is usually less than 10 nm. A structure with one well is called a single-quantum well (SQW) structure and one with several wells is called a multiple-quantum (multiquantum) well (MQW) structure. A simplified MQW structure is shown schematically in Fig. 1.49, where the effects of band-bending in the heterojunction are ignored. In the figure, E_e, E_{hh}, and E_{lh} indicate the quantized energy levels for electrons, heavy holes, and light holes, and ΔE_c and ΔE_v are the band offsets in the conduction and valence bands. The properties of the well are determined by the two kinds of band offsets, ΔE_c and ΔE_v, the thickness of well itself, L_w, and the thickness of the barrier, L_b. When the barrier layer is thinned to about 5 nm in the ordinary MQW structure, the electrons in each well combine with each other through the barrier layer by quantum-mechanical tunneling, and then the quantized (subband) energy levels also combine and create minibands. This structure having the minibands is

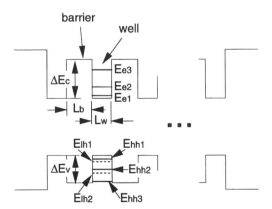

FIGURE 1.49   Simplified illustration of a multiquantum well structure.

called a (compositional) superlattice structure and behaves like a new material having an effective band-gap nearly equal to the energy separation between the electron and hole minibands.

1.3.1.1 Density of States in Quantum Wells

Electrons and holes in a bulk layer can move in any direction because the thickness of the layer is much larger than the de Broglie wavelength. In a quantum well, in contrast, their motions perpendicular to the quantum well are restricted. This results in lower-dimensional (two-dimensional) carrier confinement. The injected electrons and holes are quantized in the direction perpendicular to the well, but those carriers are not restricted in the direction parallel to the well layer and the continuum states are formed. The density of states in a quantum well is modified from the parabolic density of the states of the normal bulk semiconductors and can be calculated on a two-dimensional plane. The two-dimensional density of the states for electrons (per unit energy and unit area) near the conduction band edge is given by

$$\rho_c(E) = \sum_{\theta=1}^{\infty} m_{de}/(h^2/\pi)H[E - E_\theta], \qquad (1.179)$$

where $H[x] = 1(x \geq 0)$ and $H[x] = 0(x < 0)$, and E_θ is the quantized energy level. This equation indicates that the density of states changes in a steplike fashion with a step hight of $m_{de}/(h^2/\pi)$. This is shown in Fig. 1.50, where the density of states in the bulk semiconductors is a parabolic function of energy. The density of states in a quantum well structure is constant even as the energy increases because of this steplike shape. When the barrier layer is sufficiently high and wide, the quantized energy levels measured from the conduction band edge (minimum) can be expressed by the following equation:

$$E_\theta = [(h/\pi)^2]/2m_{de})(\theta\pi/L_w)^2 \qquad \theta = 1, 2, 3, \cdots. \qquad (1.180)$$

The energy is inversely proportional to the square of the well thickness. The quantized energy level heightens from the bottom of the well (conduction band minimum) as the well thickness decreases. For holes, a similar treatment can be performed for the heavy- and light-hole by using the density of state effective mass of holes. Actual data of the optical absorption spectra for MQW and bulk structures are shown in Fig. 1.51.[30] As the wavelength increases, the absorption intensity decreases abruptly at the wavelengths corresponding to the electron transitions between the quantized energy levels. The steplike density of states is clearly shown in the MQW structure.

In the strict sense, the expression of quantized energy levels shown in Fig. 1.50 holds only in the direction perpendicular to the well. As shown in Fig. 1.52, these levels correspond to the bottoms of the subbands because the carriers are free to move parallel to the well. In the figure, the energy levels in the parallel

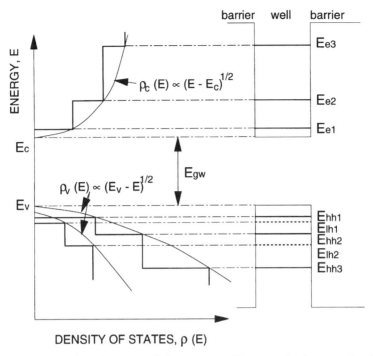

FIGURE 1.50 Density of states and the corresponding quantized energy levels in a quantum well.

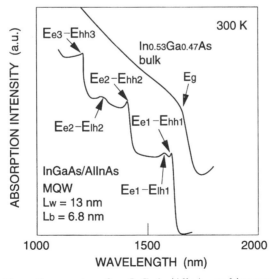

FIGURE 1.51 Absorption spectra of an InGaAs/AllnAs multiquantum well structure and an InGaAs bulk epitaxial layer. From H. Asai.[30] Reprinted by permission of the author.

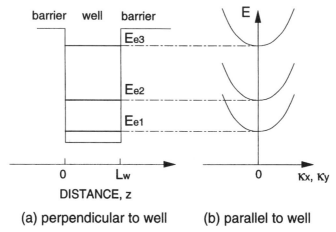

(a) perpendicular to well   (b) parallel to well

FIGURE 1.52 Simplified illustration of the electron energy subbands in a quantum well (a) perpendicular to the well and (b) parallel to the well.

direction are expressed in the **k**-space. Therefore, the energy of the carriers in the well is given by

$$E_{well} = E_\theta + (h/\pi)^2/m^*(k_x^2 + k_y^2), \tag{1.181}$$

where subscripts x and y are two perpendicular directions parallel to the well.

1.3.1.2 Optical Properties of Quantum Wells   The electron transitions in a quantum well differ from those in bulk semiconductors because of the quantized subbands formed in the well (see Figs. 1.50 and 1.52). The electrons in the bulk semiconductors transit between the conduction and valence bands, whereas the transitions that occur in the quantum well during light emission or absorption are between the quantized subbands. Consequently, the wavelength of the emitted light and the optical absorption edge in the quantum wells is shorter (or higher energy) than the corresponding wavelength for bulk layers having the same composition as the wells (see Fig. 1.51). Under excitation, the transition mainly occurs between the levels marked E_{e1} and E_{hh1} in Fig. 1.50 because most of the electrons taking part in the transition gather locally in the lower energy level and the density of states of the heavy-hole band is larger than that of the light-hole band. Consequently, from Fig. 1.50, the photon energy emitted in the transition can be roughly expressed as

$$h\nu \approx E_{gw} + E_{e1} + E_{hh1}, \tag{1.182}$$

where E_{gw} is the band-gap energy of the well material.

The optical absorption process is rather complicated. As shown in Fig. 1.51, the transitions between several subbands are observed. In addition, there is a strong influence of the exciton absorption in the quantum well. The exciton absorption in the bulk layers of very pure semiconductors can be observed at low temperature, as described in Section 1.1.3.1. In quantum wells, on the other hand, exciton absorption occurs at room temperature or higher temperatures.[31] This is because the binding energy of excitons and the exciton absorption coefficient are larger in the quantum well than in the bulk layer. The exciton absorption is inversely proportional to the well thickness and dominates the absorption spectra near the band-gap energy. This is quite important for optical modulators and optical switches because the peak energy of the exciton absorption shifts to a lower energy (or longer wavelength) under an electric field (quantum confined Stark effect; QCSE) (see Section 5.1.4).

1.3.2 Strained Quantum Wells

As noted in the discussion of heterojunctions, the lattice constants for the two different semiconductors used to form heterojunctions or heterostructures need to coincide. If there is a lattice mismatch between the two semiconductors, a large number of misfit dislocations is introduced. These dislocations severely degrade the device characteristics because they are responsible for the generation of interfacial states. If the lattice mismatch is not more than a few percent and the well is not more than a certain thickness, a quantum well without any misfit dislocations can be formed by a combination of materials that have different lattice constants. This type of quantum well is called the strained quantum well and that certain thickness at which misfit dislocations are not generated is called the critical layer thickness. In other words, the strain in the strained quantum well structures, which is introduced by the lattice mismatch between the well and the barrier, is elastically relaxed by the deformation of the lattice of the well material if the thickness is less than the critical layer thickness. The InGaAs/GaAs and Si/Ge structures are typical strained quantum well structures.[32,33] A result of calculation (by Matthew's law) for an InGaAs/GaAs structure is shown in Fig. 1.53 as a function of the mole fraction of In.[34] In the figure the misfit dislocation is generated in the region above a thick solid line. There are two kinds of strained quantum wells: those under compressive strain and those under tensile strain (see Fig. 1.54). The deformation of the lattice in the well layer induces a variation of band-gap energy as shown in Fig. 1.53, where the variation of the band gap at Γ valley is indicated. This is the special feature of the strained quantum well structures. The band-gap energy is a function of the hydrostatic and shear components of the strain. The conduction and valence band edges are shifted by the hydrostatic component and light- and heavy-hole bands are changed by the shear component. The emission wavelength and absorption edge can be adjusted by controlling the strain. Recently developed 980 nm-band InGaAs/GaAs strained quantum well lasers used for pumping erbium-doped fiber amplifiers are typical examples of devices with a lasing wavelength that cannot be obtained from lattice-matching quan-

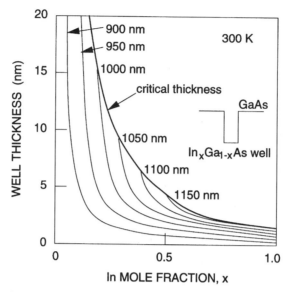

FIGURE 1.53 Correlation between lasing wavelength ($E_{e1} - E_{hh1}$ transition) and well thickness/mole fraction in an $In_xGa_{1-x}As$/GaAs strained quantum well structure. From M. Okayasu.[34] ©1991 John Wiley & Sons, Inc. Reprinted by the authors and permission of John Wiley & Sons, Inc.

tum well structures. In addition, the carrier effective mass can be reduced in strained quantum well structures, and this results in a decrease in the density of states as indicated with Eq. (1.179)—especially in the valence band. This is a very effective way to improve the device characteristics; for example, to reduce the lasing threshold current of laser diodes (see Section 3.3).

1.3.3 Other Quantum Structures

In addition to the quantum well structure and strained quantum well structure, there are lower-dimensional structures: quantum wires and quantum boxes (or dots) as shown in Fig. 1.55. Dimension, here, refers to carrier motion; the quantum wire is a one-dimensional structure and the quantum box is a zero-dimensional structure. The quantum wire has a square cross section with sides whose length is roughly the de Broglie wavelength of an electron or hole or has an arbitrary shape with cross-sectional area comparable to the square cross section. Carriers in the wire are free to move only in the direction along the wire. The density of the states for this structure is less than that for quantum wells. The quantum box is a cube having sides whose length is roughly the de Broglie wavelength of an electron or hole or has an arbitrary shape having nearly the same volume as the box. In the box, carrier motions are perfectly restricted, and the density of the states is further reduced. The lower density of the states in

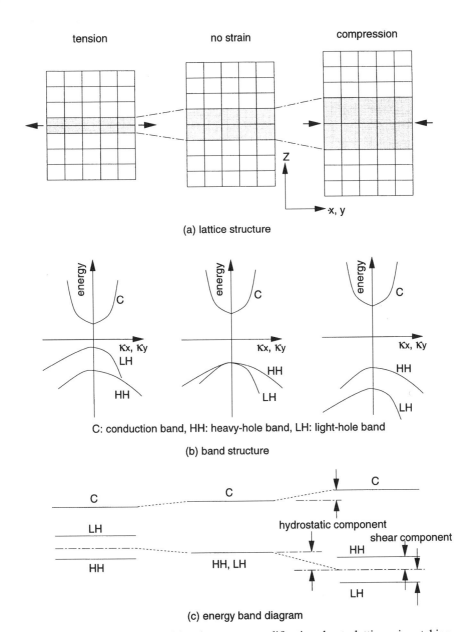

FIGURE 1.54 Strain-induced band structure modification due to lattice mismatching.

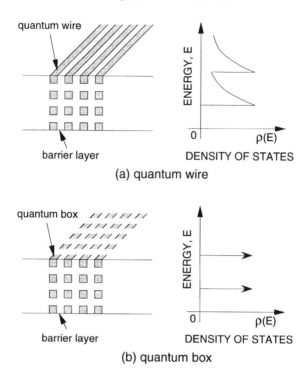

FIGURE 1.55   Schematic illustration of (a) quantum wire and (b) box, and their density of states.

the lower-dimensional structure has been calculated theoretically and various improvements in device characteristics are expected for one- or zero-dimensional structures.[35,36]

REFERENCES

1. J. C. Phillips, *Bonds and Bands in Semiconductors*, Academic Press, New York, 1973.

2. S. Wang, *Fundamentals of Semiconductor Theory and Device Physics*, Prentice Hall, Englewood Cliffs, New Jersey, 1989.

3. Many books related to the compound semiconductors and their devices refer to Vegard's law, and originate from L. Vegard, *Phys.*, **5,** 17, 1921.

4. H. Kressel and J. K. Butler, *Semiconductor Lasers and Heterojunction LEDs*, Academic Press, New York, 1977.

5. H. C. Casey, Jr. and M. B. Panish, *Heterostructure Lasers*, Academic Press, New York, 1978.

6. S. M. Sze, *Physics of Semiconductor Devices*, 2nd ed., John Wiley & Sons, New York, 1981.

7. E. O. Kane, Thomas–Fermi approach to impure semiconductor band structure, *Phys. Rev.*, **131,** 79, 1963.

8. B. I. Halperin and M. Lax, Impurity-band tails in the high-density limit. I. Minimum counting method, *Phys. Rev.*, **148,** 722, 1966.

9. Y. Suematsu and A. R. Adams, eds., *Handbook of Semiconductor Lasers and Photonic Integrated Circuits*, Ohmsha, Ltd., Tokyo and Chapman & Hall, London, 1994.

10. J. R. Hayes, A. R. Adams, and P. D. Greene, Low-field carrier mobility, in T. P. Pearsall, ed., *GaInAsP Alloy Semiconductors*, John Wiley & Sons, New York, 1982, Ch. 8.

11. R. F. Leheny, A. A. Ballman, J. C. DeWinter, R. E. Nahory, and M. A. Pollack, Compositional dependence of the electron mobility in $In_{1-x}Ga_xAs_yP_{1-y}$, *J. Electron. Mater.*, **9,** 561, 1980.

12. R. N. Hall, Electron-hole recombination in germanium, *Phys. Rev.*, **87,** 387, 1952.

13. W. Schockley and W. T. Read, Statistics of the recombination of holes and electrons, *Phys. Rev.*, **87,** 835, 1952.

14. J. I. Pankove, *Optical Processes in Semiconductors*, Prentice-Hall, Englewood Cliffs, New Jersey, 1971.

15. P. Bhattacharya, *Semiconductor Optoelectronic Devices*, Prentice Hall, Englewood Cliffs, New Jersey, 1994.

16. A. S. Grove, *Physics and Technology of Semiconductor Devices*, John Wiley & Sons, New York, 1967.

17. H. C. Casey Jr. and E. Buehler, Evidence for low surface recombination velocity on *n*-type InP, *Appl. Phys. Lett.*, **30,** 247, 1977.

18. L. R. Weisberg, Auger recombination in GaAs, *J. Appl. Phys.*, **39,** 6096, 1968.

19. G. P. Agrawal and N. K. Dutta, *Long Wavelength Semiconductor Lasers*, Van Nostrand Reinhold, New York, 1986.

20. R. A. Smith, *Semiconductors*, Cambridge University Press, 1959.

21. H. Nagai, S. Adachi, and T. Fukui, *III-V Mixed Crystals*, Corona Publishing Co. Ltd., Tokyo, 1988 (Japanese).

22. G. Lasher and F. Stern, Spantaneous and stimulated recombination radiation in semiconductors, *Phys. Rev.* **133,** A553, 1964.

23. M. G. A. Bernard and G. Duraffourg, Laser conditions in semiconductor, *Phys. Stat. Solidi*, **1,** 699, 1961.

24. Y. P. Varshni, Band-to-band radiative recombination in groups IV, VI, and III-V semiconductors (1), *Phys. Stat. Sol.*, **19,** 459, 1967.

25. F. Stern, Elementary theory of the optical properties of solids, in F. Seitz and D. Turnbull, eds., *Solid State Phys.*, vol. 15, Academic Press, New York, 1963.

26. For example, R. M. Rose, L. A. Shepard, and J. Wulff, *The Structure and Properties of Materials*, vol. VII, John Wiley & Sons, New York, 1966.

27. C. H. Henry, R. A. Logan, and F. R. Merritt, The effect of surface recombination on current in $Al_xGa_{1-x}As$ heterojunctions, *J. Appl. Phys.*, **49,** 3530, 1978.

28. M. Fukuda, Current drift associated with surface recombination current in InGaAsP/InP optical devices, *J. Appl. Phys.*, **59,** 4172, 1986.

29. A. G. Milnes and D. L. Feucht, *Heterojunctions and Metal-Semiconductor Junctions*, Academic Press, New York, 1972.

30. H. Asai, Epitaxial growth of III-V compound semiconductors and its applications, doctoral thesis, Nagoya University, Nagoya, Japan, p. 238, 1992.

31. H. Iwamura, H. Kobayashi, and H. Okamoto, Excitonic absorption spectra of GaAs-AlAs superlattice at high temperature, *Japan. J. Appl. Phys.*, **23,** L795, 1984.

32. S. L. Chuang, Efficient band-structure calculations of strained quantum wells, *Phys. Rev.*, **B.43,** 9649, 1991.

33. L. Friedman and R. A. Soref, Linear electro-optic effect in Ge_xSi_{1-x}/Si strained-layer superlattices, *Electron. Lett.*, **22,** 819, 1986.

34. M. Okayasu, T. Takeshita, O. Kogure, and S. Uehara, *Electronics and Communications in Japan*, Scripta Technica, Inc., New York, part 2, vol. 74, p. 68, 1991.

35. Y. Arakawa, K. Vahala, and A. Yariv, Quantum noise and dynamics in quantum well and quantum wire lasers, *Appl. Phys. Lett.*, **45,** 950, 1984.

36. Y. Arakawa and A. Yariv, Quantum well lasers–gain, spectra, dynamics, *IEEE J. Quantum Electron*, **QE-22,** 1887, 1986.

CHAPTER TWO

Light-Emitting Diodes

Light-emitting diodes (LEDs) are a typical *pn*-junction device used under a forward bias. The basic operating mechanisms are based on the electrical and optical properties of *pn*-junctions and of semiconductor materials. Depending on the semiconductor material used in the light-emitting layer (active layer), the wavelength of the emitted light can be anywhere within the range from visible to infrared. Most commercially available LEDs are made from III-V compound semiconductors.[1-9] Some II-VI compound semiconductors such as ZnS and ZnSe are used in a few LEDs emitting visible light, though these materials are not used frequently because of the difficulty of *pn*-junction formation.

LEDs have been widely used in various kinds of equipment and systems. LEDs composed of a combination of InGaAsP and InP cover the wavelengths from the 1300 to 1550 nm band and are the ones usually used in optical fiber communication systems. A combination of GaAs and AlGaAs or a combination of different compositions of AlGaAs has been used to make LEDs emitting in the 780 to 900 nm band, and those LEDs have been used in optical fiber communication systems, data links, remote controllers, and so forth. For display and indicator applications, LEDs emitting from blue to red light are usually applied. In this chapter, the basics of these LEDs are described.

2.1 BASIC STRUCTURE OF LEDs

The basic structures of various kinds of LEDs are illustrated in Fig. 2.1. In LEDs used in optical fiber communication systems, efficient spontaneous emission originating from the injection excitation is favorable to reduce input power, and thus a *pn*-heterojunction—usually double heterostructure—is used. The active layer in those LEDs is usually formed by bulk materials. In LEDs emitting visible light for display use, however, a variety of structures are used. A *pn*-homojunction (as well as *pn*-heterojunctions) is still used in LEDs emitting red or

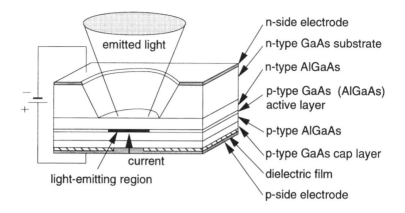

n-side electrode
n-type GaAs substrate
n-type AlGaAs
p-type GaAs (AlGaAs) active layer
p-type AlGaAs
p-type GaAs cap layer
dielectric film
p-side electrode

emitted light

current
light-emitting region

(a) surface emitting (Burrus type)

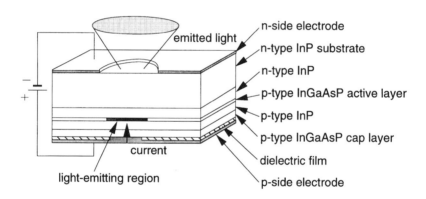

n-side electrode
n-type InP substrate
n-type InP
p-type InGaAsP active layer
p-type InP
p-type InGaAsP cap layer
dielectric film
p-side electrode

emitted light

current
light-emitting region

(b) surface emitting

FIGURE 2.1   Cross-sectional illustration of the basic surface- and edge-emitting types of LEDs: (a) Burrus type,[2] (b,c) surface-emitting type (double heterostructure type and homojunction type)[3,4], (d) edge-emitting type.[5] Reprinted with permission from *Reliability and Degradation of Semiconductor Lasers and LEDs* by Artech House, Inc., Norwood, MA, USA, http://www.artech-house.com.

yellow light because it is easy to fabricate. Some LEDs emitting blue or green light, most of which are composed of group-III nitrides, use quantum well structures, such as InGaN/GaN and GaN/AlGaN, because they produce bright light.[1] In these kinds of LEDs, however, the light originating in the spontaneous emission process is emitted from the light-emitting region (active layer) in all directions. Consequently, several structures restricting the emitted light to a certain

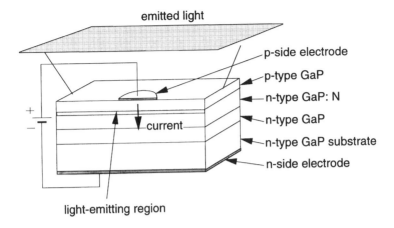

(c) surface emitting (homojunction type)

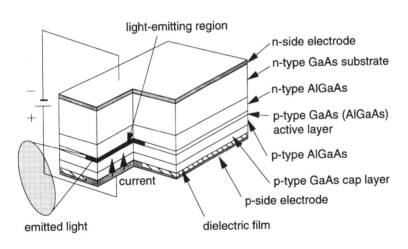

(d) edge emitting

FIGURE 2.1 (*Continued*)

direction have been developed. These LED structures are basically divided into two groups: the surface-emitting and edge-emitting types. The surface-emitting type LEDs emit the light in a direction perpendicular to the *pn*-junction plane. Figure 2.1 shows three surface-emitting types of LEDs. The Burrus type is a typical surface-emitting type of LED in which the emitted light is absorbed in the substrate.[2] If the wavelength (photon energy) of the emitted light is shorter (higher) than the fundamental absorption edge (band-gap energy) of the substrate, most of the emitted light is absorbed in the substrate. This is often the case in AlGaAs/GaAs LEDs with GaAs substrates. In such a case, a part of the substrate corresponding to the light-emitting region is etched off to prevent the absorption. Under a forward bias, high densities of electrons and holes are injected from *n*-type and *p*-type AlGaAs layers to the *p*-type GaAs (or AlGaAs) active layer, where spontaneous emission occurs. The light-emitting region is usually limited in order to keep the operating current low and, for LEDs used in optical fiber communication systems, to facilitate the coupling of the light to a fiber. The emitting region is limited by confining the injected current to a part of the electrode. The electrode is in contact with a cap layer, usually 30 to 50 μm in diameter, and is limited by a dielectric film such as SiO_2. LEDs for display use, for example, are often mounted in a junction-up configuration to avoid the absorption at the substrate. The whole active layer in such LEDs is usually excited and emits visible light. If the emitted light is transparent in the substrate, the substrates can be chemically or physically etched down to lens structure to improve the light-emitting pattern. This type of structure is employed in some AlGaAs/GaAs LEDs with AlGaAs substrates and InGaAsP/InP LEDs (see Fig. 2.2).

In the edge-emitting type, the light is emitted in a direction parallel to the *pn*-junction plane. The current is restricted, by a dielectric film, to within a stripe having a width less than several tens of micrometers, and then the light-emitting region is formed along the current injected (stripe) region. The basic operating mechanism is the same as that of the surface-emitting types. This edge-emitting type of LEDs is used in the optical sources of fiber communication systems, but not often.

Cross sections of several typical LEDs are shown in Fig. 2.2, and typical semiconductor materials used in LEDs are summarized (with typical emission wavelengths) in Table 2.1.

2.2 DEVICE CHARACTERISTICS

2.2.1 Basic Characteristics

The output light of the LEDs originates from the spontaneous emission in the active layer, which in turn is a product of the radiative recombination of the injected electrons and holes. The band structure and the distribution of injected carrier density under a forward bias are shown schematically in Fig. 2.3 for a *p*-type/*p*-type//*n*-type double heterostructure composed of direct band-gap semiconductors. When a certain density of electrons is injected into an active layer having a thickness less than the carrier diffusion length [see Eqs. (1.145a) and (1.145b)], the same density of holes is also injected because of the requirement for charge neutrality. Those injected carriers distribute uniformly in the active layer because its thickness is less than the carrier diffusion length (a few

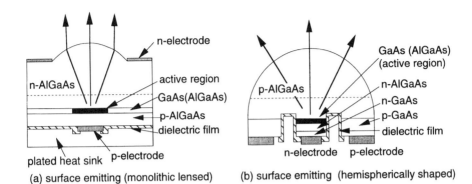

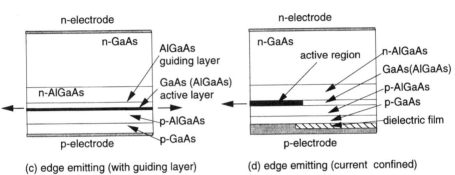

FIGURE 2.2   Cross-sectional views of some widely used LEDs: (a) monolithic-lensed type,[6] (b) hemispherically shaped type,[7] (c) edge-emitting type with a guiding layer,[8] (d) edge-emitting type with local emitting area.[9] Reprinted with permission from *Reliability and Degradation of Semiconductor Lasers and LEDs* by Artech House, Inc., Norwood, MA, USA, http://www.artech-house.com.

micron-meters). The thickness of the active layer in LEDs is ordinarily between 0.5 and 2 μm. The electron and hole densities in the active layer under the injection excitation can be expressed with the next equations:

$$n = n_0 + \Delta n, \qquad (2.1a)$$

$$p = p_0 + \Delta p, \qquad (2.1b)$$

and

$$\Delta n = \Delta p, \qquad (2.1c)$$

where Δn and Δp are the injected electron and hole densities, n_0 and p_0 are the electron and hole densities in the active layer in a thermal equilibrium state (without injection excitation), and $n_0 p_0 = n_i^2$ [see Eq. (1.25)]. The electrons and holes in the active layer radiatively recombine at a finite rate, so the injected

TABLE 2.1   Typical Semiconductor Materials, Emission Wavelength, and Examples of External Quantum Efficiency for Commercially Available LEDs

Peak Emission Wavelength (nm)	Active Layer Material	Transition	External Quantum Efficiency (example) (%)
470 (blue)	ZnS: Al	(MIS structure)	<0.1[a]
450–530 (blue/green)	InGaN/GaN (quantum well)	direct	>5[b]
565 (green)	GaP: N	indirect	<0.2
590 (yellow)	$GaAs_{0.15}P_{0.85}$: N	indirect	0.3
590–620 (yellow-orange)	AlInGaP	direct	>2
610 (orange)	$GaAs_{0.25}P_{0.75}$: N	indirect	0.3
630 (orange/red)	$GaAs_{0.35}P_{0.65}$: N	indirect	0.4
650 (red)	$GaAs_{0.6}P_{0.4}$	direct	0.2–0.5
680 (red)	$Al_{0.35}Ga_{0.65}As$	direct	3–7
700 (red)	GaP: Zn-O	indirect	1–4
850	$Al_{0.03}Ga_{0.97}As$	direct	3–7
860	GaAs	direct	2–3
1300	$In_{0.76}Ga_{0.24}As_{0.55}P_{0.45}$	direct	1–2
1550	$In_{0.65}Ga_{0.35}As_{0.79}P_{0.21}$	direct	1–2

[a]H. Katayama et al., *Appl. Phys. Lett.*, **27**, 697, 1975.
[b]S. Nakamura et al., *Japan J. Appl. Phys.*, **34**, L1332, 1995.

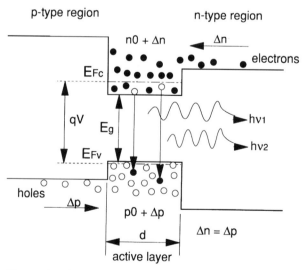

FIGURE 2.3   Energy band diagram and injected carrier distribution for a double heterostructure under a forward bias.

carrier density changes with time. The variation of the density can be expressed by a behavior of the injected minority carrier. For the injected electrons in a p-type active layer of thickness, d, the following equation holds:

$$dn/dt = J/qd - R_{\text{sp}}$$
$$= J/qd - \Delta n/\tau_{\text{e}}, \qquad (2.2)$$

where J is the injected current density, q is the electron charge, R_{sp} is the spontaneous emission rate given by Eq. (1.76), and τ_{e} is the electron lifetime in the p-type active layer. Similarly, for an n-type active layer,

$$dp/dt = J/qd - R_{\text{sp}}$$
$$= J/qd - \Delta p/\tau_{\text{h}}, \qquad (2.3)$$

where τ_{h} is the hole lifetime in the n-type active layer. The injected carrier lifetime and the spontaneous emission rate are related as specified by Eqs. (1.107) through (1.77). In Eqs. (2.2) and (2.3), the first term on the right-hand side expresses the electron and hole density injected into the active layer per unit time, and the second term corresponds to the density of electrons and holes disappearing by means of radiative recombination. The injected carrier density is more than 10^{17} cm^{-3} under the usual operating conditions in LEDs. In addition, the active layer of normal LEDs are intentionally undoped or lightly doped in order to increase light output power, since the influences of Auger recombination, intraband absorption, and nonradiative recombination become stronger with increasing impurity doping (see Sections 1.1.2 and 1.1.3). Therefore, the following relations hold for electron density in the p-type active layer and hole density in the n-type active layer: $n \approx \Delta n$ and $p \approx \Delta p$. Equations (2.2) and (2.3) can thus be rewritten as

$$dn/dt = J/qd - n/\tau_{\text{e}}, \qquad (2.4)$$

for p-type (or undoped) active layers and as

$$dp/dt = J/qd - p/\tau_{\text{h}}, \qquad (2.5)$$

for n-type (or undoped) active layers. These are basic equations describing the change in carrier density with time in LEDs. Under a stationary state, as it is during dc operation, dn/dt and dp/dt are equal to 0 and from Eqs. (2.4) and (2.5),

$$n \approx \Delta n = \tau_{\text{e}} J/qd, \qquad (2.6)$$

and

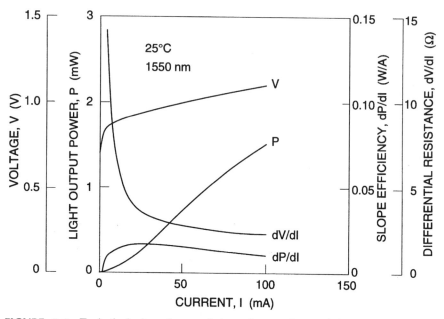

FIGURE 2.4 Typical device characteristics of a surface-emitting 1550 nm-band InGaAsP/InP LED.

$$p \approx \Delta p = \tau_h J/qd. \qquad (2.7)$$

In the ideal case, the injected carrier density depends linearly on the injected carrier lifetime.

Current–light output power and current-voltage characteristics for a 1550 nm-band InGaAsP/InP surface-emitting LED are shown in Fig. 2.4. The light output power is nearly proportional to the injected current in the relatively low-current range. The light power emitted in the active layer is given by the product of the number of photons emitted and the photon energy, $h\nu$. The number of photons created by the spontaneous recombination process in a unit volume of the active layer is given by the product of the injected carrier density, J/qd, and the internal quantum efficiency. The light power emitted from a unit volume of the active layer, P_{act}, is therefore given by

$$P_{act} = h\nu(\eta_i J/qd), \qquad (2.8)$$

where η_i is the internal quantum efficiency corresponding to the ratio of emitted photons to injected electrons and is given by Eq. (1.111). By substituting Eq. (1.111) into Eq. (2.8) and using Eq. (2.6) or (2.7),

$$P_{\text{act}} = h\nu(J/qd)B_{\text{sp}}\tau_n(p_0 + n_0 + \tau_n J/qd), \tag{2.9}$$

where τ_n is the injected carrier lifetime (τ_e for electrons in the p-type active layer and τ_h for holes in the n-type active layer). Under low-injection conditions, p_0 and $n_0 \gg \tau_n J/qd$, and Eq. (2.9) is rewritten for the p-type active layers ($p_0 > n_0$) as

$$P_{\text{act-}p} = h\nu B_{\text{sp}}p_0\tau_e J/qd, \tag{2.10}$$

and for the n-type active layers as

$$P_{\text{act-}n} = h\nu B_{\text{sp}}n_0\tau_h J/qd. \tag{2.11}$$

Under high-injection conditions, in contrast, $\tau_n J/qd \gg p_0$ and n_0 and thus

$$P_{\text{act}} = h\nu B_{\text{sp}}(\tau_n J/qd)^2. \tag{2.12}$$

In Eqs. (2.10), (2.11), and (2.12), the variables are only τ_n (or τ_e or τ_h) and J. Consequently, the light output power of the active layer is proportional to the injected current density in the low-excitation range, while in the high-excitation rate it is proportional to the square of the injected current density if the lifetimes are constant. The carrier lifetime, however, changes slowly with change in the carrier density. Other problems, such as those due to heat generation, also occur during device operation and make the relations in Eqs. (2.10), (2.11), and (2.12) imprecise for actual LEDs.

2.2.2 Radiative Transition in LEDs

The lifetime of the injected minority carrier in the active layer of an LED is determined by various radiative and nonradiative recombination processes and is basically given by Eq. (1.105). Band-to-band, band-to–impurity level (conduction band–to–acceptor level and donor level–to–valence band), and donor level–to–acceptor level transitions dominate the radiative recombinations in direct band-gap semiconductors. These radiative transmission processes in direct band-gap semiconductors have already been discussed in Section 1.1.3 from a physical point of view. The materials generally used for LEDs are AlInGaP for wavelengths from 590 to 620 nm, InGaP for wavelengths from 660 to 680 nm, GaAs for the 850 nm band, and $In_{0.76}Ga_{0.34}As_{0.55}P_{0.45}$ for the 1300 nm band (see Table 2.1).

Those types of radiative transitions in direct band-gap semiconductors are not efficient in indirect band-gap semiconductors. In LEDs composed of indirect wide band-gap semiconductors, especially in LEDs emitting visible light, the recombination of bound excitons at isoelectronic centers or at similar centers is dominant.[10] The isoelectronic centers are introduced by replacing, during crystal growth, one host atom in the crystal with another kind of atom having the same number of valence electrons. An isoelectronic center easily captures an electron and becomes negatively charged. A hole is therefore attracted to the center, and a bound exciton is formed. As a result, the radiative recombination of the exciton (electron-hole pair) at the isoelectronic center localized in the crystal occurs with a high probability without any phonon assistance (momentum is conserved) (see Table 2.1). The nitrogen atom is well known as an element forming isoelectronic centers in GaP and GaAsP. The indirect band-gap energy of GaP is 2.26 eV (corresponding wavelength: 547 nm) at room temperature. The light emission caused by the level due to the isoelectronic center formed by an N atom has a peak near 565 nm (green) and this is a wavelength slightly longer than the wavelength corresponding to the band-gap energy because of the recombination associated with the exciton. As the mole fraction, y, increases in ternary $GaAs_{1-y}P_y$, the band-gap changes from direct to indirect when the mole fraction reaches 0.45. The direct band-gap semiconductor $GaAs_{0.6}P_{0.4}$ is frequently used in LEDs emitting red light (649 nm), whereas the indirect band-gap $GaAs_{0.35}P_{0.65}$ or $GaAs_{0.15}P_{0.85}$ with N atoms is used to make LEDs emitting orange (632 nm) or yellow (589 nm) light. The quantum efficiency associated with isoelectronic centers is usually smaller than that of direct band-gap semiconductors (see Table 2.1). A Zn-O pair in GaP also forms a center similar to the isoelectronic one.[11,12] If Zn and O atoms are simultaneously doped into GaP, *Zn* substitutes at a Ga-site as an acceptor while O substitutes at a P-site as a deep donor. When the Zn and O atoms occupy adjacent sites as a nearest pair, the pair can form a neutral molecular center like an isoelectronic impurity. The radiative recombination of the bound exciton therefore occurs at the molecular center. The LEDs composed of GaP with Zn-O emit dark-red light with wavelength near 700 nm.

2.2.3 Current–Light Output Characteristics

The total light output power from an LED is never equal to the light power emitted from the active layer because the light due to a spontaneous emission process in the LED is emitted in a random direction. The light output power from a defined surface is therefore only a part of the total power emitted from the active layer. This is recognized in the very low slope efficiency (about 0.02 W/A in Fig. 2.4), which is the ratio of the output power from a defined surface to the injected current. The output power is usually only a few percent of the total power emitted from the active layer, and the slope efficiency depends on the LED structure. The light output power is expressed with several kinds of terms related to the efficiency. The conversion efficiency (or device efficiency or external power efficiency), η_{cv}, is defined as the ratio of the optical output power from LEDs, P_{out}, to the electrical input power, $P_{e\text{-}in}$, and is often expressed as a percentage:

$$\eta_{cv} \equiv (P_{out}/P_{e\text{-}in}) \times 100. \tag{2.13}$$

The conversion efficiency is ordinarily less than 5%. The external quantum efficiency, η_{ext}, which is defined by the ratio of output photon number to input electron number, is also frequently used (expressed as a percentage) in evaluating light-emitting devices:

$$\eta_{ext} = (P_{out}/h\nu)/(I_F/q)$$
$$= (P_{out}/I_F E_g) \times 100, \qquad (2.14)$$

where I_F and E_g $(= h\nu/q)$ is the injected current and the band-gap energy of the active layer (electron volts). Consequently, the following relation between the conversion efficiency and the external quantum efficiency holds:

$$\eta_{cv} = \eta_{ext}(E_g/V_j), \qquad (2.15)$$

where the relation of $P_{e-in} = I_F V_j$ is used and V_j is the bias voltage on the *pn*-junction (junction voltage) and is equal to $V_b - R I_F$ (where V_b is the applied bias voltage and R is the total series resistance of bias circuit). In normal operation V_j is usually larger than E_g because the current starts to increase when the bias voltage is over the band-gap. Consequently, the conversion efficiency is lower than the external quantum efficiency. For InGaAsP/InP LEDs, whose characteristics are shown in Fig. 2.4, the band-gap is about 0.8 eV at 300 K and the bias voltage at 50 mA is about 0.95 V. By using Eq. (2.15), at 50 mA, the conversion efficiency is about 84% lower than the external quantum efficiency. The external quantum efficiency may be given by the product of the internal quantum efficiency [see Eqs. (1.110) and (1.111)], η_i, and the extraction efficiency, η_{out}:

$$\eta_{ext} = \eta_i \eta_{out}, \qquad (2.16)$$

where the extraction efficiency is the ratio of the power emitted from the active layer to the optical power emitted from the LED. The power output from an LED can be expressed, taking the active volume into the consideration, as

$$P_{out} = \eta_{out}(S_{act}d_{act})P_{act}, \qquad (2.17)$$

where S_{act} is the area of the light-emitting region and d_{act} is the thickness of the active layer. The slope efficiency (or responsivity), η_s, shown in Fig. 2.4 is given by the first derivative of the output power from the LED with respect to the current, I_F:

$$\eta_s = dP_{out}/dI_F = \eta_{out}dP_{act}/dI_F$$
$$= \eta_{out}\eta_i h\nu(S_{act}d_{act})d(J/qd_{act})/dI_F$$
$$= \eta_{ext}h\nu/q = 1.24\eta_{ext}/\lambda(\mu m) \qquad (W/A), \qquad (2.18)$$

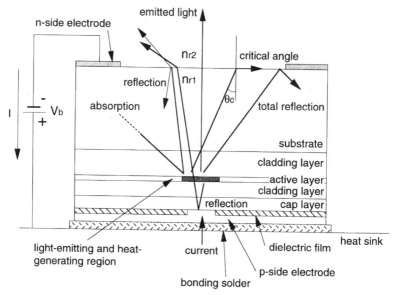

FIGURE 2.5 Geometrical factors limiting the optical output power from a surface-emitting LED.

where Eqs. (2.16) and (2.17) are used. In the case of the 1550 nm LED whose characteristics are shown in Fig. 2.4, the slope efficiency is 0.02 W/A.

The extraction efficiency, which as indicated in Eq. (2.17) strongly influences the output power, is mainly determined by the optical absorption at the substrate, the optical reflection at the interface between semiconductor surface and air (or other ambient material), and the total reflection at each interface. In addition, for surface-emitting type LEDs, there is also an effect of the optical reflection at the interface between the semiconductor cap layer and the electrode. The efficiency is given by the product of those factors, but it is difficult to calculate their exact values because the LED structures are complicated. A schematic illustration of these factors for a surface-emitting type of LED is shown in Fig. 2.5. The absorption at the substrate is the same problem described in Section 1.1.3.1 and is essentially determined by the fundamental absorption. In actual LEDs, however, this optical absorption is eliminated or suppressed to a low value by etching the substrates, using substrates transparent to the emitted light, or mounting in a junction-up configuration as shown in Fig. 2.1(c). The total reflection limits a light-emitting area at the surface because of the critical angle. The critical angle, θ_c is given by the following well-known equation:[13]

$$\theta_c = \sin^{-1}(n_{r2}/n_{r1}), \qquad (2.19)$$

where n_{r1} and n_{r2} are the refractive indexes of the semiconductor substrate and air (or other materials). The light from the active layer to the substrate with an angle larger than the critical one is not emitted outside of the LED. The critical angle is small in semiconductors because their refractive indexes are usually high. For InP the refractive index is about 3.4, and thus the critical angle between the InP and air is about 17 degrees. Most of the light emitted from the active layer is trapped within the LEDs because of the total reflection. Part of the light within the critical angle is reflected at the surface. The reflectivity is also determined by the refractive index [see Eq. (6.8)]. For InP the reflectivity is about 30%, and this means that about 30% of the light is reflected at the surface. Optical reflection also occurs at the electrode, where the reflectivity depends on the roughness of the interface between the electrode and the cap layer. In ordinary circumstances, the reflectivity at the electrode is quite high and most of the light from the active layer is reflected. About 20 to 30 percent of the total light output power from a surface-emitting type LED is due to the light reflected at the electrode. There are other mechanisms decreasing the light output power from LEDs, such as absorption at the cap layer and the active layer. They have little influence on the output power but should be considered when designing LED structures. As described above, the main factor determining the extraction efficiency is the total reflection, so some structures reducing the influence of the total reflection have been developed. The monolithic-lensed type and the hemispherically shaped type shown in Fig. 2.2 are examples of such structures. These considerations are also applicable to the edge-emitting LEDs—except for the reflection at the interface of the electrode and the cap layer, since the light is emitted in a direction parallel to the active layer.

The above discussions of efficiencies have taken into consideration only radiative recombination, but nonradiative recombination also occurs in the active layer of actual devices. Its influence on internal quantum efficiency, however, is negligible in normal LEDs composed of direct band-gap semiconductors if their doping levels are lower than 10^{18} cm^{-3} and the thickness of the active layer is set at a suitable value (described below). In such cases, internal quantum efficiency, η_i, is nearly equal to 1. The external power efficiency is decreased by nonradiative recombination current, such as generation-recombination current and surface recombination current at the perimeter of the *pn*-junction. As discussed in Section 2.2.4, the amount of the nonradiative recombination current is often between a few and a few tens of milliamperes.

The design of the active layer is obviously important in making LEDs with good characteristics. The thickness of the active layer is usually set at a value less than the injected carrier diffusion length in order to increase the carrier confinement effect, which results in a high rate of radiative recombination due to the high injected carrier density and the uniform distribution of the injected carrier. The uniform carrier distribution prevents the light emitted at a region with a high carrier density from being reabsorbed at a region with a low carrier density because the wavelength of light emitted at a region with the high carrier density is shorter than that of the light emitted at a region with a low carrier density (band-filling effect; see Section 1.1.3.1).

As indicated in Eqs. (2.6) and (2.7), the carrier density in the active layer is inversely proportional to the active layer thickness. The carrier density in a thin active layer thus increases quickly with increases in the injected current. The Auger recombination process, however, becomes dominant when the carrier density is high because the recombination rate is proportional to the cube of carrier density [see Eqs. (1.68) and (1.69)]. This Auger recombination is marked in LEDs having active layers composed of narrow band-gap materials, such as InGaAs and InGaAsP. Under such conditions, the number of carriers having high energies also increases, and these high-energy carriers flow over the heterobarrier and out of the active layer. Consequently, the thinner active layer scarcely improves the output power. From the viewpoint of active volume, a thicker active layer is favorable as indicated in Eq. (2.17). The carrier diffusion length is typically 2–3 μm in a GaAs, AlGaAs, InGaAs, or InGaAsP active layer, so the thickness of the active layer is usually set at 0.5–2 μm. Controlling active layer thickness is also important to obtaining a linear current–light output relation as indicated in Eqs. (2.10) and (2.11). When a LED having a 1 μm-thick p-type active layer and a p-side electrode of 50 μm in diameter is operated under normal operating conditions—for example, a dc current of 50 mA and an injected carrier lifetime of 3 ns—the current density is about 2.5 kA/cm^2 and the carrier density is about 5×10^{17} cm^{-3}. The hole concentration of the p-type active layer is about 10^{18} cm^{-3}. In actual devices the injected current diffuses parallel to the active layer, the light-emitting region in the active layer becomes larger, and the carrier density in the active layer is therefore smaller than the calculated value. The current–light output power relation is therefore linear under normal operating conditions (see Fig. 2.4). This linear relation is important for LEDs used in optical fiber communication systems, especially analog systems (see Section 8.1.1.1).

2.2.4   Current-Voltage Characteristics

The current-voltage characteristics of LEDs are basically given by Eq. (1.157), and several current components influence these characteristics. In the operating bias range of LEDs, however, the diffusion current is dominant and the radiative current is given by

$$I_\mathrm{F} = S_\mathrm{act} J_\mathrm{s0} \exp[q(V_\mathrm{b} - R_\mathrm{s} I_\mathrm{F})/k_\mathrm{B} T], \qquad (2.20)$$

where R_s is the series resistance (consisting of the resistances of the semiconductors, the contact resistance, and so forth). As shown in Fig. 2.4, the current increases rapidly when the applied voltage exceeds the band-gap of the active layer, which is about 0.8 V for the LED whose characteristics are shown in Fig. 2.4. The differential resistance is given by the first derivative of Eq. (2.20) with respect to I_F:

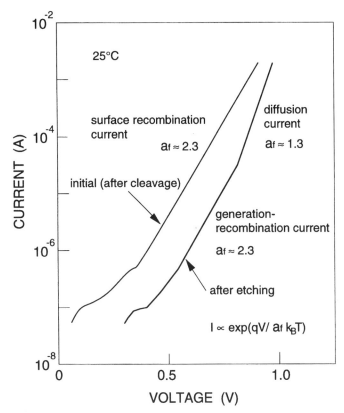

FIGURE 2.6 Current-voltage characteristics, in a low-bias range, just after cleavage and after etching for an InGaAsP/InP double-heterostructure diode, 400 μm wide and 300 μm long.[15]

$$dV_b/dI_F = (k_B T/q)(1/I_F) + R_s. \qquad (2.21)$$

As the injected current increases, the differential resistance gradually decreases and approaches the series resistance, R_s. The series resistance for the device treated in Fig. 2.4 is estimated to be 1.8 Ω.

As indicated in Eq. (1.157), the nonradiative recombination current is dominant in the low-bias range. The nonradiative recombination current, with the $2k_B T$ component, is larger than the radiative diffusion current, with a $k_B T$ component, as shown in Fig. 2.6. Just after cleavage from the wafer, the nonradiative current with the $2k_B T$ component (2.3 $k_B T$ in Fig. 2.6) is dominant in the current range over 1 mA, while after the chemical etching and passivation of the cleaved facets the diffusion current with the $k_B T$ component (1.3 $k_B T$ in Fig. 2.6) is dominant in the current range over 0.1 mA. The dominant component in the nonadiative current is different for different materials and device structures and is normally surface recombination current or generation-recombination current via deep levels in the depletion layer of the *pn*-junction. The density of deep levels in optoelectronic *pn*-junction devices is not high because of the high quality of the crystals, so the $2k_B T$ component is usually caused by the surface recombination current. This surface recombination current has been observed in GaP,[14] GaAsP,[14] AlGaAs,[14] and InGaAsP diodes.[15]

The surface recombination current is given by Eq. (1.156) and, as described in Section 1.1.2.7, is due to the surface recombination of carriers. Although in the low-excitation range surface recombination velocity is usually treated as a constant, the *pn*-junction surface is highly excited under normal operating conditions and the situation is quite different from that under low-excitation conditions. The surface recombination current at the *pn*-junction perimeter has been analyzed in detail by Henry et al.[14,16] Under a forward bias, some of the injected electrons and holes are confined as minority carriers at the surface depleted region (usually 2–5 nm deep), where in a *pn*-heterojunction only the majority carrier of the wider band-gap semiconductor is injected (see Section 1.2.4). Each minority carrier diffuses along the surface in the confined layer until it nonradiatively recombines with a majority carrier. The surface recombination current per unit length is given, exactly, by

$$J_{sur} = qs_0 L_s (n_s p_s)^{1/2}$$
$$= qs_0 L_s n_i \exp(q V_b / 2k_B T), \tag{2.22}$$

where Eq. (1.146) is used, s_0 and L_s are the intrinsic surface recombination velocity and the average surface diffusion length for minority carriers, and n_s and p_s are the electron and hole densities at the *pn*-junction surface. The concept of the average surface diffusion length is similar to that of the diffusion length of injected carriers indicated in Eqs. (1.145a) and (1.145b) and is expressed by

$$L_s = (L_{eps} L_{hns})^{1/2}$$
$$= [2(D_{eps} w_{ep} D_{hns} w_{hn})^{1/2} / s_0]^{1/2} \tag{2.23}$$

where the parameters related to surface are denoted with a subscript s and where w_{ep} and w_{hn} are the electron and hole confinement layer thicknesses at the surface. The intrinsic surface recombination velocity is mainly determined by the surface state density and the capture cross section at the surface, and it changes with bias voltage,

$$s_0 \propto (\sigma_s v_{th} N_{ts} / 2E_g) q V_b, \tag{2.24}$$

where the meaning of the symbols is similar to the meaning of those in Eqs. (1.64)–(1.67). The value of $s_0 L_s$ (in cm^2/s) is a measure of a magnitude of the surface recombination current indicated in Eq. (2.22) and has been reported for various diodes: 3–5 cm^2/s for GaP LEDs, 6–20 cm^2/s for $Ga_{0.4}As_{0.6}P$ LEDs, 1–7.5 cm^2/s for AlGaAs LEDs, 10–15 cm^2/s for GaAs LEDs, and about 0.5 cm^2/s for InGaAsP LEDs.[14,15,17,18] The recombination rate, R_s, is given by

$$R_s = s_0(n_s p_s)^{1/2}$$
$$= S_s(\text{minority carrier density}). \qquad (2.25)$$

Here S_s is the surface recombination velocity corresponding to that indicated in Eqs. (1.65) and (1.67) and is given by

$$S_s = s_0[(\text{majority carrier density})/(\text{minority carrier density})]^{1/2}, \qquad (2.26)$$

where the carrier densities are those in the outside of the depletion layer. The surface recombination velocity at the *pn*-junction perimeter is not constant and changes with bias. Compare Eq. (2.26) with Eqs. (1.65) and (1.67). The surface recombination current for LEDs as well as for laser diodes is determined by Eq. (2.22). This surface recombination current is usually at least an order of magnitude larger than the generation-recombination current in diodes consisting of III-V compound semiconductors (see Fig. 2.6).

These explanations will be applicable to dislocations at *pn*-junctions and buried heterointerfaces in buried heterostructure laser diodes (and LEDs) because a depletion region similar to that at the surface is formed in those regions. This type of current with a $2k_BT$ component has also been observed during degradation caused by generation and growth of dislocation networks in the active layer.[19]

The current drift (or voltage drift) associated with the surface recombination current is occasionally induced in LEDs and laser diodes composed of materials with low surface recombination velocities if their *pn*-junction perimeters are long and exposed to air. The remarkable drift has been observed in InGaAsP/InP double-heterostructure diodes and InP homojunction diodes. As shown in Fig. 2.7, the current at a constant bias of 0.5 V gradually decreases with time.[15] This current drift is due to a decrease in the surface recombination current, which in turn may be due to the transfer of electrons related to surface recombination current to traps located in the native oxide by means of quantum-mechanical tunneling. The native oxide rapidly grows just after cleavage. The primary cause of the trapping is a combination of electron diffusion and a low recombination rate at the semiconductor surface. Consequently, diodes composed of materials with a high surface recombination velocity, such as AlGaAs and GaAs diodes, show no current drift (or voltage drift). The intrinsic surface recombination velocity, s_0, is about 3×10^4 cm/s or less for InGaAsP and is about 4×10^5 cm/s for GaAs.[14,15] This current (or voltage) drift causes the bias point to shift in LEDs, although the shift is not large. (If devices are planer type laser diodes, their threshold currents vary with time without any degradation.)

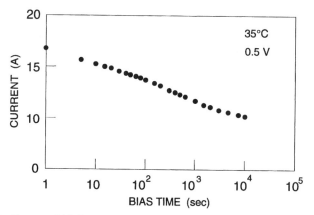

FIGURE 2.7   Current drift for an InGaAsP/InP double-heterostructure diode at 35°C. The bias is set at 0.5 V.[15]

2.2.5   Other Characteristics

2.2.5.1   *Temperature Characteristics* The performance and reliability of optoelectronic *pn*-junction devices are often limited by their temperature characteristics. The temperature dependence of the device characteristics is therefore very important from the application point of view. The optoelectronic *pn*-junction devices used under forward bias are very sensitive to the temperature of the active layer, and their current–light output power characteristics strongly depend on the junction temperature.

The junction temperature rises when the injection is carried out (Joule's heating) or the ambient (or heat sink) temperature increases. One example of the effect of Joule's heating is shown in Fig. 2.8. The light output power clearly differs between the operations at dc and pulsed bias, and this difference is due to the different junction temperatures. As the injected current increases under dc operation, the output power is gradually saturated because the heating at the active layer and the other part having ohmic resistance (such as the electrode and other semiconductor layers) becomes large. Under pulsed operation, in contrast, the output power increases linearly in proportion to the injected current because the heating is negligible under pulsed operation, if the pulse width is less than a few microseconds. The junction temperature gradually increases as the pulse width increases, and it is eventually the same as the value under dc operation. Junction temperature is not easy to measure directly because the device is small. The temperature rise can, however, be estimated (1) from the wavelength shift of emitted light, which is determined by the temperature dependence of the band-gap, (2) by comparing the output power under dc and pulsed operations, or (3) from the temperature dependence of the junction voltage (see Section 6.3.1.3). In method (2), the optical output power at a certain pulsed current is monitored as the ambient temperature increases. The junction temperature under dc operation corresponds to the value at which the output power at the certain pulsed bias at the high ambient

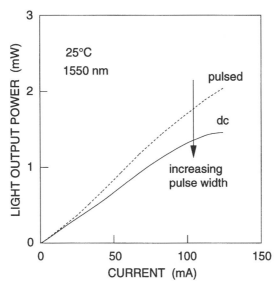

FIGURE 2.8   Current vs. light output power for a 1550 nm-band InGaAsP/InP LED under pulsed and dc biases. The active region is 1 μm thick and 80 μm in diameter.

temperature is the same as that at the certain dc bias at the initial ambient temperature. The temperature rise is often expressed by using the so-called thermal resistance in units of $^\circ$C/W. The thermal resistance indicates the temperature increase in the active layer when input power is applied and it is determined by the thermal conductivities of cladding layer, electrode, bonding part, heat sink, stem, and case (see Table 6.3). The input power heating the active layer is the forward current and the bias voltage, and the junction temperature rise is calculated by using the thermal resistance, R_{th}:

$$\Delta Tj = R_{th}I_F V_j. \qquad (2.27)$$

The thermal resistance is easily obtained by monitoring the change in junction voltage (see Section 6.3.1.3). In general, the thermal conductivity of ternary and quaternary compound semiconductors tends to be lower than that of binary or elementary semiconductors (see Table 6.3). Consequently, the thermal conductivity of the cladding layers (or wider band-gap layers) strongly influences the thermal resistance and thus the temperature rise of the active layer. The thermal resistance of AlGaAs/GaAs double-heterostructure LEDs, for example, is often more than 100°C/W, although the actual values are determined by the mounting configuration. When the injected current is set at 100 mA, the thermal resistance is 100°C/W, and the bias voltage at that current is 1.8 V at 25°C, the calculated temperature rise is 18°C. The operating junction temperature is therefore estimated to be 43°C. Suppressing and removing the heat generated in the active layer are very effective to increase output power.

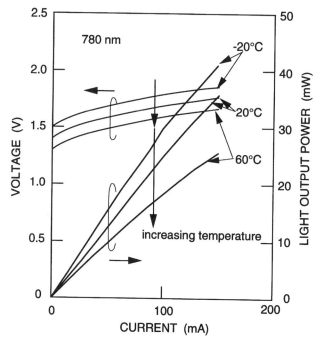

FIGURE 2.9 Effect of temperature on the relations between current and light output power and between current and voltage for a hemispherically shaped 780 nm-band AlGaAs double-heterostructure LED with a light-emitting region of 150 μm in diameter.

The ambient temperature strongly influences optical output power from LEDs (see Fig. 2.9). As the junction temperature increases, the internal quantum efficiency expressed by Eq. (1.111) decreases because of the reduction of the radiative recombination coefficient and because of the resultant increased overflow of injected carriers from the active layer.[20–22] These phenomena rapidly decrease the external power efficiency and therefore the output power. In addition, as described in Section 1.1.2.7, the influence of Auger recombination becomes large in LEDs with an active layer composed of a narrow band-gap semiconductor, such as InGaAsP and InGaAs. The change in optical output power is larger in long-wavelength-band LEDs than in short-wavelength-band LEDs as shown in Fig. 2.10. The InGaAsP double-heterostructure LED shows a large change in output power.

The temperature dependence of current–light output power characteristics is determined by several mechanisms. The overflow of the injected carrier is dominant at room temperature and the Auger recombination and intraband absorption are dominant at higher temperatures. The Auger recombination tends to be intensified as the band-gap energy of the active layer decreases (see Sections 1.1.2.7 and 1.1.3.1). These mechanisms are discussed again in Chapter 3.

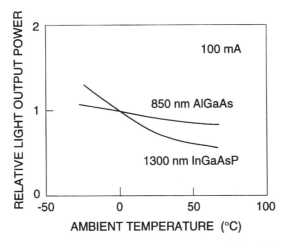

FIGURE 2.10 Temperature dependence of output power at 100 mA for a 780 nm-band AlGaAs LED with a light-emitting region about 150 μm in diameter and for a 1300 nm-band InGaAsP/InP LED with a light-emitting region about 30 μm in diameter.

The current-voltage characteristics also change with the ambient temperature as shown in Fig. 2.9. This change is due to the temperature dependence of the band-gap energy of the active layer in the double heterojunction or the active region in the homojunction or in the single heterojunction. As described in Section 1.2, forward current or diffusion current starts to increase rapidly when the bias voltage—exactly the product of junction voltage and electron charge—reaches the band-gap. At low temperatures the band-gap energy varies with the square of temperature but around room temperature decreases with increasing temperature.[23] Consequently, the temperature dependence of band-gap energy at a temperature near room temperature can be expressed, roughly, by

$$dE_g/dT = -\alpha_a \qquad (2.28)$$

where α_a (in eV/K) is the temperature coefficient of the band-gap energy. As the ambient temperature becomes higher, the forward current begins to flow at lower bias voltage (see Fig. 2.9).

2.2.5.2 *Emission Spectra* Emission spectra of LEDs correspond to the spectra of spontaneous emission in semiconductors, and the wavelength is basically determined by the band-gap energy in a direct transition process and by the energy difference between transition levels in other transition processes (indirect transition, transition related to isoelectronic traps, and so forth). Electrons and holes distribute in the conduction and valence bands or in those levels, broadening the emission spectra. For the direct transitions the peak wavelength,

λ_p, nearly coincides with the energy difference between the two quasi-Fermi levels and is given as follows [see Eq. (1.74)]:

$$\lambda_p \approx 1.24/[E_g + (E_{F_c} - E_c) + (E_v - E_{F_v})] \qquad [\mu m], \qquad (2.29)$$

where the wavelength is in microns and the energy is in electron-volts. When injected current and thus the number of injected carriers increase, the electrons in the conduction band and the holes in the valence band fill each band from the edge (bottom) in order. The quasi-Fermi levels, E_{F_c} and E_{F_v}, gradually move away from the band edge, and their energies increase according to the increase in the injected carrier density. In this case the quasi-Fermi level, E_{F_c}, moves away more quickly than the quasi-Fermi level, E_{F_v}, because the density of the states in the conduction band is generally smaller than that in the valence band. The increase in the energy difference between the transition levels makes the wavelength of the emitted light shorter (band-filling effect). This wavelength-shortening can be observed if the LEDs operate under pulsed current and the junction temperature rise is negligible. Under dc operation, however, the effect of the band-filling is covered by the effect of the band-gap reduction caused by the junction temperature rise (discussed in the previous section). The wavelength therefore usually lengthens under dc operation as the injected current increases.

The wavelength also varies with ambient temperature because of the temperature dependence of band-gap energy. The temperature dependence of the peak wavelength is given by differentiating Eq. (2.29) with respect to temperature and around room temperature is roughly given, from Eq. (1.74) and by using Eq. (2.28), by

$$\begin{aligned} d\lambda_p/dT &\approx 1.24\, d(E_g^{-1})/dT \\ &\approx -1.24\, E_g^{-2}(dE_g/dT) \\ &\approx 1.24\, \alpha_a E_g^{-2}. \end{aligned} \qquad (2.30)$$

Around room temperature, the temperature coefficient of the peak wavelength is proportional to the temperature coefficient of band-gap energy and inversely proportional to the square of band-gap energy. The wavelength variation with ambient temperature tends to be larger as the band-gap of the active layer material becomes smaller. Examples are shown in Fig. 2.11 for surface-emitting type LEDs: The peak wavelength increases linearly with temperature, and the temperature coefficients of the 850 nm AlGaAs/GaAs and the 1300 nm InGaAsP/InP LEDs are 0.28 and 0.47 nm/°C, respectively. From these data the temperature coefficients of band-gap energy, α_a, for 850 nm-band AlGaAs and 1300 nm-band InGaAsP are estimated to be 4.8×10^{-4} and 3.4×10^{-4} eV/K, respectively. Values for indirect band-gap semiconductors can be estimated similarly by using Eq. (1.75).

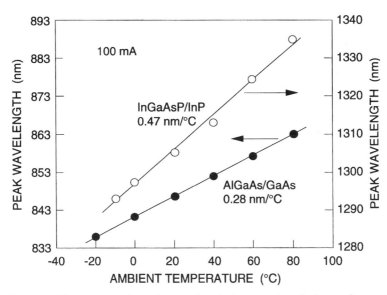

FIGURE 2.11   Temperature dependence of emission wavelength for surface-emitting 850 nm-band AlGaAs/GaAs and 1300 nm-band InGaAsP/InP LEDs.[19] Reprinted with permission from *Reliability and Degradation of Semiconductor Lasers and LEDs* by Artech House, Inc., Norwood, MA, USA, http://www.artech-house.com.

As shown in Fig. 2.3 for direct band-gap semiconductors, electrons and holes in the conduction and valence bands distribute at or around the two transition levels and recombine radiatively at random. The emission spectra therefore broaden in corresponding to the carrier distributions. The width of emission spectra is given by an integration for each radiative recombination with respect to photon energy emitted (or wavelength). In direct band-gap semiconductors the energy emitted by each transition obeys Eq. (1.74). If the width of the distribution is roughly expressed with ΔE, the spectral width, $\Delta \lambda$, can be given, by using calculus of variation and Eq. (1.74), as

$$\Delta \lambda = 1.24 \, E_g^{-2} \Delta E \qquad [\mu m]. \qquad (2.31)$$

The width of emission spectra is thus inversely proportional to the square of band-gap energy of the active layer. The spectral width increases according to the decrease in band-gap energy, and typical emission spectra for two surface-emitting LEDs are shown in Fig. 2.12. The spectral width (full width at half maximum, FWHM) is about 40 nm for 850 nm-band AlGaAs/GaAs LEDs, about 110 nm for 1300 nm-band InGaAsP/InP LEDs, and 130 nm for 1550 nm-band InGaAsP/InP LEDs.

The light-emission process in the active layer and the device characteristics for edge-emitting LEDs are the same as those for surface-emitting LEDs except

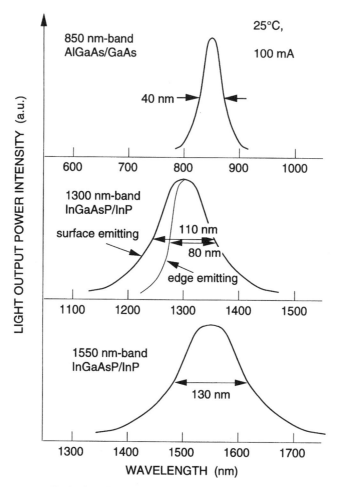

FIGURE 2.12 Typical emission spectra for surface- and edge-emitting LEDs.

for the width of the emission spectra. The emitted light propagates along the active layer from the inner region to the light-emitting surface. During the propagation, more of the light of shorter wavelengths in emission spectra is reabsorbed in the active layer because the absorption coefficients for those wavelengths are larger than those for longer wavelengths. Consequently, the emission spectra of edge-emitting type LEDs are narrow because the shorter ranges of the spectra are missing. As shown in Fig. 2.12, the spectral width decreases by about 70% from that of the surface-emitting LED made from the same material.[5]

2.2.5.3 Modulation Characteristics LEDs emitting infrared rays ranging from the 780 nm band to the 1550 nm band are used as optical sources in fiber-optic networks. They are usually made of direct band-gap semiconduc-

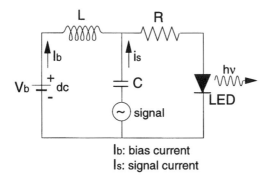

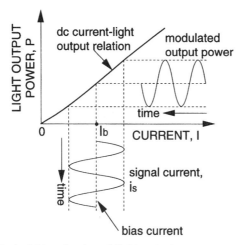

FIGURE 2.13   Typical bias circuit and light output power response under analog modulation.

tors and are often called infrared-ray-emitting diodes (IREDs). LEDs emitting visible light can also be used in these networks, if the transmission distance is short. In transmission systems, data links, and so forth, the emitted light is modulated by superimposing signals on the injected current or by operating LEDs with only signal current. The basic requirement for modulation is that output power corresponds to the modulation current signal. A simple bias circuit and the output power response are illustrated in Fig. 2.13, where the inductance and the capacitance are needed to cut the ac signal current and the dc current, respectively. The current signal is transferred to the light output power, and this is an intensity modulation: the light output power (light intensity) is modulated according to the signal current. In general, the response gradually drops as the modulation frequency increases because there are intrinsic and extrinsic factors limiting the frequency response.

The intrinsic factor is the injected carrier lifetime, which is ordinarily a few nanoseconds for spontaneous emission in LEDs. The light output power scarcely responds at all to the time variation which occurs within an interval whose length is shorter than the carrier lifetime. For the injected carrier density, the next simple rate equation holds under modulation:

$$dn/dt = [J_b + J_s \exp(i\omega t)]/qd - n/\tau_n, \tag{2.32}$$

where the bias and signal current are expressed by their densities, ω is the angular frequency ($= 2\pi f$, f = frequency) of the modulation signal, d is the active layer thickness, and τ_n is the lifetime of the injected minority carrier. Here the current signal is expressed as a complex number (or vector) and the imaginary part corresponds to the instantaneous value. The first term on the right-hand side of Eq. (2.32) indicates the injected carrier density, and the second term corresponds to the carrier density lost by recombination with a lifetime, τ_n. By solving Eq. (2.32),

$$n = (1 + \tau_n^2 \omega^2)^{-1/2}(\tau_n J_s/qd) \exp(i\omega t) + \tau_n J_b/qd. \tag{2.33}$$

In ordinary circumstances, LEDs in communication systems are used within the range where the relation between current and output power is linear, and thus the output power is proportional to the injected carrier density according to Eqs. (2.10) and (2.11). The modulated carrier density—that is, modulated light output power—changes with the injected carrier lifetime and with frequency, as indicated in Eq. (2.33). The total frequency response, $f_{re}(\omega)$, is given by the frequency response of the coefficient for the frequency term, $\exp(i\omega t)$, in Eq. (2.33):

$$f_{re}(\omega) = (1 + \tau_n^2 \omega^2)^{-1/2}. \tag{2.34}$$

The main extrinsic factor, on the other hand, is the junction capacitance of pn-junction (discussed in Section 1.2) and is a biasing circuit problem arising from RC-time constant, where R is resistance in the circuit. The capacitance is proportional to the area of the pn-junction, so the area is often reduced by chemically etching the pn-junction down to a mesa structure.

Modulation bandwidth is generally used as a measure of the frequency response and the concept of this bandwidth is illustrated in Fig. 2.14. At lower frequencies, the output response is limited by the capacitance and at higher frequencies it is limited by the lifetime of the injected minority carrier. The modulation bandwidth of LEDs, f_{3dB}, is expressed by the frequency at which the light output power decreases by 50% (or 3 dB):

$$f_{3dB} = 10 \log_{10}[P_{out}(f)/P_{out}(dc)], \tag{2.35}$$

where $P_{out}(f)$ and $P_{out}(dc)$ are the light output powers at f Hz and zero Hz

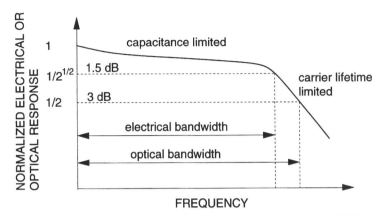

FIGURE 2.14 Conceptional illustration of modulation bandwidth.

(dc). When an optical detector operates within the range over which the optical input–current output characteristic is linear, $P_{out}(f)/P_{out}(dc)$ in Eq. (2.35) is equal to the ratio of the photocurrents, $I_{out}(f)/I_{out}(dc)$. For electrical signals, the bandwidth is defined at the point of $I_{out}(f)/I_{out}(dc) = 2^{-1/2}$ (or 1.5 dB) because the corresponding power is $[I_{out}(f)]^2/[I_{out}(dc)]^2 = 1/2$ (or 3 dB). Consequently, for the modulated carrier densities calculated using Eqs. (2.33) and (2.34), the electrical bandwidth corresponding to the 3 dB bandwidth is

$$(1 + \tau_n^2\omega^2)^{-1/2} = 2^{-1/2} \tag{2.36}$$

and the corresponding frequency, f_c, is given as

$$f_c = (2\pi\tau_n)^{-1}. \tag{2.37}$$

This frequency, f_c, is called the cut-off frequency and is determined by the minority carrier lifetime, and thus by the impurity concentration of the active layer. When the carrier lifetime is 5 ns, a reasonable value for actual LEDs, the cut-off frequency is about 32 MHz. The internal quantum efficiency is given by $\eta_i = \tau_n/\tau_r$ [see Eq. (1.110)] and thus,

$$\eta_i = (2\pi\tau_r)^{-1}/f_c. \tag{2.38}$$

The cut-off frequency and the internal quantum efficiency—that is, output power—are thus reciprocally related to each other.

2.2.5.4 *Luminous Efficiency (Eye Response) and LEDs Emitting Visible Light*
The structures and device characteristics described in the previous sections are of course applicable to LEDs emitting visible light as well as to those emitting infrared rays, though the LEDs emitting visible light are often composed

of indirect band-gap semiconductors. Major application fields for LEDs emitting visible light are displays and indicators, and a human factor becomes very important in these applications. This factor is the eye response, and it is represented by means of luminous efficiency. The optical detectors in communication systems are usually photodiodes, whereas in display and indicator applications the detectors are human eyes. The eye response is different at different wavelengths, and the luminous efficiency indicates the correlation between light output power and physiological brightness:[24]

$$V_{le} = \int_0^\infty P_{out}(\lambda)V(\lambda)d\lambda]/\int_0^\infty P_{out}(\lambda)d\lambda], \qquad (2.39)$$

where V_{le} is the luminous efficiency, $P_{out}(\lambda)$ is the LED output power [see Eq.(2.17)], and $V(\lambda)$ is the spectral luminous efficiency. An outline of the spectral luminous efficiency for the light-adapted and dark-adapted states is indicated in Fig. 2.15. The boundary between the two states is not defined clearly. The spectral response under an illumination (or to an LED output power) of more than a few candela per square meter is ordinarily used for light-adapted state, and one of less than 10^{-2} cd/m^2 is used for a dark-adapted state. The

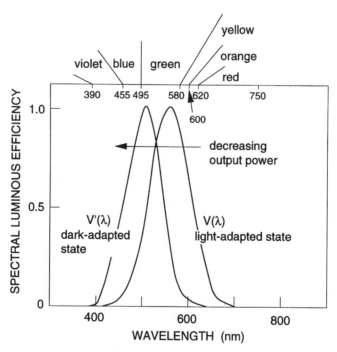

FIGURE 2.15 Spectral luminous efficiency of the human eye in light-adapted and dark-adapted states.

spectral luminous efficiency depends on the output power from LEDs and is expressed with $V(\lambda)$ and $V'(\lambda)$ for a light-adapted and a dark-adapted state in Fig. 2.15. As the output power decreases, the spectral response shifts toward shorter wavelengths as shown in Fig. 2.15. The conversion efficiency for physiological output brightness of LEDs, η'_{cv}, is therefore approximately expressed as a function of emission wavelength by rewriting Eq. (2.15) as

$$\eta'_{cv} = \eta_{ext}(E_g/V_j)V_{le}. \tag{2.40}$$

The physiological efficiency of an LED is higher in the wavelength range of yellow and green light than in the wavelength range of red and violet light (see Fig. 2.15).

REFERENCES

1. S. Nakamura, M. Senoh, N. Iwasa, S. Nagahama, T. Yamada, and T. Mukai, Superbright green InGaN single-quantum-well-structure light-emitting diodes, *Japan. J. Appl. Phys.*, part 2, **34**, L1332, 1995.

2. C. A. Burrus, and B. I. Miller, Small-area double-heterostructure aluminum-gallium-arsenide electroluminescent diode sources for optical-fiber transmission lines, *Opt. Commun.* **4**, 307, 1971.

3. S. Yamakoshi, M. Abe, O. Wada, S. Komiya, and T. Sakurai, Reliability of high radiance InGaAsP/InP LEDs operating in the 1.2–1.3 μm wavelength, *IEEE J. Quantum. Electron.*, **QE-17**, 167, 1981.

4. For example, M. Iwamoto, M. Tashiro, T. Beppu, and A. Kasami, High efficiency GaP green LEDs with double n-LPE layers, *Japan. J. Appl. Phys.*, **19**, 2157, 1980.

5. D. Botez and M. Ettenberg, Comparison of surface-and edge-emitting LEDs for use in fiber-optical communications, *IEEE Trans. Electron. Devices*, **ED-26**, 1230, 1979.

6. M. Abe, O. Hasegawa, Y. Komatsu, and Y. Toyama, A new type GaAlAs monolithic lensed LED, *Proc. 11th Int. Conf. Solid State Devices*, Tokyo, 1979; *Japan J. Appl. Phys.*, **19** (Supplement 19-1), 365, 1980.

7. K. Kurata, Y. Ono, K. Ito, M. Mori, and H. Sano, An experimental study on improvement of performance for hemispherically shaped high-power IREDs with $Ga_{1-x}Al_xAs$ grown junction, *IEEE Trans. Electron. Devices*, **ED-28**, 374, 1981.

8. Y. Seki, Light extraction efficiency of the LED with guide layers, *Japan. J. Appl. Phys.*, **15**, 327, 1976.

9. H. Kressel and M. Ettenberg, A new edge-emitting (AlGa)As heterojunction LED for fiber-optic communications, *Proc. IEEE*, **63**, 1360, 1975.

10. See for example, J. I. Pankov, ed., *Electroluminescence*, Springer-Verlag, Berlin 1977.

11. T. N. Morgan, B. Welber, R. N. Bhargava, Optical properties of Cd-O and Zn-O complexes in GaP, *Phys. Rev.*, **166**, 751, 1968.

12. C. H. Henry, P. J. Dean, J. D. Cuthbert, New red pair luminescence from GaP, *Phys. Rev.*, **166,** 754, 1968.

13. M. Born and E. Wolf, *Principles of Optics*, 5th ed., Pergamon Press, New York, 1975.

14. C. H. Henry, R. A. Logan, and F. R. Merritt, The effect of surface recombination on current in AlGaAs heterojunctions, *J. Appl. Phys.*, **49,** 3530, 1978.

15. M. Fukuda, Current drift associated with surface recombination current in InGaAsP/InP optical devices, *J. Appl. Phys.*, **59,** 4172, 1986.

16. C. H. Henry, R. A. Logan, and F. R. Merritt, Origin of $n = 2$ injection current in AlGaAs heterojunctions, *Appl. Phys. Lett.*, **31,** 454, 1977.

17. G. B. Stringfellow, Effect of structure treatment on surface recombination velocity and diode leakage current in GaP, *J. Vac. Sci. Technol.*, **13,** 908, 1976.

18. G. Leistiko Jr. and C. A. Bittman, Surface effects of $GaAs_{0.6}P_{0.4}$ light emitting diodes, *Solid State Electron.*, **16,** 1321, 1973.

19. M. Fukuda, *Reliability and Degradation of Semiconductor Lasers and LEDs*, Artech House, Norwood, MA, 1991.

20. F. Starn, Gain-current relation for GaAs lasers with n-type and undoped active layers, *IEEE J. Quantum Electron.*, **QE-9**, 290, 1973.

21. A. R. Goodwin, J. R. Peters, M. Pion, G. H. B. Thompson, and J. E. A. Whiteway, Threshold temperature characteristics of double heterostructure $Ga_{1-x}Al_xAs$ lasers, *J. Appl. Phys.*, **46,** 3126, 1975.

22. M. Yano, H. Nishi, and M. Takusagawa, Temperature characteristics of threshold current in InGaAsP/InP double heterostructure lasers, *J. Appl. Phys.*, **51,** 4022, 1980.

23. Many books related to the semiconductor physics describe the temperature dependence of band-gap energy, and originated from Y. P. Varshni, *Physica*, **34,** 149, 1967.

24. See for example, P. Bhattacharya, *Semiconductor Optoelectronic Devices*, Prentice-Hall, Englewood Cliffs, NJ 1994.

CHAPTER THREE

CHAPTER THREE

Laser Diodes

Laser diodes are, like LEDs, typical *pn*-junction devices used under a forward bias. Laser is an acronym for Light Amplification by Stimulated Emission of Radiation. The basic operating mechanisms of laser diodes and LEDs are similarly based on the electrical and optical properties of *pn*-junctions and semiconductor materials. Laser diodes are devices emitting coherent light produced in a stimulated emission process whereas LEDs under injection excitation emit light produced in a spontaneous emission process. The light emission process in laser diodes is therefore more complicated than that in LEDs. The materials used in laser diodes are quite similar to those used in LEDs, and the lasing wavelength ranges from the visible to the infrared wavelength range depending on the material of the active layer. The 850 nm-band AlGaAs/GaAs and 1300/1550 nm-band InGaAs(P)/InP material systems are commonly used in optical fiber communication systems, whereas AlGaAs/AlGaAs, group-III nitrides, and other materials are used in the visible wavelength range. Typical laser diodes emitting visible light are composed of III-V compound semiconductors, but some II-VI compound semiconductors such as (Zn, Cd)Se/ZnSe are also used in laser diodes emitting visible light. On the developmental stage of laser diodes, continuous-wave (cw) lasing at room temperature was achieved in an AlGaAs/GaAs double-heterostructure laser in 1970,[1,2] and since then the double heterostructure has been a common one for laser diodes. Consequently, this chapter's description of laser diodes is based on the double heterostructure.

3.1 BASIC STRUCTURE OF LASER DIODES: FABRY–PEROT LASERS

The basic structure of the Fabry–Perot type of double-heterostructure laser diodes is illustrated in Fig. 3.1. The structure is quite similar to that of the double-heterostructure LEDs, especially the edge-emitting types. In addition to the LED structure, it has two mirror facets—which are usually (110) cleaved

123

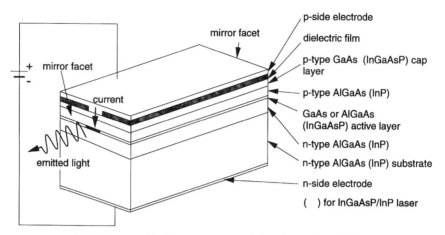

FIGURE 3.1 Basic structure of a Fabry–Perot laser diode.

facets—one at each end of the active layer. The mirror facets are parallel to each other and perpendicular to the active layer. This structure, therefore, forms a Fabry–Perot (FP) type of optical cavity for lasing, and this type of laser is called the Fabry–Perot type laser. The coherent light from the stimulated emission process in the active layer is emitted from the facets in opposite directions. The basic configuration of double-heterostructure lasers is shown in Fig. 3.2, where AlGaAs/GaAs and InGaAsP/InP lasers are used as typical examples. Under the forward bias condition the minority carriers are injected into the active layer through the *pn*-junction. Because of the requirement for electrical neutrality, what actually occurs is double injection (majority carriers also diffuse into the active layer). For stimulated emission, a very high density of injected carrier (over 10^{18} cm^{-3}) is required to form a population inversion as discussed in Section 1.1.3.1. Such a high density of the injected carriers is attained by making the active layer less than 150 nm thick. For stimulated emission to occur at a high rate leading to lasing at a low input power, a high density of injected carriers and photons (or emitted light) needs to be confined within the active layer, as indicated in Eq. (1.98). This confinement of carriers and photons can be accomplished by setting the refractive index of the active layer than that of the neighboring layers (called the cladding layers) as shown in Fig. 3.2. Emitted light propagates along the high-refractive index region, as it does in optical fibers where the refractive index of the central part is higher than that of the periphery. The combinations of GaAs (or AlGaAs) with AlGaAs and of InGaAs(P) with InP are typical examples of pairs of materials with appropriately related refractive indexes. The refractive indexes of GaAs (or AlGaAs) and InGaAs(P) active layers are a few percent higher than those of AlGaAs and InP cladding layers. The emitted light is thus confined to the active layer and the optical field is distributed around the active layer as shown in Fig. 3.2(e).

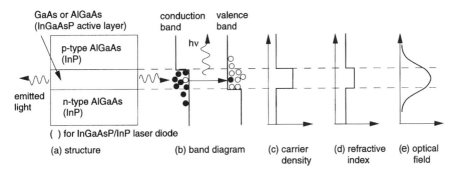

FIGURE 3.2   Basic configuration of double heterostructure laser diodes.

3.2   BASIC CHARACTERISTICS OF LASER DIODES

The operating characteristics of a typical 1300 nm-band InGaAsP/InP FP laser diode are shown in Fig. 3.3. When the forward bias is applied and gradually increased, the laser diodes initially operate like edge-emitting LEDs. In the low bias range—that is, the low excitation range—spontaneous emission is dominant because the carrier density in the active layer is not high enough for forming a population inversion (see Section 1.1.3.1). As the bias is increased, the population inversion occurs in the active layer. Stimulated emission thus

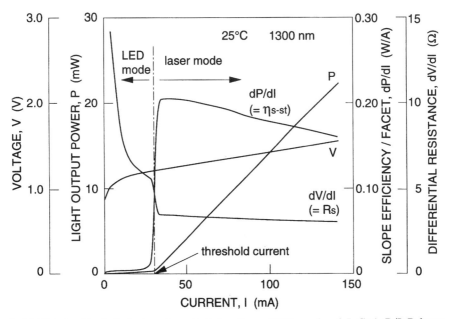

FIGURE 3.3   Typical device characteristics for a 1300 nm-band InGaAsP/InP laser diode with cleaved facets.

becomes dominant at a certain bias. This bias point is called the lasing threshold, and the corresponding current is called the threshold current. At the threshold point, the loss of light in the active region (cavity loss) is balanced by the gain from light amplification. The change from LED to laser diode occurs at the threshold current. The injected current above the threshold induces stimulated emission, and coherent light is emitted from the laser diode.

3.2.1 Lasing Conditions

3.2.1.1 Threshold Gain and Current Threshold current is the most important and basic parameter for laser diodes. It is a performance measure determined by several factors due to basic properties of materials used for structures. Below the threshold, the emitted light in the active layer propagates along the layer because the refractive index of the active layer is higher than that of the cladding layers. Part of this propagating light is reflected at the mirror facet and the rest is emitted to the outside from the facet. This propagation and reflection is repeated within the optical cavity formed by the active layer and the two mirror facets, and light is lost and gained. The cavity loss is mainly constituted of the absorption losses (see Section 1.1.3.1) in the active layer and the cladding layers. Mirror loss is caused by the mirror facet reflectivity being less than 100% (about 30% just after cleavage), and scattering loss is caused by structural inhomogeneities (such as the roughness of the heterointerface). The mirror loss can be reduced by using a highly reflective film coating (see Section 6.2.2.3), although this also decreases the light output power. The gain originates from the injection excitation. The lasing condition for an optical cavity formed by an active layer of length, L, and two mirror facets of reflectivities, R_1 and R_2, is illustrated in Fig. 3.4. The internal loss corresponding to the sum of absorption

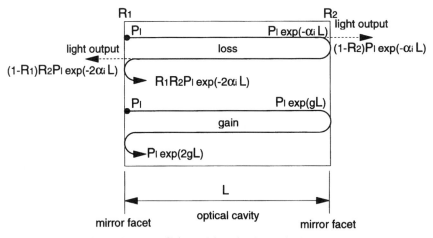

FIGURE 3.4 Gain and loss in the optical cavity.

loss and scattering loss per unit length is expressed by a loss coefficient, α_i in units of cm^{-1}, and the light power, P_i, decreases during a round trip by the factor, $R_1 R_2 \exp(-2\alpha_i L)$. The gain per unit length, on the other hand, is expressed by a gain coefficient, g in units of cm^{-1}, and the light power increases by the factor, $\exp(2gL)$, during a round trip. The propagating light power after a round trip is thus changed by the factor

$$R_1 R_2 \exp 2L(g - \alpha_i). \tag{3.1}$$

When the gain and loss balance each other, the light power after the round trip is the same as the initial power and lasing starts. The lasing condition is therefore given by

$$R_1 R_2 \exp 2L(g - \alpha_i) = 1. \tag{3.2}$$

The gain coefficient at the threshold (threshold gain) can be given by rewriting Eq. (3.2) as

$$g_{th} = \alpha_i + (1/2L) \ln (1/R_1 R_2). \tag{3.3}$$

The first-term on the right-hand side of the equation is the internal cavity loss and is dominated by free carrier absorption in the active layer. The second term on the right-hand side indicates the mirror loss and is inversely proportional to the cavity length because photons related to stimulated emission can stay within the cavity for a time proportional to the cavity length, or in other words, propagating light takes more time for a round trip as the cavity lengthens. Just after cleavage, the reflectivity (power reflectivity) of the facet is determined by the difference in refractive index between the laser material and air,

$$R_1 = R_2 = R_{cleavage} = [(n_r - 1)/(n_r + 1)]^2, \tag{3.4}$$

where n_r is the refractive index of material of the laser diodes [see Eq. (6.8)]. Consequently, Eq. (3.3) can be simplified in this case as

$$g_{th} = \alpha_i + (1/L) \ln (1/R). \tag{3.5}$$

For AlGaAs/GaAs and InGaAsP/InP lasers with a 300 μm-long cavity, a bulk active layer, and two cleaved facets, the refractive index of the active region under operation is about 3.3. Thus the power reflectivity is about 31% from Eq. (3.4), and the mirror loss in Eq. (3.3) is calculated to be about 40 cm^{-1}. The internal cavity loss in the normal bulk active layer, for example, is between 10 and 20 cm^{-1}. For lasing, therefore, a threshold gain of about from 50 to 60 cm^{-1} is needed in such laser diodes.

The gain (optical gain) changes with the injected carrier density, especially

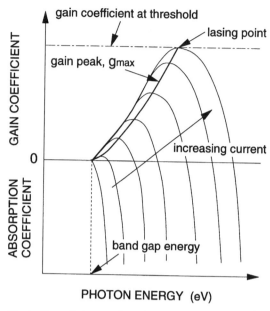

FIGURE 3.5   An illustration of changes in gain spectra with changes in injected current.

below threshold. As illustrated schematically in Fig. 3.5, gain becomes high as the injected current (carrier density) increases. According to the increase in the carrier density, the gain peak gradually shifts higher energies because of the band-filling effect. An actual case of such a change is shown in Fig. 3.6. The peak gain for this device does not coincide with the peak of spontaneous emission (or electroluminescence peak) beause the optical absorption loss at the electroluminescence peak is larger than that at longer wavelengths. The peak gain is usually at a wavelength a few tens of nanometers longer than the wavelength of the electroluminescence peak. As shown in Fig. 3.5, the threshold gain coefficient corresponds to the peak gain (maximum gain coefficient). The maximum gain coefficient, g_{max}, is also a function of injected carrier density:

$$g_{max} = \beta(J/d - J_0)^m, \tag{3.6}$$

where β is a constant called the gain constant or gain factor, J/d is the normalized current per unit active layer thickness (in $A/cm^2/\mu m$) and corresponds to the injected carrier density, and J_0 is the current (in $A/cm^2/\mu m$) required to compensate the cavity loss for a transparent cavity and is called the transparency current. The threshold gain coefficient can be combined with the threshold current, taking the optical field distribution of laser diodes into consideration. As shown in Fig. 3.2, the optical field is distributed around the active layer, and the total absorption coefficient of active region is determined by the absorption coefficients of the

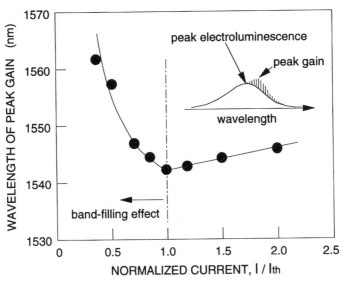

FIGURE 3.6 Peak gain as a function of injected current. The device is a 1300 nm-band Fabry–Perot strained MQW InGaAsP/InP laser diode with an active layer of 300 μm long and 1.5 μm wide. The MQW structure consists of six wells, each about 50 nm thick.

active layer and the adjacent layers. The ratio of light confined within the active layer to the total light is given by the optical confinement factor, Γ_a, and usually expressed with a numerical value from 0 to 1. This factor is a function of active layer thickness and of the refractive indexes of the active layer and the adjacent layers. The absorption coefficient can therefore be given as

$$\alpha_i = \Gamma_a\alpha_a + (1 - \Gamma_a)\alpha_{ad}, \tag{3.7}$$

where α_a and α_{ad} are the absorption coefficients of the active layer and the adjacent layers. Only the light confined within the active layer can be related to the optical gain, and Eq. (3.5) can thus be rewritten as

$$g_{th} = (1/\Gamma_a)[\alpha_i + (1/L)\ln(1/R)]. \tag{3.8}$$

Using Eqs. (3.6) and (3.8), the approximate threshold current density, J_{th} (in A/cm^2) can be expressed as

$$J_{th} = d\left\{\frac{1}{\beta\Gamma_a}[\alpha_i + (1/L)\ln(1/R)]\right\}^{1/m} + dJ_0, \tag{3.9}$$

where d is the active layer thickness in micrometers. Typical parameters for

1300 nm-band InGaAsP/InP laser diodes with bulk active layers are as follows: d, Γ_a, α_i, β, and J_0 are respectively about 0.1 μm, 0.2, 10 to 20 cm^{-1} ($L = 300$ μm and $R = 30\%$), 0.02, and 2000 A/cm^2/μm.[3] The exponent m is equal to 1 in most cases. From these parameters, the threshold current density, J_{th}, is calculated to be 1.6 kA/cm^2. The threshold current is obtained by multiplying the threshold current density by the area of the active region, which corresponds to the product of cavity length and the width of the light-emitting region. For example, if the stripe is 5 μm wide and the cavity length is 300 μm, the threshold current is about 24 mA. The value of J_0 depends on the material used in the active layer and for 850 nm-band AlGaAs/GaAs lasers is about 4000 A/cm^2/μm.[4] The calculated threshold current may be correct in an ideal case, but in actual devices there are invalid currents related to the nonradiative recombination via defects within the crystal and at the surface [see Eq. (1.157)] and to the Auger recombination process [see Eqs. (1.68) and (1.69)]. These currents increase the observed threshold current by more than a few tens of percentage points. The threshold carrier density corresponding to the threshold current, n_{th}, can be found from Eqs. (2.6) and (2.7):

$$n_{th} = \tau_s J_{th}/qd, \tag{3.10}$$

where τ_s is the lifetime of the injected minority carrier.

3.2.1.2 Phase Condition

The lasing condition is described in the previous section with regard to the relationship between gain and loss, but this relationship never completely determines the lasing conditions. The emission spectra are closely related to lasing, and only spontaneous emission corresponding to the lasing mode makes a contribution to the lasing. The phase of the propagating light after a round trip is required to coincide with the initial phase. This phase is determined by a Fabry–Perot geometry composed of the two mirror facets and the active layer. The following phase condition has to be satisfied:

$$m\lambda_m = 2L \qquad (m = 1, 2, 3, \ldots), \tag{3.11}$$

where m is called the mode index or mode number and λ_m is the wavelength of the mode in the optical cavity. If the refractive index of the active region is expressed with n_r, Eq. (3.11) can be rewritten as follows by using the wavelength in a vacuum, λ_0:

$$m(\lambda_0/n_r) = 2L. \tag{3.12}$$

This equation indicates the standing waves within the cavity. Lasing therefore occurs every wavelength (mode) space as shown in Fig. 3.7. The mode spacing, $\Delta\lambda_m$, can be obtained by using Eq. (3.12) and considering the next $(m+1)$ mode as

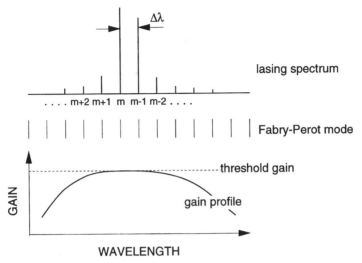

FIGURE 3.7   Illustration of gain profile and emission spectra (longitudinal modes).

$$(m + 1)[(\lambda_0 - \Delta\lambda)/n_r] = 2L, \tag{3.13}$$

and then

$$\Delta\lambda_m = \lambda_0^2/2n_{\text{eff}}L, \tag{3.14}$$

where n_r is replaced by the effective refractive index, n_{eff}. This term includes the wavelength dependence of n_r and is given by

$$n_{\text{eff}} = n_r[1 - (\lambda_0/n_r)(dn_r/d\lambda_0)]. \tag{3.15}$$

When the gain reaches the threshold level, lasing occurs at the wavelength satisfying Eq. (3.12). As shown in Fig. 3.7, the gain profile is usually broad, especially in the bulk active layer, because multiple modes often lase simultaneously. The lasing modes determined by a lightwave situation in the direction of the laser cavity are called the longitudinal modes or axial modes. The full-width at half-maximum of an enevelope of the longitudinal mode often exceeds a few nanometers. The relation between current and light output for a 1300 nm-band InGaAsP/InP FP laser is shown in Fig. 3.8, where longitudinal modes before and after lasing are also shown. Before lasing the envelope is quite broad, but it becomes narrow after lasing because only the modes near the maximum gain can lase. The mode spacing is about 0.8 nm because the cavity length is about 300 μm.

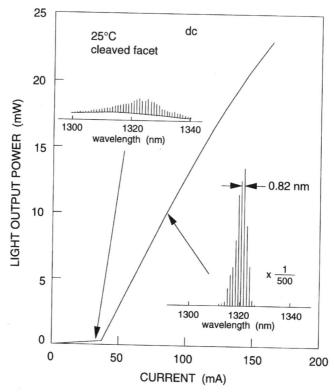

FIGURE 3.8 Emission spectra before and after lasing in a 1300 nm-band Fabry–Perot InGaAsP/InP laser diode with cleaved facets.[5] Reprinted with permission from *Reliability and Degradation of Semiconductor Lasers and LEDs* by Artech House, Inc., Norwood, MA, USA, http://www.artech-house.com.

3.2.1.3 Efficiency and Output Power Before lasing the behaviors of laser diodes are quite similar to those of edge-emitting LEDs, and a part of the spontaneous emission power in the active layer is emitted from both facets. The efficiency is quite low as described in Chapter 2, and the total output power from both facets is less than a few percent of the power emitted in the active layer. At the lasing threshold, stimulated emission exceeds spontaneous emission, and above this threshold, the efficiency—such as conversion efficiency indicated in Eq. (2.13)—increases quickly. The output power corresponds to the mirror loss from the viewpoint of the gain-loss relation shown in Fig. 3.4, and the current injected through the *pn*-junction induces gain. The efficiency after lasing is usually expressed by using the external differential quantum efficiency, which is the ratio of the increase in the number of photons emitted from the laser diode to the increase in the number of carriers injected. Taking output power from both mirror facets into consideration, the external differential quantum efficiency in units of %, η_d, is given as

$$\eta_d = \eta_{id}(2\Delta P_{\text{out-st}}/h\nu)/(\Delta l/q)$$
$$= \eta_{id}[\text{mirror loss}/(\text{internal cavity loss} + \text{mirror loss})]$$
$$= \eta_{id}\{(1/2L)\ln(1/R_1 R_2)/[\alpha_i + (1/2L)\ln(1/R_1 R_2)]. \tag{3.16}$$

Here η_{id} is called the internal differential quantum efficiency in unit of % and is given by the ratio of the increase in the number of photons to the increase in the number of injected carriers in the active layer:

$$\eta_{id} = (\Delta P_{\text{act-st}}/h\nu)/(\Delta l/q), \tag{3.17}$$

where $P_{\text{act-st}}$ is the lasing power due to the stimulated emission within the active layer. The concept of "differential" is used in the nonlinear relation, such as laser diodes, and the internal differential quantum efficiency is physically different from the internal quantum efficiency in LEDs. By using the external differential quantum efficiency, the lasing output power per facet (in W or mW) can be given as

$$P_{\text{out-st}} = (1/2)h\nu\eta_d S_{\text{act}}[(J - J_{\text{th}})/q]$$
$$= (1/2)h\nu\eta_d[(I_F - I_{\text{th}})/q], \tag{3.18}$$

where S_{act} is the area of the active layer and I_F and I_{th} are the forward injected current and the threshold current observed. In addition to the external differential quantum efficiency, the slope efficiency (or responsivity) of laser diodes, $\eta_{\text{s-st}}$, is also used very often and can be given by the first derivative of the output power with respect to the current, I_F:

$$\eta_{\text{s-st}} = dP_{\text{out-st}}/dI_F$$
$$= (1/2)\eta_d h\nu/q = 0.62\eta_d/\lambda(\mu m) \qquad [\text{W/A or mW/mA}]. \tag{3.19}$$

By using the slope efficiency, lasing output power is simplified from Eq. (3.18) as follows:

$$P_{\text{out-st}} = (1/2)h\nu\eta_d[(I_F - I_{\text{th}})/q]$$
$$= \eta_{\text{s-st}}(I_F - I_{\text{th}}). \tag{3.20}$$

Ideally, above the lasing threshold the optical output power increases according to Eqs. (3.18) and (3.20). Actually, however, the linear relationship between injected current and lasing output power is deteriorated by factors such as optical instability and Joule's heating due to current flow. These factors will be discussed in Sections 3.2.4.2 and 3.5.1.

For a 1300 nm-band laser whose characteristics are shown in Fig. 3.3, the slope efficiency, $\eta_{\text{s-st}}$, is about 0.225 mW/mA just after lasing and corresponds to the external differential quantum efficiency of about 47%. Ordinarily the

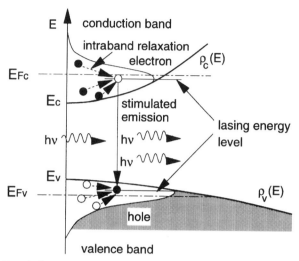

FIGURE 3.9 Band diagram of a bulk active layer under lasing conditions (stimulated emission).[5]

external differential quantum efficiency for both facets reaches a value between 40% and 60%. For the laser diode whose characteristics are shown in Fig. 3.3, the cavity length is 300 μm and the facet reflectivity is about 30% (cleavage), resulting in a mirror loss of 40 cm^{-1} and an internal loss of 10 to 20 cm^{-1}. Consequently, the internal differential quantum efficiency of the laser diode in Fig. 3.3 is 60% or more. The internal differential quantum efficiency is usually between 60% and 90%.

3.2.1.4 Injected Carriers and Lasing Some basic equations for current–light output power characteristics of laser diodes are given in the previous sections. The injected current induces gain and a lasing action starts when the injected current reaches a threshold level, above which gain saturation ideally keeps the injected carrier density constant, and the injected carriers (current) exceeding the threshold level are quickly converted into coherent light. The gain saturation indicates that the gain coefficient is kept constant due to reduction of population inversion when the excitation to laser gain medium is intensified. The stimulated emission (that is, output power) increases as the excitation increases. During these lasing processes, how do the carriers behave and take part in the lasing? As shown in Fig. 3.9, lasing occurs between the energy levels corresponding to the peaks of the electron and hole distributions in the conduction and valence bands. Just after the carriers are injected from electrodes into the active layer, they distribute randomly in the conduction and valence bands. In the active layer of an LED, the injected carriers recombine randomly after the injection, whereas the injected carriers never recombine directly in the active layer of a laser diode. In laser diodes the injected carriers have to reach the lasing levels to take part in the lasing. This process is called the intraband relaxation, and the time required for the relaxation is called the relaxation time. The relaxation time is determined by electron-electron scattering, electron-phonon scattering, and electron-impurity scattering (see Section 1.1.2.6) in each band and is typically between 10^{-12} and 10^{-13} sec. After the intraband relaxation, the carriers reach the stimulated emission levels in **k**-space (Fig. 3.9) and participate in the polarization of the semiconductor crystal for lasing. An empty state after stimulated emission is filled again immediately (within the relaxation time). The lifetime of spontaneous emission is normally about 10^{-9} sec, and thus the ratio of the relaxation time to the spontaneous emission lifetime is on the order of 10^{-3} or 10^{-4}. In addition, the carrier lifetime of nonradiative recombination, such as recombination via deep levels or surface states and Auger recombination, is on the order of 10^{-9} or 10^{-10} sec. Consequently, most of the injected carriers take part in the stimulated emission because of the short relaxation time. The internal differential quantum efficiency is determined by the ratio of the relaxation time in conduction and valence bands to the carrier lifetime of spontaneous emission and becomes high as described in the previous section, ideally 100%.

3.2.2 Current-Voltage Characteristics

The current-voltage characteristics of laser diodes, like those of LEDs, are basically given by Eq. (1.157)—except, of course, for those due to lasing. In the bias range above the voltage corresponding to the band-gap of the active layer (about 0.95 V on current-voltage characteristics shown in Fig. 3.3 for the 1300 nm-band laser diode), the current increases rapidly and then LED mode operation starts. In the bias range below that in which lasing occurs, the diffusion current is dominant and the current is approximately given by Eq. (2.20):

$$I_F = S_{act} J_{s0} \exp[q(V_b - R_s I_F)/k_B T]. \tag{2.20}$$

The injected carrier density in the active layer increases rapidly according to the relation expressed in Eq. (2.6) or (2.7) and shown in Fig. 3.10. The differential resistance is given by the first derivative of Eq. (2.20) with respect to I_F and is already given by Eq. (2.21):

$$dV_b/dI_F = (k_B T/q)(1/I_F) + R_s. \tag{2.21}$$

In the LED-mode operation before lasing, the differential resistance decreases rapidly with injected current and approaches the series resistance, R_s. At the lasing threshold current, however, the differential resistance corresponding to the slope of current-voltage characteristics becomes constant as shown in Fig. 3.3. This is due to the influence of the pinning of the quasi-Fermi level.

The injected carriers recombine and stimulated emission occurs with high

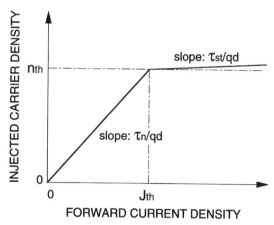

FIGURE 3.10   Injected carrier density as a function of injected (forward) current.

efficiency. The lifetime for the stimulated emission is three or four orders of magnitude shorter than those for nonradiative recombination and for spontaneous emission. Most of the injected carrier (current) is immediately converted to lasing output power. As a result of such rapid carrier recombination, the carrier density in the active layer becomes constant (see Fig. 3.10). This constancy of injected carrier density means that the quasi-Fermi levels are pinned at the lasing threshold position because the carrier distributions in the conduction and valence bands are fixed after lasing, even under further increases in the forward bias. This quasi-Fermi level pinning results in the saturation of the junction voltage because the junction voltage corresponds to the quasi-Fermi level difference, $qV_j = E_{F_c} - E_{F_v}$, as discussed in Section 1.2. This saturation of the junction voltage changes the current-voltage characteristics drastically. The junction voltage is expressed as

$$V_j = V_b - R_s I_F, \tag{3.21}$$

so by using $dV_j/dI_F \approx 0$, the first derivative with respect to the forward current gives the equation:

$$dV_b/dI_F = R_s. \tag{3.22}$$

The differential resistance changes rapidly from the value given by Eq. (2.21) to that given by Eq. (3.22) at the lasing threshold. Consequently, a steplike variation appears in the relation between current and differential resistance (see Fig. 3.3). Above the lasing threshold the slope of the current-voltage characteristic assumes a constant value determined by the series resistance. Actually, the injected carrier density still increases with the slope determined by the lifetime of stimulated emission (see Fig. 3.10), but because of the very short lifetime for stimulated emission, τ_{st}, the slope of this increase is about three orders of magnitude smaller than that of the slope before lasing. The above discussion is therefore approximately correct and is sufficient for consideration when discussing actual laser diodes.

When the bias is low, laser diodes operate in a manner similar to LEDs, and the characteristics of that operation have already been discussed in detail in Section 2.2.4.

3.2.3 Emission Spectra (Longitudinal Modes or Axial Modes)

The lasing mode and mode spacing are given by Eqs. (3.12) and (3.14). This longitudinal mode reflects the situation of the emitted light propagating in the direction of the cavity length. The mode spacing depends on the cavity length, and the number of lasing longitudinal modes, which corresponds to the number of modes within the gain width (see Fig. 3.7), decreases as the cavity length decreases. Consequently, a single-longitudinal mode operation is often observed in the laser diodes with a short cavity (less than 100 μm long) under dc bias, although, as shown in Eq. (3.9), the threshold current is high because of the large mirror loss. The conventional Fabry–Perot type of laser diode, however, normally shows multiple longitudinal mode operation as shown in Fig. 3.8.

When the dc bias on a Fabry–Perot laser diode increases, the peak wavelength, which means the highest mode in the wavelength spectrum, shifts to a shorter wavelength before lasing because of the band-filling effect (see Fig. 3.6). This is due to the shift of the peak gain or gain profile. In addition, each mode also shifts to a shorter wavelength because of the change in refractive index indicated in Eq. (3.12). The size of the wavelength shift is given by using Eq. (3.12) and calculus of variation:

$$\delta\lambda_m(n) = (\lambda_0/n_r)(dn_r/dn)\Delta n. \tag{3.23}$$

The refractive index change here is induced by the so-called plasma effect and is determined by the injected carrier density.[6] The size of the refractive index change due to the plasma effect, $\Delta n_{r\text{-plasma}}$, is expressed as

$$\Delta n_{r\text{-plasma}} = -(n_{r0}q^2/2m^*\omega_0^2\epsilon_s)\Delta n, \tag{3.24}$$

where n_{r0} is the refractive index when the carrier density is equal to zero, m^* is the effective mass of the injected carriers, and ϵ_s is the dielectric constant. The effective mass is nearly determined by the electron effective mass because the electron effective mass is smaller than hole effective masses.

The term Δn is the change in carrier density, and ω_0 is the angular frequency of the light and is equal to $2\pi\nu$ ($\nu = c_0/\lambda_0$).

The value of dn_r/dn in Eq. (3.23) is between -1×10^{-21} and -5×10^{-21} cm^3 for laser diodes lasing at an infrared range and for typical lasers is given as follows:[6]

$dn_r/dn \approx -1.3 \times 10^{-21} cm^3$ for 850 nm-band AlGaAs/GaAs lasers (3.25)

$dn_r/dn \approx -4 \times 10^{-21} cm^3$ for 1300 nm-band InGaAsP/InP lasers (3.26)

$dn_r/dn \approx -6 \times 10^{-21} cm^3$ for 1550 nm-band InGaAsP/InP lasers. (3.27)

After lasing begins, the band-filling effect and the plasma effect are nearly constant because the injected carrier density is nearly constant. The peak wavelength (or lasing peak wavelength) gradually lengthens as the injected current increases because of the Joule's heating due to the injected current. The lasing wavelength changes with temperature, and its temperature dependence is basically the same as that of LEDs. The dependence is largely determined by the temperature dependence of the band-gap energy as indicated in Eq. (2.30). The refractive index also changes with temperature. In a manner similar to that used to express the carrier density dependence of wavelength, the size of the wavelength shift for each mode is given as follows:

$$\delta\lambda_m(T_j) = (\lambda_0/n_r)(dn_r/dT_j)\Delta T_j, \tag{3.28}$$

where T_j is the junction temperature. The value of dn_r/T_j is between 2×10^{-4} and 5×10^{-4} K^{-1} for laser diodes lasing in the infrared range. The temperature dependence of each mode is about 0.08 nm/°C for 850 nm-band AlGaAs/GaAs, 0.1 nm/°C for 1300 nm-band InGaAsP/InP, and 0.12 nm/°C for 1550 nm-band InGaAsP/InP laser diodes. The average wavelength shift due to the temperature dependence of the band-gap is about 0.25 nm/°C for 850 nm-band AlGaAs/GaAs, 0.4 nm/°C for 1300 nm-band InGaAsP/InP, and 0.6 nm/°C for 1550 nm-band InGaAsP/InP laser diodes. The typical change in lasing peak wavelength for 1300 nm-band InGaAsP/InP lasers is shown in Fig. 3.11. There are several steps corresponding to the mode jumping (or hopping) at which the shift of peak gain causes the highest mode to change to an adjacent mode or longer mode. The step position often differs between cases in which the injected current increases and cases in which it decreases. The peak gain shifts smoothly according to the temperature dependence of the band-gap energy, while the gain near the lasing mode is coupled to the mode and then the mode hopping is suppressed until the gain at the adjacent range is higher than that at lasing. This results in the hysteresis loops in the wavelength change during bias current variation. The wavelength also shifts under constant current operation when the ambient temperature changes.

From Eqs. (3.23) and (3.28), the influences of the refractive index change due to carrier density change and of the temperature change on wavelength shift are opposite each other under dc operation. The influence of these factors on the direction of wavelength change is illustrated schematically in Fig. 3.12.

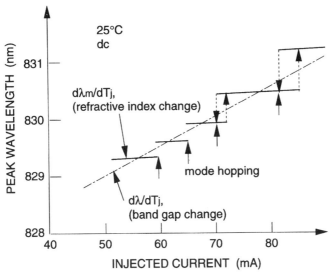

FIGURE 3.11   Wavelength shift as a function of injected dc current for a planar-type 850 nm-band Fabry–Perot AlGaAs/GaAs laser diode.

3.2.4   Transverse Modes

3.2.4.1   Modes, Optical Confinement, and Structure   Within the cavity of laser diodes the emitted light propagates along the active layer. This propagating situation reflects the optical field distribution of light emitted through the mirror facets. The behavior of propagating light has been analyzed in detail according to waveguide theory,[7] and the propagating fields are generally separated into two independent sets: TE (transverse electric) and TM (transverse magnetic) modes. The transverse mode in laser diodes is subdivided into a vertical mode and a lateral mode. The vertical transverse mode is in a direction perpendicular to the active layer, and the lateral transverse mode is in a direction parallel

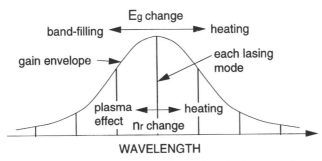

FIGURE 3.12   Conceptional diagram of wavelength change under operation.

to the active layer. The vertical transverse and the lateral transverse modes in laser diodes are conveniently called the TM- and TE-modes, and a laser diode usually lases in the TE-mode because the threshold gain for the TE-mode is lower than that for the TM-mode.

The vertical transverse mode reflects the standing wave between the two heterojunctions. In the direction perpendicular to the active layer, the confinement of optical fields to the active layer is determined by the thickness of the active layer, the difference between the refractive indexes of the active layer and the adjacent layers, and lasing wavelength. The active layer is usually less than 0.2 μm thick and the refractive index difference is less than 10%. Under these conditions laser diodes lase only in their fundamental mode. Operation in higher-order modes is possible only if the active layer is more than 1 μm thick, and thus normal laser diodes operate only in the fundamental mode perpendicular to the active layer.

The optical confinement perpendicular to the active layer, however, is low because of the thin active layer. As shown in Fig. 3.13, the optical confinement factor is usually less than 0.5 (50%) and is typically 0.2 (20%) or less. Some structures improving the optical confinement are often used to reduce the threshold current in a laser diode with a thin active layer, especially those with a quantum well structure. Those structures will be discussed in Section 3.3.

A lateral transverse mode reflecting the standing wave in the direction parallel to the active layer is complicated and influences device characteristics markedly when the active region is more than 1 μm wide. This lateral transverse mode is strongly influenced by the structure of the active region. To stabilize the

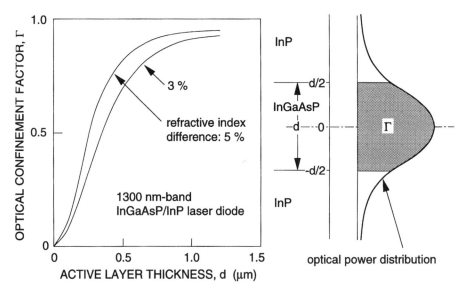

FIGURE 3.13   Calculated optical confinement in double heterostructures.

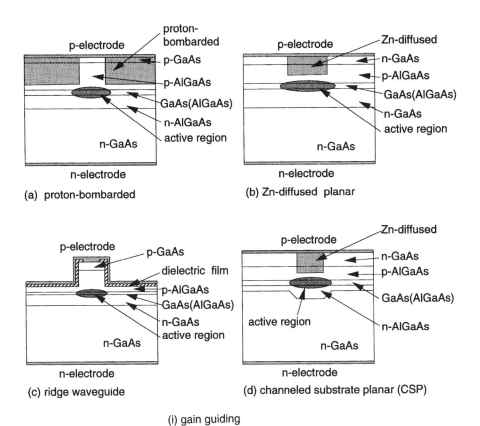

(i) gain guiding

FIGURE 3.14 Some popular structures for laser diodes: (a) proton-bombarded laser diode,[8] (b) Zn-diffused planar laser diode,[9] (c) ridge-waveguide laser diode,[10] (d) channeled substrate planar (CSP) laser diode,[11] (e) buried heterostructure (BH) laser diode,[12] (f) transverse junction stripe (TJS) laser diode,[13] (g) v-grooved substrate (VSB) BH laser diode,[14] (h) double-channel planar BH (DC-PBH) laser diode[15]. Reprinted with permission from *Reliability and Degradation of Semiconductor Lasers and LEDs* by Artech House, Inc., Norwood, MA, USA, http://www.artech-house.com.

lateral transverse mode (and, of course, to improve other characteristics) various kinds of structures have been developed (see Fig. 3.14). In those structures the lateral transverse mode is controlled by gain guiding or refractive index guiding. Gain guiding is generated by carrier injection, as shown in Fig. 3.15. Light amplification due to stimulated emission occurs only in the gain region. Under relatively high injection conditions, the carrier density of the center part of the light-emitting region decreases because of the high rate of stimulated emission. This phenomenon is called spatial hole-burning.[16] This reduction of the injected carrier density leads to an increase in refractive index, and the emitted light is confined within the center part (called self-focusing). This confinement, however, is unstable. The light-emitting region easily shifts toward

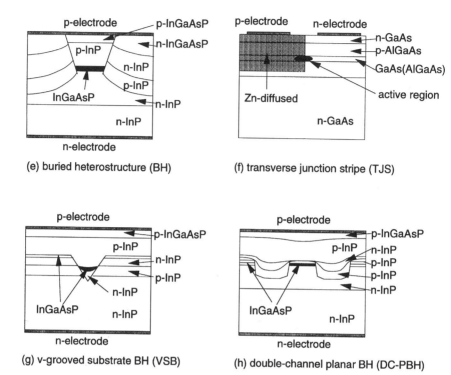

(e) buried heterostructure (BH)

(f) transverse junction stripe (TJS)

(g) v-grooved substrate BH (VSB)

(h) double-channel planar BH (DC-PBH)

(ii) refractive-index guiding

FIGURE 3.14 (*Continued*)

the higher refractive index (lower injected carrier density) side if there is any asymmetry in the refractive index distribution. Then the optical output power from laser diodes slightly or scarcely increases with increased current injection because the light confinement region does not correspond geometrically to the high gain region. These phenomena result in the kink on the current-light output power characteristics as shown in Fig. 3.16. At relatively high optical output powers, the lateral transverse mode tends to be unstable in laser diodes with gain guiding, although some planar type structures are still used in fields other than optical fiber communication because they are easy to fabricate and are reliable (see Chapter 7).

The buried heterostructure (BH) laser diodes show some excellent characteristics when compared with the planar type laser diodes. The current path and optical confinement region are strictly limited by the structure. The active layer is surrounded by another kind of material and the refractive index distribution is similar to that in an optical fiber. Consequently, if the dimensions of the active layer are appropriate, the lateral transverse mode is stable and fundamental even at high output powers. The stripe width of the active layer is

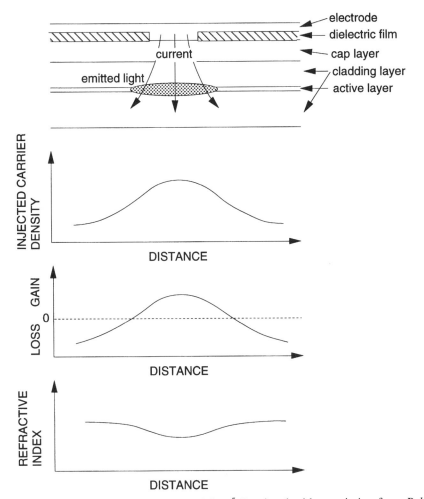

FIGURE 3.15 Basic concept of gain guiding.[5] Reprinted with permission from *Reliability and Degradation of Semiconductor Lasers and LEDs* by Artech House, Inc., Norwood, MA, USA, http://www.artech-house.com.

usually set at a value so small that only the fundamental mode can propagate within the layer: for example, less than 1.8 μm and 1.5 μm, respectively, for InGaAsP/InP laser diodes in the 1300 nm- and 1550 nm-bands. Those active layer widths for the fundamental mode operation were calculated as a function of the thickness of the active layer and the difference between the refractive indexes of the active layer and the adjacent cladding layers and burying layers. The current flow in buried heterostructure laser diodes is limited to the active layer by current blocking layers (burying layers). The *pn*-junction in the blocking layers is reversely biased under forward bias operation (see Fig. 3.14), and thus the injected current is confined to the active layer. A highly resistive layer,

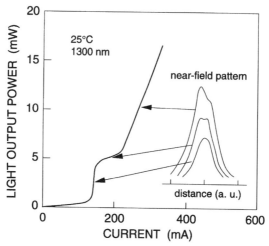

FIGURE 3.16 Current–light output power characteristics for a gain guiding laser diode. The laser diode is a 1300 nm-band InGaAsP/InP planar type laser diode with a 15 μm-wide stripe.

such as Fe-doped InP for InGaAsP/InP laser diodes, is also used as the blocking layer instead of the layers with *pn*-junctions.[17] As a result of the strict current confinement, the threshold current for a buried heterostructure laser diode is much less than that of a planar type laser diode (compare Figs. 3.3 and 3.16), although the threshold current density of a buried heterostructure laser diode is higher.

The buried heterostructure is the structure used most commonly in laser diodes, especially in the laser diodes used in optical fiber communication, because this structure results in excellent characteristics, low threshold current, and stable mode operation. The buried heterostructure, however, introduces an additional degradation mode: degradation of the buried heterointerface. This reliability problem will be discussed in detail in Section 7.2.1.4.

3.2.4.2 Near-Field Pattern and Far-Field Pattern The transverse modes are closely related to the light emission field. Two field patterns are used to describe the emitted optical field: the near-field pattern (NFP) and the far-field pattern (FFP). As shown in Fig. 3.17, the near-field pattern indicates the optical intensity distribution of emitted light at the mirror facet. The far-field pattern is important for applications (e.g., coupling to an optical fiber).

The vertical transverse mode is the fundamental mode in ordinary laser diodes, for which the near-field and far-field patterns are thus also fundamental (single-peak) configurations in the direction perpendicular to the active layer. The width of a far-field pattern is inversely proportional to that of a near-field pattern. This can be understood as a diffraction of light. The far-field pattern results from the diffraction of light through a slit having a width determined

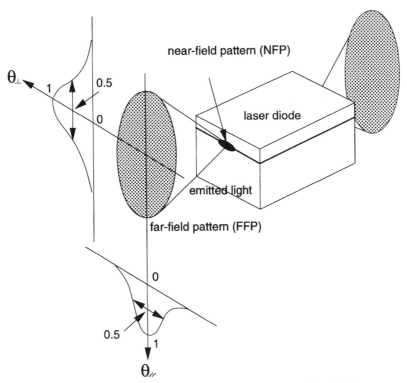

FIGURE 3.17 Light emission field: near-field pattern and far-field pattern.

by the near-field pattern. As the slit becomes narrower, the diffraction becomes more severe. The width of far-field pattern is expressed by the angle at the half-point of the intensity peak (full-width at half maximum: FWHM) in the emitted pattern. In actual laser diodes the angle in the direction perpendicular to the active layer is typically about 40 degrees. In gain guiding, the carrier injection region is often limited to within a stripe region about 5 μm wide by controlling current flow to obtain the far-field pattern with a single peak. If the stripe is more than about 10 μm wide, lasing occurs at several points within the stripe (filamentary lasing), and the near-field and far-field patterns have peaks corresponding to each lasing point. For a single peak, the width of the far-field pattern is about 20 degrees. In refractive index guiding, the gain region and high refractive index regions coincide. Therefore, the instability which occurs in the gain guiding structures scarcely appears when the refractive index guiding structures are used. The buried heterostructure laser diodes show a stable transverse mode. The width of the far-field pattern in buried heterostructure laser diodes is about 30 degrees. The smoothness of the pattern differs between gain guiding and refractive index guiding. The edge of a gain-guided beam varies gradually as shown in Fig. 3.15, whereas that of a refractive index guided beam is

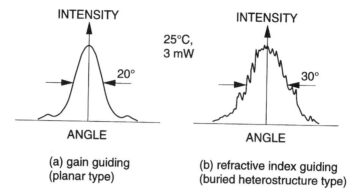

(a) gain guiding (planar type)

(b) refractive index guiding (buried heterostructure type)

FIGURE 3.18 Examples of far-field patterns in the direction parallel to the active layer for (a) gain-guiding and (b) refractive index guiding structures. The gain guiding is for a 1300 nm-band planar type laser diode with a 10 μm-wide oxide stripe and the refractive index guiding is for a 1300 nm-band buried heterostructure laser diode with 1.5 μm-wide active layer.

often irregular—especially a beam from laser diodes in which the active layer was physically and chemically etched during the formation of the buried heterostructure. The far-field pattern reflects this difference as shown in Fig. 3.18. In gain guiding, the envelope is quite smooth, and under high excitation conditions small peaks corresponding to the high refractive index region outside the active region appear on both sides of the envelope. There are many ripples on the envelope in the far-field pattern of the buried heterostructure laser diode. These are caused by interference between output light and light scattered at the irregular points on the side walls of the active layer. These ripples can be eliminated by improving the process used to fabricate the buried heterostructure.

Within the cavity, the situation of the propagating light differs for different guiding mechanisms. In refractive index-guiding, such as the guiding in buried heterostructure laser diodes, there is a refractive index difference in the direction perpendicular and parallel to the active layer. The light propagates under total reflection at the interface between the active layer and the adjacent layers, which are cladding layers and burying layers, and then the wave front is plane as shown in Fig. 3.19. The positions of the beam waist, in the direction perpendicular and parallel to the active layer nearly coincides with the mirror facet. Here, the beam waist is the minimum point of the spot size for the emitted beam and corresponds to the apparent source point of the emitted light. This means that the emitted light can be easily reformed to a parallel beam by using a lens and can be focused to a small area. The wave front in gain-guiding, in contrast, is not plane in the direction parallel to the active layer,[18] and the apparent beam waist is located within the cavity far from the mirror facet (see Fig. 3.19). The wave front in the direction perpendicular to the active layer is still plane because of the refractive index guiding, and the beam waist coincides with the mirror facet. This difference in the beam waist position in the direction perpen-

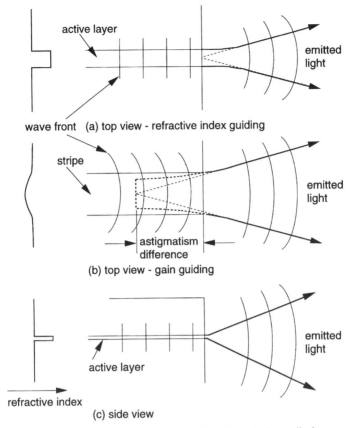

FIGURE 3.19   Beam waist and astigmatism in laser diodes.

dicular and parallel to the active layer is called astigmatism, and the distance between the positions of the two beam waists—that is, the distance from the mirror facet and the waist of the beam in the direction parallel to the active layer—is called the astigmatism difference. The astigmatism difference in normal gain guiding laser diodes is about 30 μm. The light emitted from laser diodes with the astigmatism cannot be reformed to a parallel beam by using normal lenses and cannot be focused to a small area. Laser diodes with severe astigmatism therefore cannot be used in laser printers, compact disk systems, and so forth (see Section 8.2).

3.3   ACTIVE LAYERS: BULK, QUANTUM WELL, AND STRAINED QUANTUM WELL TYPES

The basic characteristics described above are those of all kinds of laser diodes, even those with quantum well and strained quantum well structures. The exact characteristics of laser diodes, however, are slightly different for laser diodes whose active layers have different structures. Consequently, the active layer structures are described here before the device characteristics are discussed in detail.

From the standpoint of device performance, the most important part of a laser diode is its active layer, where injected carriers are present at very high densities and stimulated emission occurs as a result of their radiative recombination. Direct band-gap semiconductors are generally used as the active layers of laser diodes. In the initial developmental stage of laser diodes, in the 1970s and 1980s, the active layer was usually formed by a bulk material, such as GaAs, AlGaAs, and InGaAsP. At the end of the 1970s, lasing was achieved at 300 K under injection excitation in an AlGaAs/GaAs laser diode with quantum well structure.[19] Since then many kinds of quantum well structures have been developed for the active layers of laser diodes because of their excellent properties, such as a density of the states lower than that in bulk layers. In the latter half of the 1980s, the characteristics of laser diodes were further improved by the development of strained quantum well structures.[20,21] These structures also made it possible to produce laser diodes emitting wavelengths unattainable in lattice-matched material systems.

The thickness of a bulk active layer is usually set at a value between 0.08 and 0.2 μm to confine carriers and photons at a high density. If the layer is less than 0.05 μm thick, threshold current increases quickly, although the slope efficiency increases because the reduction of optical confinement causes absorption loss to decrease (see Fig. 3.20). The devices whose data are plotted in Fig. 3.20

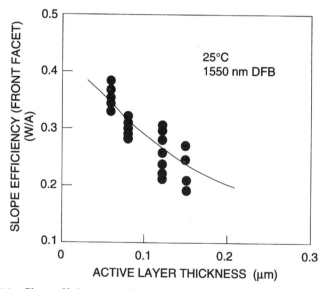

FIGURE 3.20   Slope efficiency as a function of bulk active layer thickness in 1550 nm-band DFB laser diodes (each with a 300 μm-long cavity and 0.12 μm-thick waveguide layer having a composition corresponding to a wavelength of 1300 nm).[22]

are distributed feedback laser diodes with κL values of 1.0 to 1.5 (see Section 3.4.1).

The properties of quantum wells and strained quantum wells are described in Section 1.3. In the (strained) quantum well structures, stimulated emission is due to transmissions between the quantized energy levels in the conduction and valence bands. The density of the states given by the steplike functions is reduced and the injected carrier distribution is localized near the quantized energy levels. These quantum-sized effects narrow the optical gain spectra (see Fig. 3.5). The gain, therefore, increases quickly under the injection excitation and the lasing threshold current density decreases. This corresponds to the large differential gain,

$$dg/dn = g_{max}/n, \tag{3.29}$$

where g and n are the gain and the carrier density, and g_{max} is given in Eq. (3.6). The (strained) quantum well structure can improve various characteristics of laser diodes.

The thickness of wells is typically between 5 and 10 nm. Some typical quantum well structures used in the active layer of laser diodes are illustrated in Fig. 3.21. The structures of strained quantum wells are basically the same as normal quantum well structures except for the lattice mismatching at the heterointerface. In the simple single quantum well (SQW) structure in Fig. 3.21(b), the optical confinement to the well is small and a large amount of injected current is therefore required to attain sufficient optical gain for lasing. In addition, the injected carriers overflow from the well under high-injection conditions. These effects counteract the advantage of quantum well structures (that the density of the states is reduced and lasing occurs at low-injection excitation). The simple quantum well structure is therefore not frequently used in the active layer of laser diodes, and other structures have been used to improve the optical confinement and the carrier confinement.[23] They are the separate-confinement heterostructure (SCH), the graded-refractive index separate-confinement heterostructure (GRIN-SCH), and the multiple quantum well (MQW) structure. In SCH and GRIN-SCH structures, the refractive index of the layers adjacent to the well is set at a value higher than that of the outside layers (see Fig. 3.22 for a SCH). The refractive index in a GRIN-SCH changes corresponding to the composition of the semiconductor, that is, to the band-gap [see Fig. 3.21(b)]. The emitted light is confined in the SCH and GRIN-SCH regions, and the threshold current is reduced. The multiple quantum well laser diodes show low threshold current densities even without such SCH and GRIN-SCH structures because of the strong optical confinement in the MQW structure. (Those structures are, of course, often used to further improve the device characteristics.) The number of wells is usually set between three and eight.

Strained quantum well structures result in even better device characteristics of laser diodes. As shown in Figs. 1.51 and 1.52, the subbands are formed in

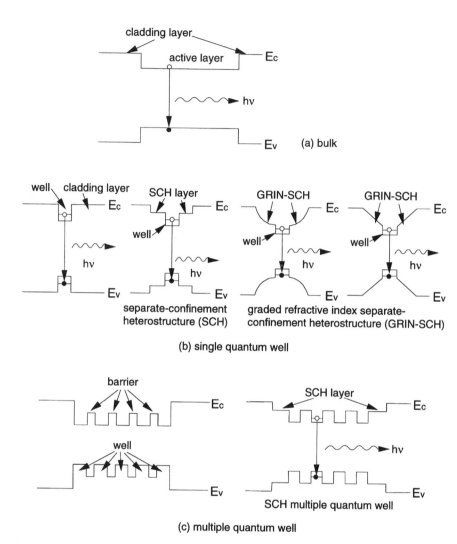

FIGURE 3.21 Typical active layer structures of laser diodes: band diagrams of (a) bulk, (b) single quantum well, and (c) multiple quantum well structures.

a parallel direction to the well. The light-hole and heavy-hole subbands in the valence band are separated by the strain as shown in Fig. 1.54. This separation of the bands results in a further reduction of the density of states and thus of hole effective mass. The reduced density of states introduces some advantages. As shown in Fig. 3.23, the influence of the strain on the threshold current clearly appears.[24] The gradual decrease in threshold current density according to the increase in compressive strain, from that of the lattice-matching quantum well (strain = 0 in Fig. 3.23), is due to the reduced density of states. The

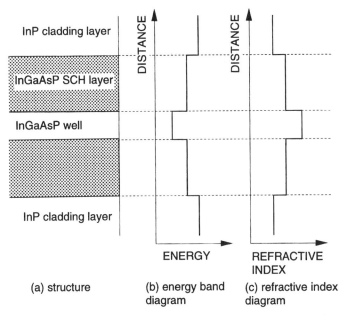

FIGURE 3.22   Schematic diagram showing the refractive index profile in a InGaAsP separate confinement heterostructure (SCH).

threshold current density also decreases when tensile stress increases as shown in Fig. 3.23. This decrease is introduced by the separation of the light-hole and heavy-hole bands. After the separation, the transition for lasing occurs between the subbands in the conduction band and one of the bands separated, and the transition between the subbands in the conduction band and the other band can

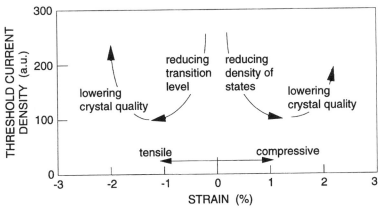

FIGURE 3.23   Schematic diagram of influence of strain on threshold current in strained quantum well laser diodes.

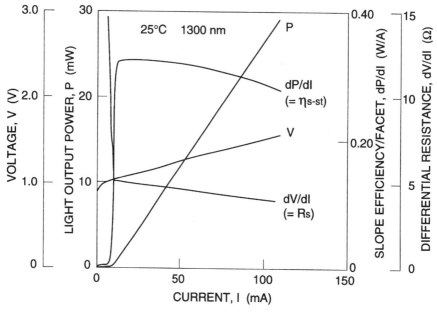

FIGURE 3.24   Device characteristics of a 1300 nm-band buried heterostructure compressively strained quantum well InGaAsP/InP laser diode with six wells and 0.8% compressive strain. The cavity is 300 μm long and both facets are cleaved.

be suppressed.[25] The actual characteristics of a 1300 nm-band compressively strained quantum well laser diode are shown in Fig. 3.24. The device parameters, except for the active layer structure, are the same as those of the device whose characteristics are shown in Fig. 3.3. Though the exact comparison cannot be carried out for the two laser diodes, the strained quantum well laser diode has a much lower threshold current than does a similar laser diode with a bulk active layer and also has a higher slope efficiency. These improvements of device characteristics in quantum well and strained quantum well laser diodes, over those of bulk laser diodes, are due to the lower density of state and smaller hole effective mass.

Other influences of the strained quantum well structures appear in the Auger recombination processes and the intraband absorption (see Sections 1.1.2.7 and 1.1.3.1) because the separation between split-off, light-hole, and heavy-hole bands changes with the strain. The temperature characteristics discussed later of laser diodes with the strained quantum well structures are better than those of laser diodes with bulk active layers. In addition, the polarization of output light is also influenced by the valence band modification in the strained quantum well structure. The laser diodes with bulk active layer or lattice-matched quantum well active layer emit light with the mixture of TE- and TM-modes, although the lasing in TE mode usually predominates because of the threshold gain difference. The heavy-hole bands are coupled to the TE modes, and the light-hole bands are coupled to the TM modes. The lasing mode is TE-like in the compressively strained quantum well laser diodes and is TM-like in tensilely strained quantum well laser diodes (see Fig. 1.54). The mode selectivity is intensified in the strained quantum well structures.

3.4 SINGLE LONGITUDINAL MODE LASER DIODES

Several kinds of single mode laser diodes that lase in the single longitudinal mode have been developed, and some of them are used in optical communication systems, such as coherent systems and high-dense wavelength-division-multiplexing (WDM) systems. The use of these laser diodes in actual equipment and systems requires stable operation with a single longitudinal mode under modulation and in severe environment conditions. To meet this requirement, distributed feedback (DFB) laser diodes and distributed Bragg reflector (DBR) laser diodes are commonly used. Their operating principles and characteristics are described in this section. Wavelength tunability for multisection DFB laser diodes and DBR laser diodes is also discussed.

3.4.1 Distributed Feedback Laser Diodes

The structure of DFB laser diodes is quite similar to that of the Fabry–Perot laser diodes shown in Fig. 3.14 except for the grating in the active region (see Fig. 3.25). If the grating is inserted into the active layer of a Fabry–Perot laser diode while satisfying some conditions related to the grating, the laser diode can operate as a DFB laser.

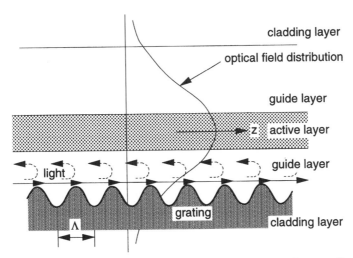

FIGURE 3.25 Cross-sectional view of a waveguide with a Bragg reflector.

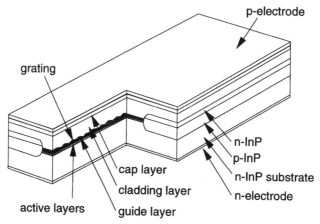

FIGURE 3.26   Cross-sectional view of a buried heterostructure InGaAsP/InP DFB laser diode.

3.4.1.1 Gratings and Lasing Wavelength

The structure of a typical buried heterostructure InGaAsP/InP DFB laser diode is shown in Fig. 3.26. The grating introduces a periodic refractive index change in the active region. If the grating shape is that of a sinusoidal wave which is used in most cases, the refractive index change is given as

$$n_r(z) = n_{eq} + \Delta n_r \cos(2\beta_0 z + \Omega_g), \qquad (3.30)$$

where n_{eq} is the equivalent refractive index (an overall refractive index of the multilayers influencing the propagation of the emitted light and a function of the layers and structures), and Δn_r is the difference in refractive index between the top and bottom of the grating. The term Ω_g is the phase of the grating at the center of the laser cavity and is equal to zero in a homogeneous grating, and β_0 is given by

$$\beta_0 = 2\pi n_{eq}/\lambda_B \equiv m\pi/\Lambda \qquad m = 1, 2, 3, \ldots \qquad (3.31)$$

where Λ is the grating pitch. The refractive index periodically changes by the grating pitch. This grating selectivity reflects a certain wavelength (Bragg wavelength) in the gain spectrum. Roughly speaking, this selectivity corresponds to taking one of the multiple longitudinal (Fabry–Perot) modes. The lasing wavelength of a DFB laser diode can be set at an arbitrary wavelength within the gain spectrum by adjusting the grating pitch and the equivalent refractive index. The shift of the lasing wavelength in the DFB mode from the gain peak is called the detuning. The wavelength selected by the grating—which corresponds to the lasing wavelength of DFB laser diodes—λ_{DFB}, is given by modifying Eq. (3.31) and by replacing λ_B with λ_{DFB} as[26,27]

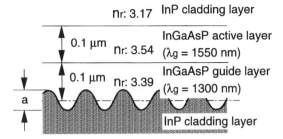

FIGURE 3.27 Typical active layer structure for a 1550 nm-band InGaAsP/InP laser diode.

$$\lambda_{\text{DFB}} = 2n_{\text{eq}}\Lambda/m \qquad m = 1, 2, 3, \ldots . \tag{3.32}$$

The grating pitches corresponding to $m = 1$, 2, and 3 are called the first-order grating, second-order grating, and third-order grating. The gratings commonly used in 1300 nm-band and 1550 nm-band InGaAsP/InP DFB laser diodes are the first- and second-order gratings, and the gratings used in 850 nm-band AlGaAs/GaAs laser diodes are usually second-order or third-order gratings because of the grating pitch and the fabrication difficulty. For a 1550 nm-band planar type of InGaAsP/InP laser diode shown in Fig. 3.27, the equivalent refractive index is calculated by using waveguide theory and each layer's refractive index at 1550 nm. The calculated equivalent refractive index is about 3.24, and the pitches for the first-order and second-order gratings are about 239 nm and 478 nm.

The intensity of reflected light is determined by the grating height and shape and by the distance between the grating and the active layer (optical field distribution). The grating height and shape are associated with the magnitude of refractive index change and with the spatial gradient of this change. The intensity of reflected light gradually becomes high as the gradient of the refractive index increases. The distance of the grating from the active layer—that is, the active layer and guide layer thicknesses—is a measure of how the optical field induced in the active layer is influenced by the grating (see Fig. 3.25). The strength of the influence is calculated by using the optical field distribution and optical confinement shown in Fig. 3.13 and can be expressed with the so-called coupling constant (in cm^{-1}). For a grating whose shape is like that of a sinusoidal wave, the coupling constant, κ is given by

$$\kappa = \pi\Delta n_{\text{r}}/\lambda_{\text{B}}. \tag{3.33}$$

The coupling constant for a 1550 nm-band InGaAsP/InP laser diode with the structure shown in Fig. 3.27 is shown in Fig. 3.28. The coupling constant for the first-order grating is about an order of magnitude larger than that for the

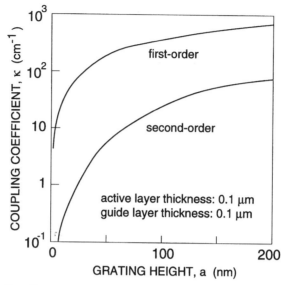

FIGURE 3.28   Coupling constant as a function of grating height in the sinusoidal wave-shape grating. The structure is shown in Fig. 3.27.

second-order grating because the wavelength of the propagating light coincides with the grating pitch of the first-order grating. The first-order grating also has advantages from the viewpoint of fabrication and crystal growth. It is not easy to fabricate a high grating while keeping crystal quality high. The total reflecting light is also influenced by the length of the distributed reflecting mirror. The magnitude of optical feedback is therefore expressed by the product of the coupling constant, κ, and laser cavity length, L. The κL is ordinarily designed to be about 1.0 for a homogeneous grating and to be about 2.0 for a grating with a phase-shifter (see next section). These values are determined from the viewpoint of lasing characteristics, including spectral stability.

3.4.1.2   Gratings and Phase-Shift   The optical reflection behaviors need to be solved in the structure shown in Fig. 3.25 by using coupled-mode theory for light propagating in opposite directions (right to left and left to right in Fig. 3.25).[28–30] As shown in Fig. 3.25, the phases of the two propagating lights in a homogeneous grating never coincide because the phase changes by $\pi/2$ during the reflection at the grating, and the two modes on both sides of the Bragg wavelength lase under injection excitation. The region between the two modes is called the stop band, and in the stop band there is no mode. The width of stop band is strongly influenced by the value of κ. This two-mode operation is corrected by changing the phase of the grating. If Ω_g in Eq. (3.30) is set at $\pi/2$, the two lights propagating in the opposite directions are coupled to each other, and lasing at the Bragg wavelength occurs. This means that a

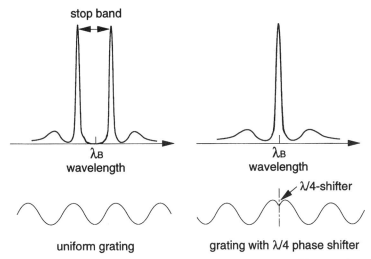

FIGURE 3.29   Emission spectra for DFB laser diodes with and without a phase shifter.

single longitudinal mode is operating at the Bragg wavelength. The phase shift of $\pi/2$ corresponds to the shift of $\lambda_0/4n_{eq}$ (λ_0 = wavelength in the vacuum) and of $\Lambda/2$ for the first-order grating. Laser diodes of this type are called phase-shifted DFB laser diodes. For both cases, lasing wavelength is illustrated in Fig. 3.29. The threshold of a phase-shifted DFB laser diode is lower than that of a DFB laser with a homogeneous grating because the two propagating lights are coupled.

The above discussions holds when both facets have no reflectivity, but the mirror facet is usually formed by a cleavage process that is carried out without consideration of the grating. The phase of the grating at a cleaved facet is determined by the cleaved position of the grating. This facet phase of the grating and reflectivity (due to the refractive index difference between laser material and air) influences the lasing characteristics.[31] The degree of the influence is different with the facet phase, and both facets are scarcely equal for the phase of the grating just after cleavage. Laser diodes with a phase-shift have a high probability of showing multiple longitudinal mode operation. Laser diodes with a homogeneous grating, on the contrary, show single-longitudinal mode operations with a certain probability. Consequently, both facets of laser diodes with phase-shifters are usually coated with antireflecting films, and the facet reflectivity is controlled to be less than 1%. If the quantity of the phase shift is more than $\pi/15$ at the center part of the cavity, a single mode operation is maintained, though the magnitude of the threshold gain is different with the phase shift.[32] For laser diodes with homogeneous gratings, the probability of single mode operation is increased by using a dielectric film coating to adjust the facet reflectivity. A front facet reflectivity of 1% combined with a rear facet reflectivity of 90% results in a single mode operation probability greater than 90%

for laser diodes with a κL value of about 1.[28] This structure with one antireflective facet and one highly reflective facet is nearly equivalent to the structure in which the cavity is doubled and symmetric with respect to the highly reflective facet. In the symmetric cavity, the highly reflective facet is located at the center position and acts as a phase-shifter.

3.4.1.3   Current–Light Output Characteristics and Spectral Characteristics

The gain at the threshold is decreased, when compared with that of Fabry–Perot laser diodes, by the contribution of the distributed mirror, so the lasing in the DFB mode begins before the lasing in the Fabry–Perot mode begins. Consequently, the spectral characteristics are more tightly related to gain-loss relation for lasing in DFB laser diodes than in Fabry–Perot laser diodes. The lasing wavelength—that is, the grating pitch—cannot be set arbitrarily. The Bragg wavelength indicated in Eqs. (3.31) and (3.32) has to be adjusted to the gain peak (or to the neighborhood of the gain peak) by controlling the grating pitch or composition of the active and guide layers. The simultaneously monitored DFB and Fabry–Perot modes in the emission spectra are shown in Fig. 3.30. Below the threshold current the intensities of the DFB mode and the Fabry–Perot mode are comparable. At and above the threshold, the DFB mode is dominant and a single longitudinal mode is intensified as the injected current increases. The description of the threshold current in DFB laser diodes is rather complicated because of the grating, but the basic concept for lasing is of course the same as that of Fabry–Perot laser diodes. It is based on the gain-loss relation and the threshold current density, for example, is expressed by the next equation:[33]

$$J_{\text{th-DFB}} = A d (\alpha_{\text{loss}} + \Gamma_{\text{a}}\alpha_{\text{in}})^2 / \Gamma_{\text{a}}^2, \tag{3.34}$$

where A is a constant, d is thickness of the active layer, and Γ_{a} is the optical confinement factor of the active layer. The term of $\Gamma_{\text{a}}\alpha_{\text{in}}$ represents the loss until the population inversion is built or the cavity is transparent at the lasing wavelength, and α_{loss} is the total loss:

$$\alpha_{\text{loss}} = \Gamma_{\text{a}}\alpha_{\text{a}} + \Gamma_{\text{g}}\alpha_{\text{g}} + (1 - \Gamma_{\text{a}} - \Gamma_{\text{g}})\alpha_{\text{clad}} + \alpha_{\text{th}}, \tag{3.35}$$

where α_{a}, α_{g}, and α_{clad} are the absorption coefficients of the active, guide, and cladding layers, and Γ_{g} is the optical confinement factor of the guide layer. α_{th} is the required minimum threshold gain and is related to the grating, that is, to the coupling constant and the facet phase.[28] The threshold current gradually decreases as the coupling strength, κL, increases because of the distributed optical reflection mechanism. This is similar to the case in which the facets of Fabry–Perot laser diodes are coated with highly reflective films. The slope (or external differential quantum) efficiency decreases in DFB laser diodes with high κL values. The external differential quantum efficiency, $\eta_{\text{d-DFB}}$, is given by the following equation:[28]

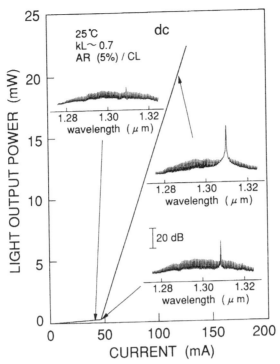

FIGURE 3.30 Change in emission spectra before and after lasing in 1300 nm-band buried heterostructure DFB laser diode.[5] Reprinted with permission from *Reliability and Degradation of Semiconductor Lasers and LEDs* by Artech House, Inc., Norwood, MA, USA, http://www.artech-house.com.

$$\eta_{\text{d-DFB}} = \eta_{\text{id}} P_R \alpha_{\text{th}} / \alpha_{\text{loss}}, \tag{3.36}$$

where P_R is the ratio of output power from a front facet to total output power from the both facets. This ratio is given as follows:

$$P_R = [(1 - R_F^2)E_F^2]/[(1 - R_F^2)E_F^2 + (1 - R_R^2)E_R^2], \tag{3.37}$$

where R_F and R_R are the reflectivities of the front and rear facets, and E_F and E_R are the optical field intensities at the front facet for the mode propagating toward the front facet and at the rear facet for the mode propagating toward the rear facet, respectively. Equation (3.36) is basically the same as Eq. (3.16).

The current-voltage characteristics of DFB laser diodes are the same as for Fabry–Perot laser diodes, and the discussions in Section 3.2.2 are also applicable to DFB laser diodes.

The behaviors of the DFB mode and the gain profile differ from each other

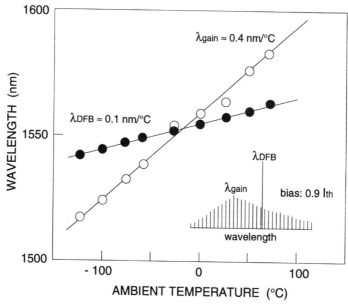

FIGURE 3.31 Temperature coefficients of the DFB mode and of the gain peak (Fabry–Perot mode) in a 1550 nm-band InGaAsP/InP DFB laser diode. The bias is set at 0.9 I_{th}.[5] Reprinted with permission from *Reliability and Degradation of Semiconductor Lasers and LEDs* by Artech House, Inc., Norwood, MA, USA, http://www.artech-house.com.

under operation because the DFB mode is determined by the effective grating pitch while the gain profile is associated with the band-gap energy (exactly the difference between two quasi-Fermi levels). A typical example of the different behaviors is observed under temperature variation. The temperature coefficient of the gain peak shift is about 0.25 nm/°C for 850 nm-band AlGaAs/GaAs laser diodes and is 0.4 to 0.6 nm/°C for long-wavelength InGaAsP/InP laser diodes. The coefficient of DFB mode, in contrast, is determined by the shift of the effective grating pitch due to the refractive index change of the active region. This difference in temperature coefficient is clearly recognized in Fig. 3.31. For a 1550 nm-band InGaAsP/InP laser diode with a 300 μm-long cavity, the temperature coefficients of the DFB mode and the gain peak (Fabry–Perot mode) are about 0.1 and 0.4 nm/°C, respectively. The temperature coefficient of the DFB mode is close to the value of the shift for each mode in Fabry–Perot laser diodes (see Fig. 3.12). This difference in temperature coefficients often limits the range of DFB mode operation under temperature variation. This problem can be understood schematically in Fig. 3.32. The gain peak usually moves 3 to 5 times more than the DFB mode does when temperature changes. If the DFB mode coincides with the gain peak, the DFB mode separates from the gain peak at lower and higher temperatures. In a severe case the Fabry–Perot mode starts

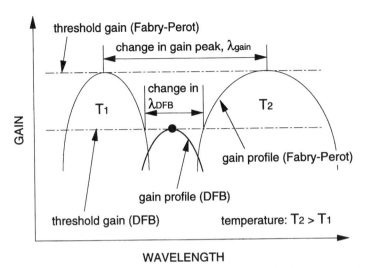

FIGURE 3.32   Temperature range of DFB mode operation in laser diodes. From T. Matsuoka.[34] ©1986 *Japan. J. Appl. Phys.* Reprinted by permission of the author and *Japan. J. Appl. Phys.*

to lase because the gain resulting from the injection excitation scarcely couples to the DFB mode. The temperature range of DFB mode operation is a function of the coupling strength, κL, and tends to increase as the coupling strength increases. A temperature range greater than 100 K around room temperature can be obtained in DFB laser diodes with a coupling strength greater than 1. A large coupling strength seems to be desirable for keeping the threshold current density low and the temperature range of DFB mode operation wide. As discussed in the next section, however, the coupling strength has to be kept small to obtain stable operation in the DFB mode.

3.4.1.4 Optical Field Distribution and Emission Spectra   When the coupling strength is set at high values, DFB laser diodes show unstable current-light output power characteristics and spectral characteristics. An inhomogeneous optical field distribution along the cavity is easily generated in DFB laser diodes because of the distributed feedback mechanisms as shown in Fig. 3.33. This inhomogeneity is enhanced by increasing the coupling strength, κL, because when the κL value is high the photons emitted from the active layer are stored within the cavity. This inhomogeneous optical field introduces a nonlinearity on the current-output characteristics as shown in Fig. 3.34(a). A DFB laser diode with a homogeneous grating and a κL value of 2 has a superlinear current-light output relation and a convexly upward current-slope efficiency curve. The spectral behavior is somewhat anomalous, as shown in Fig. 3.34(b). Even after lasing, the wavelength becomes shorter as the injected current increases. This behavior results from the threshold carrier density not being clamped at the

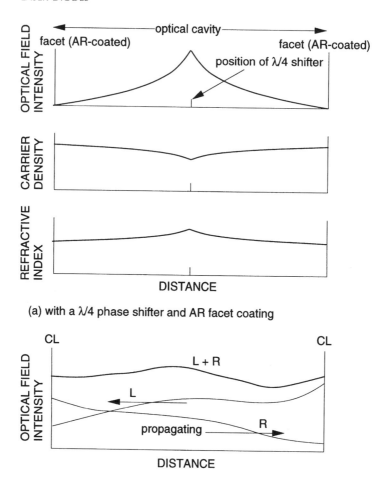

(a) with a λ/4 phase shifter and AR facet coating

(b) one example for a uniform grating and cleaved (CL) facets

FIGURE 3.33 Optical field, carrier density, and refractive index distribution in a DFB laser diode.

threshold value and increasing with increased injected current even after lasing. The wavelength shortening is caused by the decrease in the grating pitch that is due to the reduction of the refractive index induced by the increase in the injected carrier density (plasma effect). In severe cases, spatial hole-burning occurs along the cavity.[35] Spatial hole-burning is a phenomenon in which the injected carrier density at a certain position is spatially decreased by the localized reduction of carrier lifetime as shown in Fig. 3.33(a). This reduction of carrier lifetime is caused by the increases in stimulated emission rate due to the increased photon density (or optical field). The hole burning occurs at the position corresponding to the peak of optical intensity in the laser cavity. After the reduction of the carrier lifetime, the optical field of the lasing mode

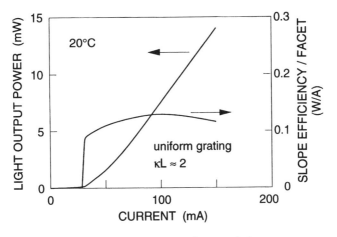

(a) current - light output characteristics

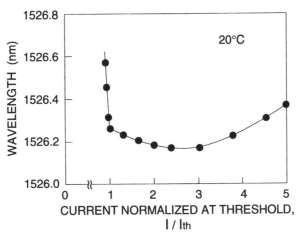

(b) lasing wavelength

FIGURE 3.34 (a) A nonlinear current-light output power characteristics and (b) longitudinal mode for a DFB laser diode with a uniform grating of high κL value (≈ 2).[5] Reprinted with permission from *Reliability and Degradation of Semiconductor Lasers and LEDs* by Artech House, Inc., Norwood, MA, USA, http://www.artech-house.com.

also decreases. This inhomogeneous distribution of carriers and the optical field results in unstable lasing in the longitudinal mode. Under a generation of the hole burning, the optical field of the lasing mode gradually decreases and the field of the next mode (a side-mode on the shorter-wavelength side of the initial lasing mode) increases. When the two-mode operation begins, a kink in the current-light output power relation is generated. The kink resulting from the

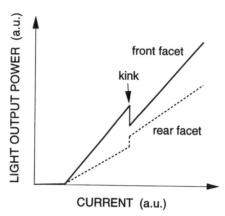

FIGURE 3.35 Relation between current and light output power from the front and rear facets of a DFB laser diode without a phase shifter.

instability of the longitudinal mode is sharper than that resulting from the instability of the transverse mode. In DFB lasers with phase shifters, the peaks of optical field coincide with the position of the shifters and, as shown in Fig. 3.33, there are no peaks near mirror facets coated with antireflecting films. The peak position in DFB lasers without phase-shifters depends on the facet phase of the grating and on the inhomogeneity of the active region along the cavity. The intensity of the peak of the optical field depends on the coupling constant and cavity length. Consequently, linear current-light output power relations without kinks are usually obtained by setting the coupling constant of a DFB laser without a phase-shifter to a value between 0.7 and 1.2 and by setting the coupling constant of a DFB laser with a phase-shifter to a value near 2.0. As discussed above, the intensity of optical field is given as a summation of the two optical fields for lightwaves propagating in opposite directions in the laser cavity. The inhomogeneities of the two optical fields influence the current–light output power relation in different ways because each field is individually determined by some structural factors, such as facet phase of grating. In DFB laser diodes with high coupling constants the opposite kinks in the current–light output power relations for the front and rear facets are often observed (see Fig. 3.35). In the figure the output power from the front facet decreases just after the kink, whereas that from the rear facet increases. These different behaviors of kinks are caused by the different change in each optical field within the cavity.

3.4.1.5 Wavelength Tuning and Multisection DFB Laser Diodes The wavelength of a DFB laser diode can be tuned if the electrode is divided into several sections as shown in Fig. 3.36. The lasing wavelength changes according to the variation of the effective grating pitch, which is a function of the refractive index of the active region. The refractive index can be changed by varying the carrier density. The refractive index variation of multisection DFB laser diodes

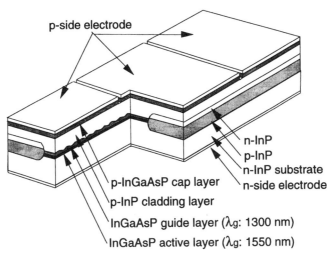

p-side electrode

n-InP
p-InP
n-InP substrate
n-side electrode

p-InGaAsP cap layer

p-InP cladding layer

InGaAsP guide layer (λ_g: 1300 nm)

InGaAsP active layer (λ_g: 1550 nm)

FIGURE 3.36   Schematic diagram of a multisection (three-section) DFB laser diode.

results mainly from the threshold carrier density variation produced by changing the fractions of current injected by each of the electrodes. An example of
wavelength tuning for a two-electrode DFB laser diode under a constant output
power is shown in Fig. 3.37. Because the amount of injected current required
to keep output power constant increases when the current ratio becomes small
or large, the influence of Joule's heating is also large at the ends of the tuning range, and the tuning range due to the effective grating pitch modulation
is about one-half of the tuning range shown in Fig. 3.37.[36,37] In multisection
DFB laser diodes the change in refractive index along the cavity is small even
under spatially inhomogeneous current injection because the whole region of
the cavity lases and the change in threshold carrier density is not large. This
results in a small tuning range. A larger range can be obtained using distributed
Bragg reflector laser diodes.

3.4.2   Distributed Bragg Reflector Laser Diodes

A schematic diagram of DBR laser diode is shown in Fig. 3.38. The structure
of a DBR laser diode is one in which one or both of the mirror facets in a
Fabry–Perot laser diode has been replaced with a DBR which reflects the light
emitted in the active region. The distributed reflectivity can be controlled in a
manner similar to that used to control the coupling constant, κ, of DFB laser
diodes. The threshold gain of the DBR laser diodes can basically be given in a
manner similar to that giving the threshold gain of a Fabry–Perot or DFB laser
diode by means of the gain-loss relation and phase condition. The threshold
gain is expressed by the following equation:[38]

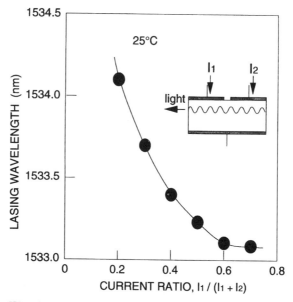

FIGURE 3.37 Wavelength tuning characteristics of a two-section DFB laser diode under nonuniform current injection and a constant light output power of 9 mW from the front facet.

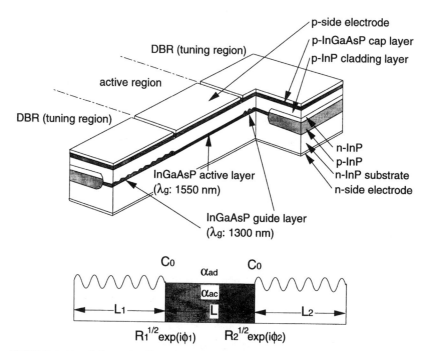

FIGURE 3.38 Schematic diagram of a DBR laser diode with two DBR regions.

$$g_{th} = \alpha_{ac} + [(1 - \Gamma)/\Gamma]\alpha_{ad} + (1/\Gamma L) \ln[1/(R_1 R_2)^{1/2} C_0^2], \qquad (3.38)$$

where α_{ac} and α_{ad} are the absorption coefficients of the active layer and the adjacent cladding layer, R_1 and R_2 are the reflectivities of DBRs, and C_0 is the coupling efficiency of the lightwave between the active region and the DBR region (see Fig. 3.38). The reflectivities of DBRs are given by $R_1^{1/2} \exp(i\phi_1)$ and $R_2^{1/2} \exp(i\phi_2)$, respectively. The phase condition is given by the following equation:[38]

$$\phi_1 + \phi_2 = 2\beta L + 2m\phi \qquad m = 1, 2, 3, \ldots. \qquad (3.39)$$

Among the modes in which the phase change of lightwave after a round trip within the cavity is $2m\pi$, the lasing occurs at the nearest mode to the Bragg condition. In the DBR region there is no gain, and thus the effective grating pitch can be changed a great deal by the current injection. The wavelength tuning range is normally more than a few nanometers, but it is not continuous and, as shown in Fig. 3.39, the continuous tuning range is similar to that of multisection DFB laser diodes. The tuning range has been increased by developing structures based on the DBR laser diodes, such as that of the superstructure-grating (SSG) laser diodes.[39,40]

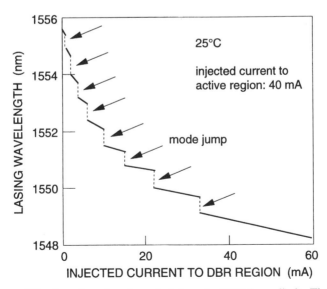

FIGURE 3.39 Wavelength tuning characteristics of a DBR laser diode. The active and DBR regions are 300 and 500 μm long, respectively.

3.5 OPERATING CHARACTERISTICS

The previous sections have described the principles of operation of Fabry–Perot, DFB, and DBR laser diodes with bulk or quantum well active layers. This section discusses the actual operating characteristics of those laser diodes (except for multisection DFB lasers and DBR laser diodes, because those laser diodes are mainly used as tunable laser diodes and their characteristics have been discussed in the previous section).

3.5.1 Temperature Characteristics

The temperature dependence of device characteristics is very important from the application point of view. When the temperature changes, the optical output power of laser diode decreases under a constant current or the injected current increases to keep an output power constant under a constant output power. Consequently, the temperature characteristics govern device reliability as well as device performance. As the ambient (or junction) temperature increases, the output power from a laser diode gradually decreases because of the increase in threshold current and decrease in the slope efficiency (or external differential quantum efficiency). These temperature dependences are shown in Fig. 3.40 for 1300 nm-band normal type and strained quantum well InGaAsP/InP laser diodes. The characteristic temperature, T_0, is often used as a measure for the temperature dependence of lasing characteristics and is empirically expressed for lasing threshold as[41]

$$I_{th} = I_{th0} \exp(\Delta T_j / T_0), \tag{3.40}$$

where I_{th0} and I_{th} are the threshold currents before and after the temperature change, ΔT_j. The characteristic temperature decreases as junction (ambient) temperature increases, and a large characteristic temperature indicates a small change in lasing characteristics with temperature. The temperature characteristics near room temperature for several kinds of laser diodes are shown in Fig. 3.41. The characteristic temperature is not a physical parameter but is associated with some physical mechanisms determining the temperature dependence of the lasing characteristics. Physical mechanisms thought to be the main factors determining the temperature characteristics are the following: overflow of injected carriers,[42–45] Auger recombination,[42,46–49] and intravalence band absorption.[33,50] The free carrier absorption (see Section 1.1.3.1) also increases according to the threshold current increase. The characteristic temperatures of AlGaAs/GaAs laser diodes are usually between 100 and 200 K and are determined primarily by the overflow of injected carriers from the active layer. In InGaAsP/InP laser diodes the characteristic temperature is greatest for the active layer lasing in the 1300 nm band and gradually decreases as the lasing wavelength moves toward shorter or longer range. When the lasing wavelength is shorter and the

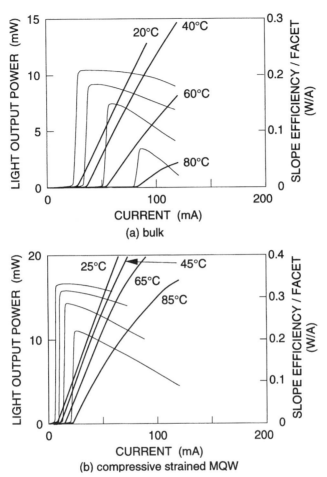

FIGURE 3.40   Temperature dependence of current-light output power characteristics for 1300 nm-band (a) bulk type and (b) strained quantum well InGaAsP/InP laser diodes with cleaved facets and 300 μm-long cavities.

composition of the InGaAsP active layer approaches the InP cladding layer, the characteristic temperature decreases primarily by the overflow of injected carriers, especially electrons because of their small effective mass, over the heterobarrier. This is the main reason that the characteristic temperature of 1200 nm-band laser diodes is smaller than that of 1300 nm-band laser diodes. In contrast, Auger recombination (see Section 1.1.2.7) and intravalence band absorption (see Section 1.1.3.1) strongly reduce the characteristic temperature when the band-gap energy decreases. The Auger recombination also increases the overflow of injected carriers because hot electrons having energies greater than the band offset of the heterobarrier are produced as a result of the recombination. These factors are responsible for the low characteristic temperature for

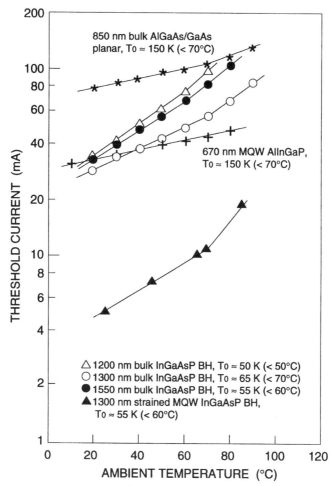

FIGURE 3.41   Temperature dependence of lasing threshold current for various laser diodes.

1550 nm-band InGaAsP/InP laser diodes. The slope efficiency (external differential quantum efficiency) can also be expressed by an equation similar to Eq. (3.40). The main factors influencing the temperature characteristics of the efficiency, however, are intravalence band absorption and free carrier absorption in the active and cladding layers. These behaviors result from the nearly constant Auger recombination rate and constant carrier overflow because the carrier density is nearly kept constant after lasing. The temperature dependence of free carrier absorption can be derived by replacing the threshold current with the threshold carrier density in Eq. (3.40) and using Eq. (1.84).

The temperature characteristics of quantum well and strained quantum well laser diodes are better than those of laser diodes with bulk active layers. This improvement of the temperature characteristics is mainly due to the reduction of threshold carrier density and of carrier overflow (or escape) from wells to barriers, optical confinement layers, and clading layers.[51] As the threshold current increases, the lasing energy levels increase because of the increase in injected carrier density and lead to the increase in the quasi-Fermi levels. Under such a situation a large amount of carriers exist in the higher-order subband in the well (see Section 1.3.1). These carriers, especially electrons, having higher energies easily escape from wells and move into the adjacent layers.

The low internal absorption loss (free carrier absorption) due to the low optical confinement factor also improves the temperature characteristics of quantum well laser diodes. In addition, the Auger recombination and intravalence band absorption can be reduced by modifying the valence band in strained quantum well structures (see Section 3.3).

The temperature characteristic for a 1300 nm-band InGaAsP/InP strained quantum well laser is plotted in Fig. 3.41. The apparent value of the characteristic temperature is comparable to or lower than that of the 1300 nm-band InGaAsP/InP laser diode with a bulk active layer. As described above, the characteristic temperature is not a physical parameter and it varies with other parameters, such as resistive leakage current and nonradiative recombination current via surface states and deep levels. For example, if a resistance is connected in parallel to a laser diode and resistive leakage current is introduced, the characteristic temperature is always larger after the connection of the resistance than it is before. The characteristic temperature tends to increase with the threshold current if the active layers are of the same composition. The apparent effects of quantum well and strained quantum well structures on the temperature characteristics are therefore not so drastic as shown in Fig. 3.41. The improved temperature characteristics of quantum well and strained quantum well laser diodes are more evident in the difference between the highest temperatures at which lasing occurs, as shown in Fig. 3.40. The maximum lasing temperature for strained quantum well InGaAsP/InP laser diodes is often more than $100°C$,[52,53] whereas that of InGaAsP/InP laser diodes with bulk active layers is usually less than $100°C$. The main factor determining the temperature characteristics of these strained quantum well InGaAsP/InP laser diodes around room temperature is thought to be the Auger recombination and that determining the temperature characteristics at temperatures above $70°C$ is thought to be the overflow of the injected carriers.[54]

The temperature characteristics of the laser diode itself are reflected in the different current–light output power characteristics under pulsed and dc operations (see Fig. 3.42). This difference is quite similar to that in the case of the LED shown in Fig. 2.8. Under dc operation the junction temperature rises as specified by Eq. (2.27). The same temperature rise occurs under pulsed operation, but the influence of the temperature rise is ordinarily negligible because of the intermittent operation. The junction temperature rise is strongly influenced by the thermal resistance determined by the configuration of the bonding to the heat sink and package stem and by the materials used (see Section 6.3.1).

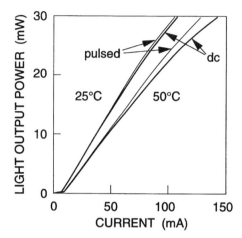

FIGURE 3.42 Current–light output power characteristics of a 1300 nm-band strained quantum well InGaAsP/InP laser diode under pulsed and dc biases.

3.5.2 Transient Phenomena and Modulation

Modulation characteristics are important when laser diodes are used in optical fiber transmission systems, and a variety of techniques are used to modulate the output power from laser diodes. A simple way to modulate the light is a direct modulation in which signal current is biased to the laser diodes directly and the output power varies according to the signal. Direct modulation is also used in LEDs, but laser diodes can be modulated at much higher frequencies—gigahertz or more—because of the properties of stimulated emission. Analog and digital modulations are used as the direct modulation techniques, and the modulation scheme is basically the same for both (see Fig. 3.43). The current modulation circuit is the same as that in the case of the LEDs shown in Fig. 2.13. Actually, as shown in Fig. 3.43, transistors or field effect transistors (FETs) are often used to superimpose the electrical signal on the operating current. Some *n-p-n* type transistors are usually used for high frequency modulation because that type of transitor can operate more quickly than *p-n-p* type transistors. The different operating speed in those transistors is due to the mobility difference between electrons and holes. The dc bias indicated with bias current in Fig. 3.43 is usually set below the lasing threshold current under digital modulation and is usually set above the threshold under analog modulation. In the high frequency range (10 GHz or more), the dc bias point is often set above the threshold current even under digital modulation. These direct modulation characteristics are closely associated with the lasing behavior of laser diodes.

3.5.2.1 *Transient Phenomena* The behavior of laser diodes under modulation is strongly related to the phenomena induced by the variations of injected carrier density and of emitted photon density. If only the gain-loss relation is

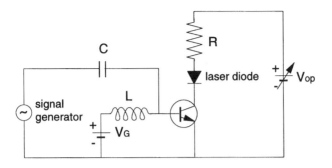

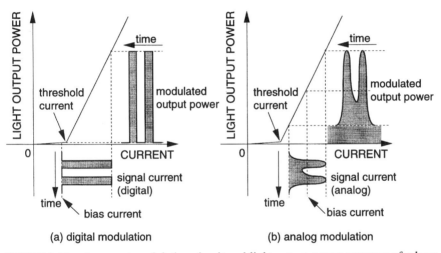

(a) digital modulation   (b) analog modulation

FIGURE 3.43 A current modulation circuit and light output power response of a laser diode under (a) digital and (b) analog modulations.

considered, the injected carrier lifetime and photon lifetime in the lasing condition can be connected by the rate equations.[55,56] For the injected carrier density,

$$dn/dt = J/qd - n/\tau_s - gn_{ph}, \qquad (3.41)$$

where τ_s is the injected carrier lifetime. The first term on the right-hand side is the number of carriers injected per unit time and unit volume, the second term indicates the spontaneous emission, and the third term corresponds to the stimulated emission. In the third term, g is the gain indicated in Eq. (3.6), and n_{ph} is the number of photons in the lasing mode. For photons,

$$dn_{ph}/dt = gn_{ph} - n_{ph}/\tau_{ph} + \beta_{sp}n/\tau_s, \qquad (3.42)$$

where τ_{ph} is the photon lifetime. The first term on the right-hand side is the number of photons generated by stimulated emission, and the second term is the number of photons lost from the cavity. The third term indicates the contribution of spontaneous emission to the lasing mode and expresses the origin of lasing. The term β_{sp} is called the spontaneous emission factor and is given by

$$\beta_{sp} = \Gamma_a \lambda_0^4 / 4\pi^2 n_r^3 \Delta\lambda_{sp} V_{act}, \qquad (3.43)$$

where Γ_a and λ_0 are the optical confinement factor and lasing wavelength and $\Delta\lambda_{sp}$ and V_{act} are the spectral width (FWHM) of spontaneous emission and the volume of the active layer. The value of β_{sp} thus depends on the structure of the laser diode and is between 10^{-4} and 10^{-6}. In Eqs. (3.41) and (3.42), the unit volume is considered. Photons generated by stimulated emission are lost for photon lifetime through the mirror facets and are absorbed within the cavity. Consequently, the photon lifetime is given by

$$\tau_{ph} = 1/\{(c_0/n_r)[\alpha_i + (1/2L)\ln(1/R_1 R_2)]\}, \qquad (3.44)$$

where c_0/n_r is the velocity of light in the cavity. When the refractive index, absorption coefficient, and cavity length are 3.5, 20 cm^{-1}, and 300 μm, and both mirror facets are cleaved ($R_1 = R_2 \approx 30\%$), the photon lifetime is about 2×10^{-12} sec. This value is about three orders of magnitude smaller than the injected carrier lifetime. In a stationary state $dn/dt = 0$ and $dn_{ph}/dt = 0$, and the following equations can thus be derived from Eqs. (3.41) and (3.42) by neglecting the influence of spontaneous emission:

$$J_{th} = qdn_{th}/\tau_s \qquad (3.45)$$
$$g(n_{th}) = 1/\tau_{ph} \qquad (3.46)$$

where J and n are replaced with J_{th} and n_{th} because the carrier density is nearly kept constant at n_{th} after lasing begins. Equations (3.45) and (3.46) are equivalent to Eqs. (3.10) and (3.8).

The rate equations, Eqs. (3.41) and (3.42), cannot be solved analytically because they are the nonlinear equations. They are usually solved by neglecting the influence of spontaneous emission [the third term on the right-hand side of Eq. (3.42)]. Responses to a step-change in injected current is illustrated schematically in Fig. 3.44, and actual data for a 1300 nm-band InGaAsP/InP buried heterostructure laser diode with bulk active layer are shown in Fig. 3.45. Damping oscillation is observed when a step-current is applied to laser diodes, and this transient phenomenon is called relaxation oscillation. This oscillation originates from the different times required for the carrier and photon populations to reach the equilibrium state, where the difference between the injected carrier lifetime and the photon lifetime is quite large (three or more orders of magnitude). As illustrated in Fig. 3.44, when a step-current, J_p, is applied to a laser diode based at $J_b < J_{th}$ (see Fig. 3.43), lasing starts after a time delay, t_d. Relaxation oscillation then occurs

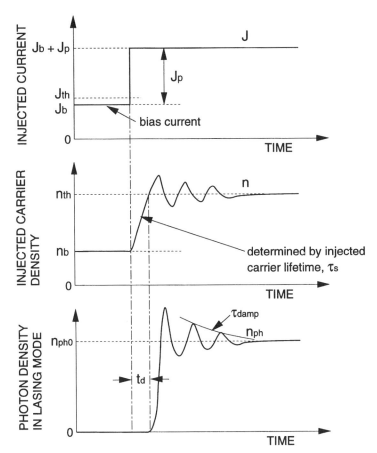

FIGURE 3.44 Change in injected carrier density and photon density in response to a step-change in injected current.

until the injected carrier density and photon density become constant. The lasing delay time is expressed by the following equation:[57]

$$t_d = \tau_s \ln[I_p/(I_b + I_p - I_{th})]. \qquad (3.47)$$

The injected carrier lifetime can be calculated from Eq. (3.47) if the delay time is monitored. As the pulsed current, I_p, or the bias current, I_b, increases, the lasing duration lengthens and then the delay time becomes small as shown in Fig. 3.46. The output response produced when the pulse current increases is shown in Fig. 3.46. The lasing delay time is ordinarily a few nanoseconds and the calculated carrier lifetime ranges from less than a subnanosecond to a few nanoseconds.

The relaxation oscillation can also be analyzed by using the rate equations. Under small signal modulation, by substituting $J = J_0 + J_m(\omega) \exp(i\omega t)$, $n = n_0$

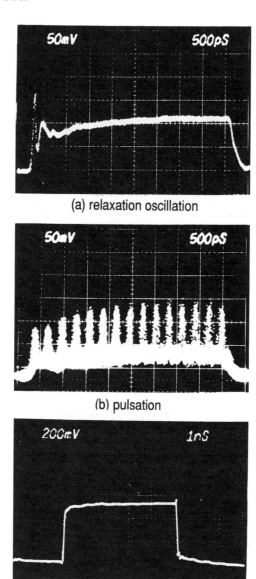

(a) relaxation oscillation

(b) pulsation

(c) input current

FIGURE 3.45 Step-current response, damping (relaxation) oscillation and pulsation in laser diode. The device is a 1300 nm-band buried heterostructure type InGaAsP/InP laser diode with a bulk active layer.[5] Reprinted with permission from *Reliability and Degradation of Semiconductor Lasers and LEDs* by Artech House, Inc., Norwood, MA, USA, http://www.artech-house.com.

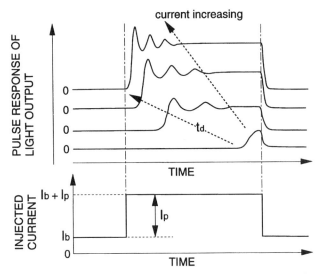

FIGURE 3.46 A schematic illustration of pulse response as a function of pulsed current in laser diode.

$+ n_m(\omega)\exp(i\omega t)$, and $n_{\mathrm{ph}} = n_{\mathrm{ph0}} + n_{\mathrm{ph}m}(\omega)\exp(i\omega t)$ into Eqs. (3.41) and (3.42) and solving the equations, the oscillation frequency is given as follows:[58]

$$f_{\mathrm{r}} = (1/2\pi)[n_{\mathrm{ph0}}(dg/dn)/\tau_{\mathrm{ph}}]^{1/2} \qquad (3.48)$$

or, approximately, as

$$f_{\mathrm{r}} \approx (1/2\pi)(\tau_{\mathrm{s}}\tau_{\mathrm{ph}})^{-1/2}[(I - I_{\mathrm{th}})/I_{\mathrm{th}}]^{1/2}, \qquad (3.49)$$

where dg/dn is called the differential gain [see Eq. (3.29)]. The damping constant, τ_{damp}, (see Fig. 3.44) is given approximately by $2\tau_{\mathrm{s}}$. The relaxation oscillation frequency corresponds to the resonance frequency (see next section) and can be monitored as a peak in the noise spectrum of a laser diode under dc operation (see Section 3.5.4). The relaxation oscillation in planar type laser diodes is influenced by the lateral diffusion of the injected carriers. After injected carriers at the lasing region are lost due to stimulated emission (see Fig. 3.44), lateral diffusion of the carriers existing outside the lasing region occurs very quickly within the active layer in the planar type laser diodes. This rapid diffusion suppresses the change in carrier density in the lasing region and thus suppresses the relaxation oscillation. During the relaxation oscillation, the injected carrier density changes as shown in Fig. 3.44. The deviation of the carrier density from the equilibrium value is ordinarily more than a few percent and less than 10 percent. This carrier change induces the wavelength shift due to the plasma effect as indicated Eq. (3.23). If the longitudinal mode of a laser diode is monitored over period longer than the oscillation span, the broadening of the emission spectrum is therefore generated as an overlapping of the momentary spectra. This wavelength broadening is called chirping (see Section 3.5.3.3). In distributed feedback lasers the width of a single longitudinal mode increases because of the chirping, and Fabry–Perot mode (multiple longitudinal mode) lasing occurs—in addition to the distributed feedback mode (a single-longitudinal mode) of lasing—if the relaxation oscillation is severe and the injected carrier density reaches the threshold level of the Fabry–Perot mode. In addition, if the lasing wavelength of DFB mode is set at the edge of the longer wavelength side in the gain spectrum, the Fabry–Perot mode operation often occurs under modulation even though the laser diode operates in a single longitudinal mode under dc bias. This is because the band-filling effect causes gain spectrum shifts to shorter wavelengths under an additional carrier injection (modulation), and the DFB mode comes out of the gain spectrum. These transient phenomena severely degrade the spectral aspect of modulation characteristics of DFB laser diodes.

A self-sustained oscillation (pulsation) shown in Fig. 3.45 occasionally appears in the step-current response. The main cause of the pulsation is a saturable absorber induced in the cavity by defects, impurities, inhomogeneous gain regions, and so forth.[59,60] A saturable absorber is one whose absorption coefficient varies with the quantity of absorbed light. For example, photoinduced electrons and holes progressively fill higher-level states in the conduction and valence bands (Burstein–Moss shift; see Section 1.1.3.1), and thus the absorption edge shifts to higher energies. This shift results in the decrease in the absorption coefficient. After the optical absorption decreases, the number of photons in the cavity increases repidly. This increase in the number of photons quickly leads to a decrease in the number of carriers because of stimulated emission, and this decrease in the number of carriers again increases the absorption coefficient of the saturable absorber. As a result, the lasing threshold increases and the number of injected carriers increases. After that the absorption coefficient of the saturable absorber decreases again, and this cycle is repeated. This results in the similar changes in photon and carrier density shown in Fig. 3.44. This is the main mechanism of pulsation (generally called Q-switching). Sometimes the lasing under unstable transverse modes also introduces pulsation in planar type laser diodes. The pulsation can be considered as an undamped relaxation oscillation. The frequency is roughly expressed by Eq. (3.49) and is linearly proportional to the square root of the magnitude of injection current (see Fig. 3.47). Extrapolation of the linear relation (dashed line) gives the threshold current density calculated from Eq. (3.49).[61]

When the bias of the step-current is suddenly removed, the number of photons and injected carriers decreases rapidly. After the number of photons in lasing mode becomes zero, the number of carriers decreases gradually according to the carrier lifetime in the same way as in LEDs. These behaviors can also be analyzed by using the rate equations.

3.5.2.2 *Factors Limiting Modulation*   By using the circuit shown in Fig. 3.43, a modulation response can be monitored as shown in Fig. 3.48 under a small signal

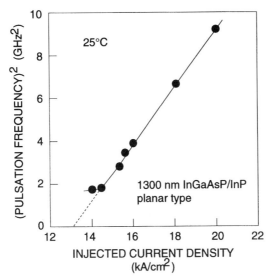

FIGURE 3.47 Pulsation frequency as a function of injected current density for a 1300 nm-band planar type InGaAsP/InP laser diode with a bulk active layer.[60]

modulation—that is, one in which the signal current is so small that the change in lasing conditions is negligible. As the modulation frequency becomes high, the modulation response (which corresponds to the modulated output power in Fig. 3.43) gradually becomes small (roll-off) because of the additional factors, such as the parasitic capacitance of the device and the circuit, and then a peak appears. The frequency of the peaking is called the resonance frequency and corresponds to the relaxation frequency. At still higher frequencies, the modulation response decreases rapidly because stimulated emission cannot respond to the change in the carrier density. These two factors, the resonance frequency and the parasitic effects, determine the modulation characteristics of laser diodes.

Relaxation Frequency   The relaxation frequency fixes the limit of the modulation bandwidth of a laser diode.[62] According to Eq. (3.48) the following methods might be used to increase the relaxation frequency and extend the modulation bandwidth: (1) increasing the photon density within a laser cavity, (2) decreasing the photon lifetime, and (3) increasing the differential gain, dg/dn. The photon density can be increased simply by increasing the injected current. Except for long-wavelength InGaAsP/InP lasers, however, the bias is limited by catastrophic optical damage (see Section 7.2.1.3).[63] Photon lifetime can be reduced by shortening the optical cavity. The relaxation frequency of a laser diode with a cavity about 100 μm-long is twice that of a laser diode with a cavity about 300 μm long.[63] Another way to reduce photon lifetime can be inferred from Eq. (3.44): increasing the optical cavity loss and decreasing the mirror reflectivity, although these degrade the device characteristics. The use

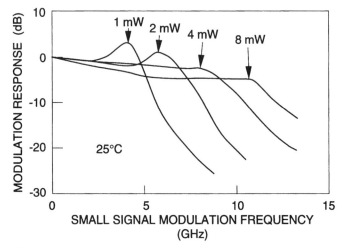

FIGURE 3.48 Small-signal modulation response of a 1550 nm-band buried heterostructure InGaAsP/InP laser diode with bulk active layer.[5] The modulation signal is a −10 dBm sinusoidal wave. The arrows indicate resonance peak frequency at each bias level (light output power). Reprinted with permission from *Reliability and Degradation of Semiconductor Lasers and LEDs* by Artech House, Inc., Norwood, MA, USA, http://www.artech-house.com.

of a quantum well or strained quantum well structure in the active layer is an effective way to increase the differential gain. The relaxation frequency of quantum well laser diode is clearly higher at a given output power as shown in Fig. 3.49. Doping acceptors is also an effective way to increase differential gain,[64] but in a bulk layer heavy doping induces the band-tailing effect (see Section 1.1.2.5), a decrease in injected carrier lifetime, and an increase in absorption coefficient. These lead to an increase in threshold current density and a decrease in quantum efficiency. In quantum well and strained quantum well structures, the band-tailing can be eliminated and the impurity doping is increased by a modulation doping in which a *p*-type impurity is doped into only the barrier layers of a quantum well structure. A quantum well laser diode with this modulation doping shows a high relaxation frequency, about 30 GHz.[65]

In Fig. 3.49 the linear relation indicated by Eq. (3.48) is broken at high output powers, where the resonance frequency scarcely increases as the output power increases. This is due to the nonlinear gain, g_{nl}, empirically expressed by the next equation:

$$g_{nl} = g(1 - \epsilon n_{ph}),$$ (3.50)

where ϵ is called the gain compression factor. This nonlinearity is more pronounced in (strained) quantum wells than in bulk layers.[66–68] Some models, such as spectral hole-burning[69,70] and well-barrier hole-burning in quantum well laser

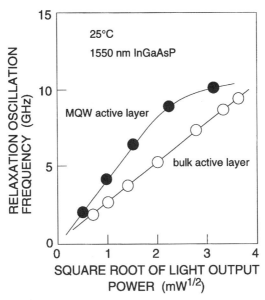

FIGURE 3.49   Relaxation oscillation frequency as a function of the square root of optical output power for a 1550 nm-band InGaAsP/InP laser diodes with bulk and MQW active layers. Each laser diode has a cavity 300 μm long, and the well width and number of wells for the MQW structure are 50 nm and 4.

diodes,[71] have been proposed to explain the nonlinear gain. Among the models, the spectral hole-burning explains the experimental data well. As discussed in the lasing mechanisms (see Section 3.2.1.4 and Fig. 3.9), the injected carrier has to reach the lasing level (intraband relaxation). As the injection excitation is intensified, the stimulated emission rate becomes high, and then the intraband relaxation process cannot respond gradually to the stimulated emission. As a result, the carrier at the lasing level is depleted and a hole in the carrier distribution in conduction and valence bands appears around the level (see Fig. 3.9), although the influence of the spectral hole-burning in energy space is not observed clearly in longitudinal modes of emission spectra. By introducing the nonlinear gain, the resonance frequency is rewritten as

$$f_{\mathrm{r}} = (1/2\pi)[(dg/dn)n_{\mathrm{ph}}(1 - \epsilon n_{\mathrm{ph}})/\tau_{\mathrm{ph}}]^{1/2}. \qquad (3.51)$$

The final limit of the relaxation frequency, that is, modulation bandwidth, is determined by the influence of the nonlinear gain. In distributed feedback laser diodes, the spatial hole-burning within the cavity (see Section 3.4.1.4) is also a major factor influencing the nonlinear gain. Distributed feedback laser diodes with high κL values often show the saturation of the relaxation frequency at relatively low output powers of less than 10 mW.

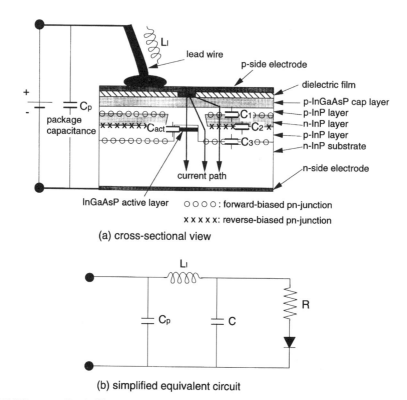

FIGURE 3.50 Buried heterostructure type of InGaAsP/InP laser diode under a forward dc bias: (a) cross-sectional view and (b) simplified equivalent circuit.

Parasitic Factors The parasitic capacitance is mainly a result of the junction capacitances at *pn*-junctions (see Sections 1.2.2 and 1.2.4.1). A cross-sectional view of an InGaAsP/InP buried heterostructure laser diode is illustrated, with the *pn*-junctions, in Fig. 3.50. Under dc bias, current flows ideally through the active layer because of the current confinement due to the dielectric film and reversely biased *pn*-junctions, whereas under ac bias or modulation some current leakage paths appear. The capacitances, C_1, C_2, and C_3 exist in the vicinity of the active layer. From Eq. (1.137), the capacitance of reverse-biased *pn*-junctions is smaller than that of forward-biased *pn*-junctions because of the difference in the widths of their depletion layers. As shown in Fig. 3.50(b), the total capacitance—including C_1, C_2, and C_3—can be expressed as a capacitance, C, between the *p*- and *n*-side electrodes in the simplified equivalent circuit.[72] The resistance R includes the resistances due to ohmic contact and to the semiconductor material. The inductance, L_1, and the capacitance, C_p, correspond to the inductance of the lead wire and package capacitance. The capacitance, C, strongly depends on the structure of the laser diode. The total impedance of the equivalent circuit, Z_{laser}, is therefore given by the next equation, which takes into consideration the impedance of driving circuit (or matching resistance, ordinarily 50 Ω):

$$Z_{\text{laser}} = V/I = (50 + R - \omega^2 L_1 CR) + i\omega(L_1 + 50CR), \qquad (3.52)$$

where the package capacitance is neglected. For a small signal modulation at a bias above the threshold current, the cut-off frequency determined by the parasitic factors, $f_{\text{c-para}}$, can be obtained by using Eq. (3.52). The cut-off frequency corresponds to the frequency satisfying the relation, $I(\omega)/I(\omega = 0) = 1/2$ (see Fig. 2.14):

$$f_{\text{c-para}} = \omega_{\text{c-para}}/2\pi = (1/2\pi)[(50 + R)/L_1 CR]^{1/2}. \qquad (3.53)$$

For example, the cut-off frequency calculated for a buried heterostructure laser having a capacitance of 100 pF, a resistance, R, of 5 Ω, and an inductance, L_1, of 1.5 nH is about 1.4 GHz. The same equation is applicable for planar type laser diodes.

The parasitic capacitance is often reduced by forming a groove on each side of the active region, along the stripe, in order to separate the *pn*-junction of the active region from those of the other areas. Burying the active region with a semi-insulator material, such as Fe-doped InP, eliminates the capacitances in the case of buried heterostructure laser diodes. For a buried heterostructure InGaAsP/InP laser diode with a 300 μm-long cavity, the capacitance can be reduced from a few tens of picofarads to less than 1 pF by replacing the *p*- and *n*-type burying layers with an Fe-doped InP.[73]

Another parasitic factor is related to the heterostructure and is primarily a carrier transport problem. As shown in Fig. 3.51, carrier transport, capture, and tunneling occur within the double heterostructure. The time delay due to carrier transport in a separate confinement layer is remarkable if the confinement layer is thick (see Figs. 3.21 and 3.22). Similar to the roll-off caused by the parasitic capacitance, the response under modulation gradually decreases as the modulation frequency increases. In single quantum wells with separate confinement heterostructures, the influence of the time delay due to carrier diffusion in the separate confinement layer, $\tau_{\text{d-sch}}$, appears in the modulation bandwidth when the layer is more than 50 nm thick. In InGaAs/GaAs single quantum well laser diodes, the modulation bandwidth is determined by the relaxation frequency when the confinement layer is 75 nm thick but is limited by the carrier transport time when the confinement layer is 300 nm thick.[74] Thin separate confinement layers are better from the modulation point of view, although they result in higher threshold currents (see Section 3.3).

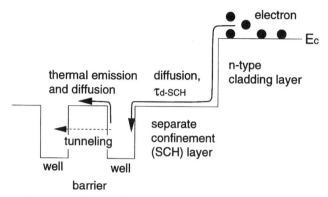

FIGURE 3.51 Schematic illustration of carrier (electron) transport in a separate confinement heterostructure (SCH).

3.5.3 Coherence, Spectral Linewidth, and the Linewidth Enhancement Factor

3.5.3.1 Coherent Light There are two kinds of coherence: temporal and spatial. Perfectly temporal coherent light is a continuous lightwave with a single-line spectrum (or a single frequency), and perfectly spatial coherent light is a plane wave. The spatial coherent light in such a plane wave can be focused to a small area by a lens. The light emitted from laser diodes is coherent light similar to that emitted from gas and solid state lasers, whereas the light emitted from LEDs is incoherent light. The light from laser diodes with a single longitudinal mode, such as DFB and DBR laser diodes, shows high temporal and spatial coherence. The light from laser diodes with multiple longitudinal mode also shows reasonably high spatial coherence, although its temporal coherence is not high. In addition, the light emitted from the actual laser diodes is not perfectly coherent light and has amplitude and phase fluctuations even under operation in a single longitudinal mode.

The coherence length, l_c, is often used as a measure of the coherence of the emitted light. It is given by the following equation:

$$l_c = \pi c_0 / n_r \Delta f = \pi \lambda^2 / n_r \Delta \lambda, \tag{3.54}$$

where c_0 is the velocity of light in the vacuum, n_r is the refractive index of the medium, and λ is the wavelength. The terms Δf and $\Delta \lambda$ are the spectral widths in units of frequency and of wavelength and

$$\Delta f = (c_0 / \lambda^2) \Delta \lambda. \tag{3.55}$$

For a 1300 nm laser diode lasing in multiple longitudinal modes occupying a spectral width of 5 nm ($\approx$ 888 GHz) (FWHM of the envelope of the spectrum), the coherence length in a silica fiber ($n_r \approx 1.45$) is calculated to be about 0.73 mm. For a 1300 nm DFB laser diode lasing in a single longitudinal mode with 1 MHz wide, the coherence length in a silica fiber is about 650 m. For an 850 nm LED with a spectral width of 30 nm, the coherence length is about 0.052 mm.

3.5.3.2 *Spectral Lindwidth*

The coherence length related to the spectral width is specified by Eq. (3.54) and is especially important in laser diodes lasing in a single longitudinal mode, such as DFB and DBR laser diodes. For those laser diodes the spectral width is called the spectral linewidth. The actual spectral linewidth of laser diodes broadens because the light generated by spontaneous emission is mixed with lasing output produced by stimulated emission. The spontaneous emission process occurs randomly and introduces phase noise as well as optical intensity noise to the light output power. In addition, the fluctuation of light intensity induces a fluctuation of a carrier density within the laser cavity, and then the lasing frequency fluctuates because of the refractive index fluctuation due to the carrier density variation. As a result, the spectral linewidth of a laser diode, $\Delta\nu$ [Δf in Eqs. (3.54) and (3.55)], is given by multiplying the Schawlow–Townes equation by the constant term $(1 + \alpha^2)$:[75]

$$\Delta\nu = (gn_{sp}/4\pi I_P)(1 + \alpha^2), \tag{3.56}$$

where g and I_P are the gain coefficient and the number of photons in the lasing mode within the laser cavity. The term n_{sp} is the spontaneous emission coefficient and is given by

$$n_{sp} = 1/\{1 - \exp\{[h\nu - (E_{F_c} - E_{F_v})]/k_B T\}\}. \tag{3.57}$$

For ordinary DFB and DBR laser diodes this coefficient is between 1 and 2. The term α is called the linewidth enhancement factor.[75] The spectral linewidth can ideally be expressed by the Lorenzian as shown in Fig. 3.52. The subpeaks on each side of the main peak is due to the relaxation oscillation, and they separate from the main peak as the bias increases (see Fig. 3.49). By adding a constant term, C_{rl}, the Eq. (3.56) is rewritten for actual laser diodes as

$$\Delta\nu = (h\nu n_{sp}/4\pi P_{out\text{-}st})v_g^2(\alpha_i + \alpha_m)\alpha_m(1 + \alpha^2) + C_{rl}, \tag{3.58}$$

where $h\nu$ and $P_{out\text{-}st}$ are the photon energy of the lasing mode and the lasing output power from both facets, v_g is the group velocity, and α_i and α_m are the internal cavity (absorption) loss and the equivalent mirror loss including an influence of the grating. The constant C_{rl} is called the residual linewidth corresponding to the linewidth at $1/P_{out\text{-}st} = 0$. As indicated in Eq. (3.58), the spectral linewidth gradually decreases in inverse proportion to the light output

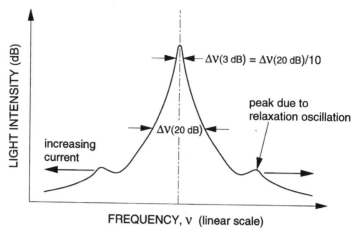

FIGURE 3.52 Spectral linewidth of DFB laser diodes.

power. The actual data for two DFB laser diodes with 1.2 mm-long cavities and bulk active layers are shown in Fig. 3.53. When the active layer is thin, the internal cavity loss becomes low, and thus the spectral linewidth narrows.[76] The spectral linewidth of quantum well laser diodes, therefore, further decreases because of the low internal cavity loss. Another way to decrease the linewidth

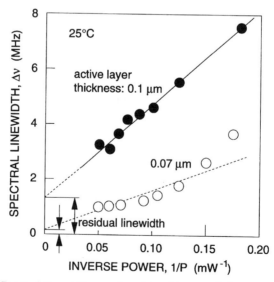

FIGURE 3.53 Spectral linewidth as a function of inverse light output power for 1550 nm-band buried heterostructure type InGaAsP/InP DFB laser diodes with 1.2 mm-long cavities and bulk active layers.[76]

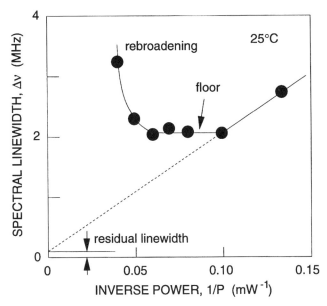

FIGURE 3.54 Typical behavior of spectral linewidth at high light output power range for a 1550 μm-band buried heterostructure InGaAsP/InP DFB laser diode with a 1.2 mm-long cavity and 0.08 μm-thick active layer.

is to increase cavity length, which is equivalent to reducing the effective mirror loss. The residual linewidth, C_{rl}, is mainly caused by $1/f$ noise, which is thought to originate from a *pn*-junction perimeter of the active layer.[77,78]

The typical behavior of spectral linewidth in a high output power range is shown in Fig. 3.54 for a 1550 nm DFB laser diode with a bulk active layer. The decrease in spectral linewidth saturates and a floor appears in the relation between spectral linewidth and inverse light output power. At higher output powers, rebroadening of the linewidth often occurs. The causes of the floor and rebroadening are thought to be the spatial hole-burning along the cavity,[79] spectral hole burning,[80] and side-mode instability.[81] Controlling the κL value is therefore very important for suppressing the spatial hole-burning and the side-mode instability (see Section 3.4.1.4).

In DBR laser diodes, the characteristics related to the spectral linewidth depend on structure as shown in Fig. 3.55. As the laser cavity lengthens [compare structures (a) and (b)] and the reflectivity of the DBR reflector increases [compare structures (b) and (c)], the spectral linewidth can be greatly reduced. When the current is injected into the DBR region and the wavelength is tuned, the spectral linewidth often increases tremendously and rebroadening and floor appear in a manner similar to DFB laser diodes. This instability of spectral linewidth is thought to be due to the lasing wavelength fluctuation induced by variation of effective grating pitch (refractive index change) in the DBR

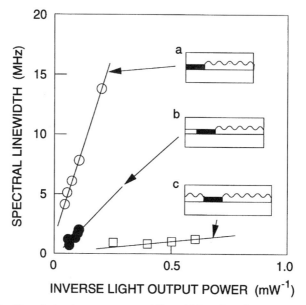

FIGURE 3.55   Correlation between spectral linewidth and cavity structure in DBR laser diodes.[82] The active layers are 200 μm long for structure (a) and 300 μm long for structures (b) and (c). The length of DBR with a κ of about 200 cm^{-1} is 500 μm for rear parts in all structures and is 100 μm for front part in structure (c).

region.[83] The refractive index change is caused by leakage of injected carriers from the active region to the DBR region.

As discussed above, the spectral linewidth can basically be reduced by improving the quality of the laser cavity (called the Q-value). Those improvements are lengthening laser cavity, increasing reflectivity, decreasing internal cavity loss, and so forth. A spectral linewidth less than a few hundred kilohertz can be obtained relatively easily in a strained quantum well DFB laser diode with a long cavity and a suitable κL value.

3.5.3.3   Linewidth Enhancement Factor: α Parameter   In Eqs. (3.56) and (3.58) the term of $(1 + \alpha^2)$ indicates the broadening of spectral linewidth due to the fluctuation of injected carrier density. The linewidth enhancement factor (α parameter) is quite important in describing the spectral behaviors associated with carrier density variation in laser diodes. The α parameter is mainly determined by the structure of the active layer and is given by the following equation:[75]

$$\alpha = (\partial n_r / \partial n)/(\partial g / \partial n), \qquad (3.59)$$

where n_r is the refractive index, n is the carrier density, and g is the optical gain. The n_r and g respectively correspond to the real and imaginary parts of the complex refractive index of the active layer [see Eq. (1.112)]. Equation (3.59) indicates that the refractive index and the gain fluctuate simultaneously as a result of the carrier density fluctuation induced by spontaneous emission within the laser cavity. When the α parameter is small, the spectral linewidth decreases, and then the coherence becomes high because the refractive index change is small. In InGaAsP/InP DFB laser diodes, the α parameter is between -4 and -10 for bulk active layers and between -1.5 and -4 for quantum well active layers. The value can be further reduced by using a strained quantum well structure, suitable detuning (that is, a setting of lasing wavelength at a wavelength shorter than the gain peak), and other techniques. The value of -1 or less can be obtained as α parameter by using those techniques.[84]

The α-parameter is also useful in analyzing the spectral characteristics of laser diodes with single longitudinal modes (such as DFB and DBR laser (diodes) under modulation; that is, for analyzing the chirping (see Section 3.5.2.1) or frequency modulation (FM) response (see Section 8.1.1.4). Each of the multiple longitudinal modes in Fabry–Perot laser diodes is in the same situation as the single mode of the DFB and DBR laser diodes (see Fig. 3.12) even though there is no grating. Under current modulation, intensity and frequency modulation of $P_{\text{out-st}}(t)$ and $\Delta \nu(t)$ occur because gain and refractive index are modulated simultaneously and are combined as follows:[85]

$$\Delta \nu(t) = (-\alpha/4\pi)\{[1/P_{\text{out-st}}(t)][dP_{\text{out-st}}(t)/dt] + (2\Gamma\epsilon/V_{\text{act}}h\nu\eta_d)P_{\text{out-st}}(t)\}, \tag{3.60}$$

where Γ is the optical confinement factor, ϵ is the gain compression factor [see Eq. (3.50)], V_{act} is the active layer volume, and η_d is the total external differential quantum efficiency. The chirping for a single mode (that is, FM-response), which is observed as a spectral broadening, results from an overlapping of instantaneous lasing wavelength (frequency). Chirping is a harmful aspect of modulation for laser diodes used in optical communication systems employing intensity modulation techniques. Frequency chirping, in contrast, is indispensable in laser diodes used in communication systems employing FM techniques. The α parameter is therefore required to be small from the viewpoint of chirping, but large from the viewpoint of FM-response.

3.5.4 Noise

The noise in laser diodes has an important factor effect on signal or data transmission in optical fiber communication systems, compact disks, and so forth. Most of the noises are associated with the principle of the device operation and degrade the performance of equipment and systems. The basic noises are quantum noise, mode-partition noise, and mode-hopping noise. A noise induced by optical feedback from a distant point is also important from an application point of view. In the following sections, these noises are discussed for laser diodes with a single transverse mode.

3.5.4.1 Quantum (Shot) Noise   The most basic noise in laser diodes as well as LEDs (and photodiodes) originates from the behavior of injected carriers and emitted photons as particles. It is therefore called quantum (shot) noise because electrons, holes, and photons are quantized particles. The quantum noise in LEDs is due only to spontaneous emission (that is, it is only shot noise), whereas the quantum noise in laser diodes is due to the mixture of spontaneous emission in lasing light as discussed in Section 3.5.3. Although there are several ways to analyze the quantum noise in laser diodes,[86-89] the rate equations indicated in Eqs. (3.41) and (3.42) are basically solved after adding Langevin force (a shot noise term), $F(t)$, to their right-hand sides. Here the photon density is replaced by the amplitude of a lightwave with a phase because the number of photons is proportional to the value (amplitude of lightwave)^2/photon energy. Intensity noise (AM noise) and phase noise (FM noise) can be analyzed by solving those equations. Schematic illustrations of these noises are shown in Fig. 3.56 as a function of frequency. In the spectrum of the lasing light, a subpeak corresponding to the relaxation frequency appears on each side of the lasing frequency (wavelength). Plots of the intensity noise and phase noise against frequency are similar in that a peak corresponding to the relaxation frequency (or resonance frequency) indicated in Eqs. (3.48) or (3.49) appears. At very low frequency range less than about 100 kHz, $1/f$ noise is dominant and the noise level increases. The phase noise for laser diodes with single longitudinal modes increases the spectral linewidth as shown in Fig. 3.52 (see Section 3.5.3.2). An example of the intensity noise for a 1550 nm buried heterostructure InGaAsP/InP Fabry–Perot laser diode with a bulk active layer is shown in Fig. 3.57. The peak at each output power corresponds to the relaxation frequency. The position of the noise peak shifts to higher frequencies as output power (injected current) increases (see Fig. 3.48).

The intensity noise is often expressed as the relative intensity noise (RIN), which is defined by the ratio of the fluctuation of light intensity per unit frequency to the average output power. It is usually monitored by using a circuit consisting of a photodetector, an amplifier, and a spectrum analyzer (Fig. 3.58). The relative intensity noise (in Hz^{-1} or dB/Hz) is given by

$$RIN = (P_{noise} - N_N)/GR_L I_R^2 \Delta f, \qquad (3.61)$$

where P_{noise} is the measured noise power, N_N is the noise power from photodetector (shot noise) and amplifier (thermal noise), G is the gain of the amplifier, R_L and I_R are the load resistance and the photocurrent of the photodetector, and Δf is the measuring frequency bandwidth. Here LEDs are often used to check a background noise level of the circuit because the intensity noise of LEDs

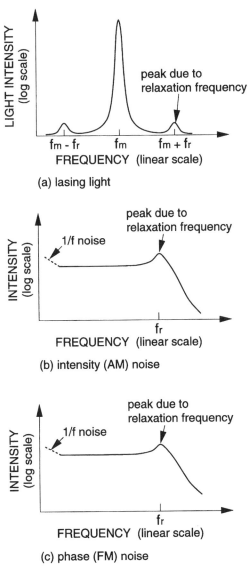

FIGURE 3.56 Schematic illustrations of (a) spectral line shape for lasing light, (b) intensity noise, and (c) phase noise.

is due only to shot noise. For a 1300 nm buried heterostructure InGaAsP/InP Fabry–Perot type laser diode with a bulk active layer, the relative intensity noise is shown as a function of light output power in Fig. 3.59. In the low-excited (output power) range, the quantum noise is very large because the random spontaneous emission is amplified and mixed in each longitudinal mode. As the excitation (output power) increases, more of the injected carrier is transformed

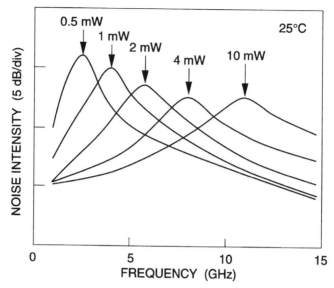

FIGURE 3.57   Noise spectra of a 1550 nm-band buried heterostructure InGaAsP/InP laser diode (shown in Fig. 3.48) under dc operation. The arrows indicate the noise peak at each bias.

into stimulated emission and therefore the intensity noise decreases. The similar behaviors are generated in the spectral linewidth (phase noise) shown in Fig. 3.53. The intensity and phase noises are strongly influenced by the behaviors of the longitudinal and transverse modes. When the transverse mode—that is, the near-field and far-field patterns–changes, the intensity noise and phase noises become large because of the increase in quantum noise.[90] The intensity noise also increases because of the instability of the longitudinal mode (see Fig. 3.59). In laser diodes with single modes, therefore, a relative intensity noise below −140 dB/Hz is easily obtained, whereas it is difficult to decrease the noise to below −140 dB/Hz in laser diodes with multiple longitudinal modes.

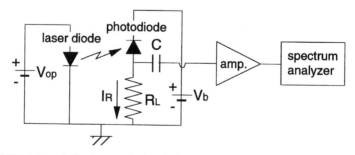

FIGURE 3.58   A fundamental circuit for measuring a relative intensity noise.

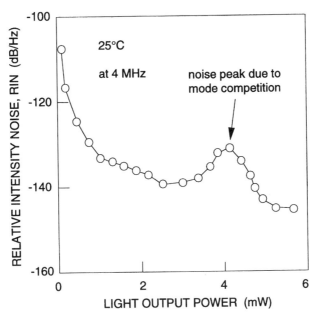

FIGURE 3.59   Relative intensity noise as a function of light output power for a 1300 nm-band buried heterostructure InGaAsP/InP Fabry–Perot laser diode. The measurement is carried out at 4 MHz under a bandwidth of 30 kHz.

3.5.4.2 Mode-Partition Noise and Mode-Hopping Noise   The mode-partition and mode-hopping noises are the intensity noise observed in laser diodes. In laser diodes operating in multiple longitudinal modes, the intensity noise for each longitudinal mode is larger than that for all the longitudinal modes under a constant output power. Therefore, when only one mode is selected from the multiple longitudinal modes by using a filter, the intensity noise of the mode is quite large. This kind of intensity noise is called mode-partition noise,[91] and it originates from the random partitioning of the total output power into each lasing mode even under a constant output power operation.

For laser diodes lasing in single mode or in only a few longitudinal modes, the lasing mode hops to another mode or the peak mode (peak wavelength) changes when the injection current or the temperature changes (see Fig. 3.11). The mode-competition occurs at the point of the mode-hopping as shown in Fig. 3.60. There the two modes lase alternately at random. The intensity fluctuation due to the random lasing between the modes is called the mode-hopping noise. The alternate lasing is due to the fluctuation of spontaneous emission. Small kinks corresponding to the mode-hopping appear on the current-light output power characteristics and the current-voltage characteristics. The mode-hopping noise is quite large when the laser diode operates in the single mode because the total output power fluctuates corresponding to the two competing wavelengths. This mode-hopping noise is usually induced below 50 MHz. At mode-hopping

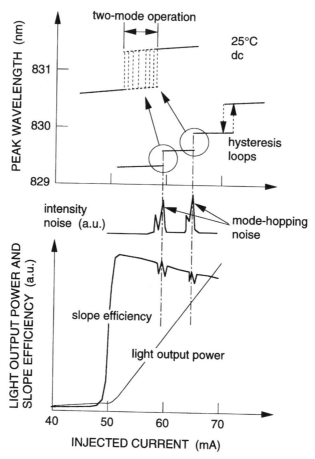

FIGURE 3.60 Mode-hopping noise and lasing characteristics for a planar type AlGaAs/GaAs laser diode with a bulk active layer.

with a hysteresis loop, no mode-hopping noise is generated because the mode changes instantaneously and no two-mode operation occurs (see Fig. 3.60). In Fig. 3.59, the noise peak indicated with an arrow is an example of the intensity noise increase due to the mode-hopping.

Laser diodes lasing in many longitudinal modes show no mode-hopping noise, although their intensity noise level is higher than that of laser diodes operating in single or a few longitudinal modes. The mode-hopping noise is suppressed by superimposing a sinusoidal wave (see Fig. 3.43) to increase the lasing longitudinal modes.[92] To avoid a single longitudinal mode operation at any moment, the minimum current level of the sinusoidal wave is set below the threshold current and the frequency is usually set higher than 600 MHz. Laser diodes with self-sustained pulsation (discussed in Section 3.5.2.1) are also effective for suppressing the mode-hopping noise, because of their multilongitudinal mode operations. As shown in Fig. 3.60, the mode-competition scarcely occurs at the mode-hopping with a hysteresis loop. This hysteresis loop is intensified by introducing an saturable absorber (see Section 3.5.2.1) to the laser cavity, and the mode-hopping noise can also be suppressed.[93]

3.5.4.3 Optical Feedback Noise When laser diodes are set in some kinds of equipment, such as compact disks and the transmitters in communication systems, a small percent of the emitted light is reflected at distant points, such as the ends of optical fibers. The reflected light couples with the lasing modes within the laser cavities if optical isolators are not inserted between the distant points and laser diodes, and this coupling tremendously perturbs the lasing situations.[94] The noise induced by the reflected light is often called optical feedback noise. As shown in Fig. 3.61, the change in lasing mode induced by

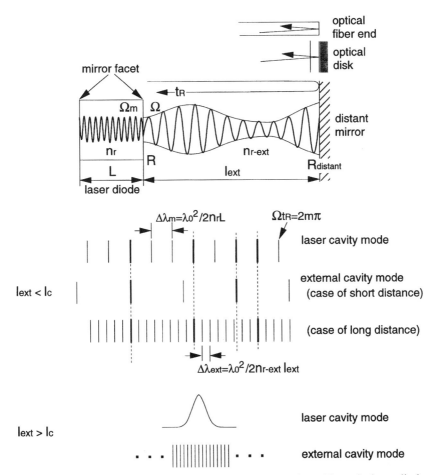

FIGURE 3.61 Schematic illustration of optical feedback problems in laser diode.

the optical feeback from a distant mirror can be analyzed by investigating the behaviors of a complex resonator laser consisting of the cavity of a laser diode and an external resonator between the mirror of the laser diode and the distant mirror. The behaviors depend on the distance between the mirror facet on the laser diode and the distant mirror and on the reflectivity of each mirror.

Some of the light emitted from the laser diode is reflected at the distant mirror and returns to the mirror facet of the laser diode after a round trip time, t_R:

$$t_R = 2n_{\text{r-ext}}l_{\text{ext}}/c_0, \tag{3.62}$$

where $n_{\text{r-ext}}$, l_{ext} and c_0 are respectively the refractive index of the medium, the distance between the laser mirror and the distant mirror, and the velocity of light in a vacuum. The phase difference of the reflected light is given by the product of the lasing angular frequency and the round trip time (Ωt_R in Fig. 3.61). When the phase difference corresponds to $2m\pi$ ($m = 1, 2, 3 \ldots$), the phases of the laser cavity mode and the external cavity mode coincide with each other and a standing wave is formed within the complex cavity. This leads to an increase in the effective reflectivity of the mirror facet and thus decreases the threshold current and increases output power. If the phase difference is $2(m - 1)\pi$, which gives the maximum phase mismatch, the threshold current increases and the output power decreases. These phase conditions change greatly with changes in the lasing wavelength and the distance, l_{ext}, and thus with changes in the injected current, ambient temperature, and so forth. As shown in Fig. 3.62, ripples are, therefore, often observed on the current–light output power characteristics and

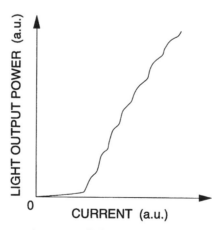

FIGURE 3.62   Ripples on the current-light output power characteristics of a 1300 nm-band buried heterostructure InGaAsP/InP Fabry–Perot laser diode under optical feedback.

thus on the current-voltage characteristics. As a result, light intensity fluctuation is induced. Optical feedback induces these phenomena in laser diodes but not in LEDs because the light output from an LED is not coherent.

The problems on the phase matching between the laser cavity and external cavity modes can be understood by inspecting Fig. 3.61. When the distance to the external mirror is less than the coherence length, l_c [see Eq. (3.54)], the phase condition discussed above is important. In such a situation, the lasing mode in the phase matching condition, at which the laser cavity and external cavity modes coincide, tends to be intensified. In addition, the mode (lasing wavelength) jump and light power fluctuation occur corresponding to the change in the phase condition. When the distance to the external mirror is greater than the coherence length of the laser diode, the phase condition no longer has meaning.

The optical feedback noises are mainly related to the quantum shot noise and mode-hopping noise. The noise related to the quantum noise typically appears in the light intensity and phase noise, especially at the relaxation oscillation frequency. The mode-hopping noise is, of course, initially induced by the mode coupling between the two cavities and increases noise at a relatively low frequency range as discussed in the previous section. In addition, the beat noise originating from each longitudinal mode is generated in laser diodes showing multiple longitudinal mode operation under the optical feedback.[95] These optical feedback noises tend to be intensified as the optical feedback increases. The measure of the optical coupling between the two cavities, κ_{of}, is given by the following equation:[94]

$$\kappa_{of} = (1 - R)(R_{distant}/R)^{1/2} c_0/2 n_r L. \qquad (3.63)$$

In the following part of this section the behavior of DFB laser diodes under conditions in which there is optical feedback is treated in order to discuss simply the optical feedback noise related to the quantum shot noise. (The mode-hopping noise described in the previous section can be neglected.) Typical changes in relative intensity noise as a function of optical feedback level are shown in Fig. 3.63 for a 1550 nm InGaAsP/InP DFB laser diode with a bulk active layer. As shown in Fig. 3.64, sharp periodic noise peaks appear and are intensified at the frequencies corresponding to those satisfying the phase matching condition, $\Omega t_R = 2m\pi$, as the level of optical feedback increases. The frequencies, f_{of} ($= \Omega/2\pi$), are given as

$$f_{of} = m/t_R = mc_0/2n_{r\text{-}ext}l_{ext} \qquad m = 1, 2, 3, \ldots. \qquad (3.64)$$

The spacing between the noise peaks is therefore equal to $c_0/2n_{r\text{-}ext}l_{ext}$, and the position of the external distant mirror can be determined from this spacing. As the level of optical feedback further increases, the base level of the noise spectrum (indicated by a closed circle in Fig. 3.63) arises from the output

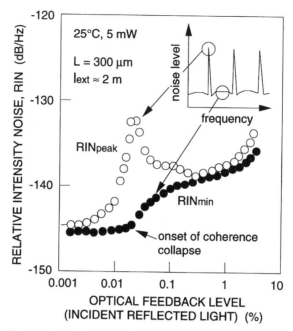

FIGURE 3.63 Change in relative intensity noise at 100 MHz in the dc noise spectrum of a 1550 nm-band buried heterostructure DFB laser diode with a bulk active layer.[96]

power fluctuation. The noise spectra under dc operation and pulse responses are shown in Fig. 3.65 for the same laser diode whose changes in relative intensity noise are shown in Fig. 3.63. In the range of optical feedback over which the noise peak is small, the external cavity mode is strongly enhanced around the relaxation oscillation frequency (at 7.6 GHz in Fig. 3.65). A small oscillation

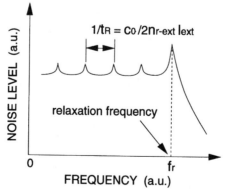

FIGURE 3.64 Illustration of intensity noise as a function of frequency for laser diodes under optical feedback.

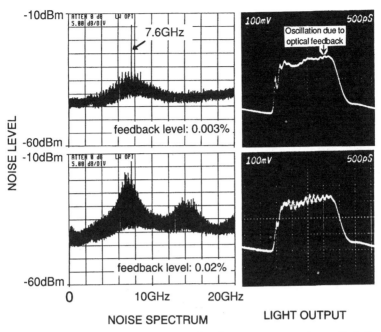

NOISE SPECTRUM   LIGHT OUTPUT

FIGURE 3.65 Change in noise spectra and pulse responses under optical feedback levels of 0.003% and 0.02% for the same laser diode shown in Fig. 3.63. For the pulse response, dc current was biased at $0.9\,I_{th}$ and the pulsed peak light output power was set at 5 mW.[96]

having the relaxation oscillation frequency appears in the pulse response. Just before the base level of the noise spectrum increases, the external cavity mode is further enhanced and subpeaks at the frequencies corresponding to the integral multiples of the relaxation oscillation frequency appear in the noise spectrum. In the pulse response, the oscillation becomes clear. This phenomenon, that the laser diode is modulated at the relaxation oscillation frequency, results from the change in the quantum shot noise spectra due to the optical feedback. After the base level of the noise spectrum increases, the coherence deteriorates tremendously even under a single longitudinal mode operation, and a self-sustained pulsation at the relaxation oscillation frequency starts. This deterioration of coherence is called the coherence collapse.[97] The onset of the coherence collapse is determined by the relaxation oscillation damping factor and the spectral linewidth enhancement factor (see Section 3.5.3.3).[98] If the level of optical feedback exceeds 10%, the noise decreases again and the coherence is recovered because the whole cavity formed by the distant mirror and the laser diode starts to lase. The phase noise behaves similarly because the intensity noise couples with the phase noise through the fluctuations of carrier density and refractive index.

Some methods have been used to suppress the optical feedback noise in laser diodes. This feedback noise is essentially a result of the light being coherent. The coherence is usually lowered by superimposing a high-frequency signal or by using laser diodes with self-sustained pulsation, although the intensity noise increases to some extent under such modulation or pulsation. As indicated in Eq. (3.63), another way to decrease the influence of the optical feedback is to increase the facet reflectivity and to lengthen the laser cavity—that is, to heighten the coherence. Laser diodes with low linewidth enhancement factors (α-parameters), such as (strained) quantum well laser diodes, tend to be resistant to optical feedback.[99] Optical feedback does not always produce noise. The spectral linewidth of DFB laser diodes changes according to the amount of the optical feedback, and a very narrow linewidth can be obtained under an appropriate level of optical feedback.[100]

3.6 HIGH-POWER LASER DIODES AND SURFACE-EMITTING LASER DIODES

High-power laser diodes and surface-emitting laser diodes are based on the (edge-emitting) laser diodes discussed in this chapter. These high-power and surface-emitting types of laser diodes are summarized in this section because their structures and operating characteristics are slightly different from those of the edge-emitting laser diodes.

3.6.1 High-Power Laser Diodes

Most high-power laser diodes are edge-emitting types. To obtain a stable high optical output power, several conditions are required in addition to the use of antireflective and highly reflective dielectric films on the mirror facets (see Section 6.2.2.3). For laser diodes themselves, the following are mainly important: (1) low internal loss (low threshold and high quantum efficiency), (2) stable transverse mode (no kink on current–light output power characteristics), (3) high temperature characteristics, (4) no catastrophic optical damage or high generation level of catastrophic optical damage.

1. *Low Internal Loss.* Loss in the optical cavity is usually reduced by using (strained) quantum well structures so that the active layer can be thin and the optical confinement factor therefore is lowered. In addition, a long cavity (for example, about 1 mm) is often used to reduce operating current density (carrier density) and thermal resistance, although a long cavity also decreases slope efficiency.

2. *Stable Transverse Mode.* Transverse modes can be stabilized by using buried heterostructures or ridge waveguide structures to provide refractive index guiding because the transverse modes are unstable in gain guiding. If there are kinks, the optical output power decreases or is saturated.

3. *High Temperature Characteristics.* High output power is a result of high current injection. The high current injection also results in a large amount of Joule's heating, which in turn increases the temperature of the active region. If the temperature dependence of the current–light output power characteristics is large (low temperature characteristics), the output power saturation occurs at a relatively low output power (injected current).

4. *No Catastrophic Optical Damage (see Section 7.2.1.3).* Catastrophic optical damage occurs suddenly as a result of melting of facets because of the heat produced by the nonradiative recombination of photoinduced carriers at facets. These phenomena are triggered by the absorption of emitted light at the facet. The damage level limits the maximum output power. Consequently, either a nonabsorbing mirror (NAM) structure is used or the optical power density at the facet is decreased in order to increase the damage level.

The bonding configuration should be one with high thermal conductivity (low thermal resistance), so a junction-down configuration is used in most cases. Large packages, such as those used for power transistors, are also used to release the heat generated.

In addition to the solitary high-power laser diodes mentioned above, laser diodes integrated with semiconductor amplifiers are also used as high-power lasers.[101] One-dimensional or two-dimensional laser diode arrays are also used as high-power optical sources. In those arrays, laser diodes with gain-guiding are often used. The AlGaAs/GaAs quantum well laser diode arrays easily emit more than 1 W.[102] These arrays are used, for example, as the pumping sources of solid state lasers. The patterns of emitted light (far-field patterns) are not controlled in those arrays. The phased array laser diodes show high spatial coherence, and the lateral far-field pattern is less than a few degrees in one-dimensional arrays.[103] Active (light-emitting) regions a few micrometers apart are built in a single active layer as shown in Fig. 3.66. The light emitted in the active region is propagated along each active region and leaks in the lateral direction. Each lasing mode is coupled by the light leaked, and the phase of each optical field is synchronized with those of the others. This results in high temporal and spatial coherence. When the array operates under the fundamental mode, a very sharp emission pattern can be obtained as illustrated in Fig. 3.66. The optical power density at the facet can be reduced in this type of array and then the level of catastrophic optical damage increases when compared with solitary laser diodes. These phased arrays will be candidate optical sources for satellite communication because of their high power and high directivity.

3.6.2 Surface-Emitting Laser Diodes

The surface-emitting laser diodes, as shown in Fig. 3.67, lase in the direction perpendicular to the *pn*-junction planes (active layers). The active vol-

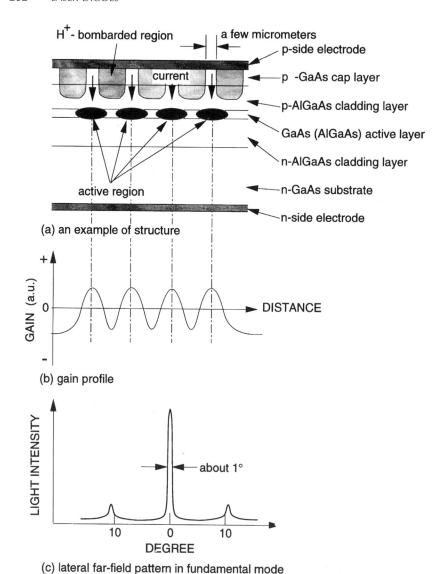

FIGURE 3.66   Conceptual illustration of a phased array laser diode and its far-field pattern.

ume is very small when compared with that of the normal edge-emitting type of laser diodes. The lasing area can be less than 10 μm in diameter and the thickness (cavity length) can be less than a micrometer. Such a short cavity increases the threshold current because the threshold gain required for lasing is inversely proportional to the cavity length [see Eq. (3.3)]. The problem related to the threshold current is solved by increasing facet reflectivity (or mirror

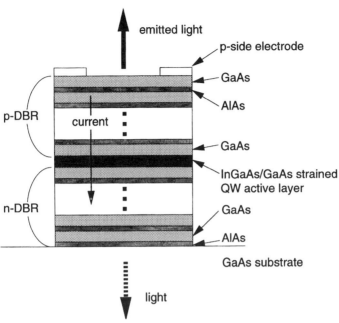

FIGURE 3.67   Cross-sectional view of a surface-emitting type of laser diode.

reflectivity). Metal coatings and highly reflective dielectric films on the surface increase the reflectivity to some extent. A distributed Bragg reflector (DBR) composed of semiconductor layers can also increase the reflectivity.[104,105] The multilayer structure is similar to a high-reflecting dielectric film and consists of two kinds of semiconductors, such as GaAs/AlAs, each with a thickness of $\lambda_0/4n_r$ (see Section 6.2.2.3). Those semiconductor DBR mirrors can be successively grown on both sides of the active layer, so highly reflective mirrors (with an optical power reflectivity of more than 99%) can easily be set in the vicinity of the active region. The small active volume and the high reflectivity of the DBR result in a very low lasing threshold and operation at a low input power.[106-109] The strained quantum well InGaAs/GaAs laser diodes with the DBRs, for example, show very low threshold current below 100 μA, where the current injected (active) region is surrounded by the oxide of the host semiconductors.[108,109] The surface-emitting laser diodes operate in a single longitudinal mode because of the very large mode spacing due to the short cavity [see Eq. (3.14)] and the DBR mirrors. The electrical resistance, however, is high (over 10 Ω) because of the electrical resistance of the DBR region.

The surface-emitting type of laser diodes operate under a low input power, and one-dimensional and two-dimensional arrays of them can be formed easily. This type of device is suitable for parallel processing of information, parallel interconnection between computers, and so forth.

Surface-emitting laser diodes can also be fabricated by using the normal edge-emitting type of laser diodes: the light emitted from the mirror facet can be reflected by an external mirror set so that the emitted light is reflected perpendicular to the *pn*-junction plane. Such surface-emitting laser diodes, however, lack the advantage of a very low threshold.

REFERENCES

1. I. Hayashi, M. B. Panish, P. W. Foy, and S. Sumski, Junction lasers which operate continuously at room temperature, *Appl. Phys. Lett.*, **17**, 109, 1970.

2. Zh. I. Alferov, V. M. Andreef, D. Z. Garbutov, Yu. V. Zhilyaev, E. P. Morozov, E. L. Portnoi, and V. G. Trofim, Investigation of the influence of the AlAs-GaAs heterostructure parameters on the laser threshold current and the realization of continuous emission at room temperature, *Sov. Phys. Semicond.*, **4**, 1573, 1971. (Translated from *Fiz. Tekh. Poluprovodn.*, **4**, 1826, 1970.)

3. R. E. Naholy and M. A. Pollack, Threshold dependence on active layer thickness in InGaAsP/InP DH lasers, *Electron Lett.*, **14**, 727, 1978.

4. F. Stern, Gain-current relation for GaAs lasers with *n*-type and undoped active layers, *IEEE J. Quantum Electron.*, **QE-9**, 290, 1973.

5. M. Fukuda, *Reliability and Degradation of Semiconductor Lasers and LEDs*, Artech House, Boston, 1991.

6. F. R. Nash, Mode guidance parallel to the junction plane of double heterostructure GaAs lasers, *J. Appl. Phys.*, **44**, 4696, 1973.

7. For example, H. C. Casey, Jr., and M. B. Panish, *Heterostructure Lasers*, Academic Press, New York, 1978.

8. J. C. Dyment, L. A. D'Asaro, J. C. North, B. I. Miller, and J. E. Ripper, Proton-bombardment formation of stripe-geometry heterostructure lasers for 300 K cw operation, *Proc. IEEE*, **60**, 726, 1972.

9. H. Yonezu, I. Sakuma, K. Kobayashi, T. Kamejima, M. Ueno, and Y. Nannichi, A GaAs-$Al_xGa_{1-x}As$ double heterostructure planar stripe laser, *Japan. J. Appl. Phys.*, **12**, 1585, 1973.

10. N. Chinone, R. Ito, and O. Nakada, Limitations of power outputs from continuously operating GaAs-$Ga_{1-x}Al_xAs$ double heterostructure lasers, *J. Appl. Phys.*, **47**, 785, 1976.

11. K. Aiki, M. Nakamura, T. Kuroda, J. Umeda, R. Ito, N. Chinone, and M. Maeda, Transverse mode stabilized $Al_xGa_{1-x}As$ injection lasers with channeled-substrate-planar structure, *IEEE J. Quantum Electron.*, **QE-14**, 89, 1978.

12. H. Kano and K. Sugiyama, Operation characteristics of buried-stripe GaInAsP/InP DH lasers made by melt-back method, *J. Appl. Phys.*, **50**, 7934, 1979.

13. H. Namizaki, H. Kan, M. Ishii, and A. Ito, Transverse-junction-stripe-geometry double heterostructure lasers with very low threshold current, *J. Appl. Phys.*, **45**, 2785, 1974.

14. H. Ishikawa, H. Imai, T. Tanahashi, Y. Nishitani, M. Takusagawa, and K. Takahei, V-grooved substrate buried heterostructure InGaAsP/InP laser, *Electron. Lett.*, **17**, 465, 1981.

15. I. Mito, M. Kitamura, K. Kobayashi, and K. Kobayashi, Double channel planar buried heterostructure laser diode with effective current confinement, *Electron Lett.*, **18,** 953, 1982.

16. H. Statz, C. L. Tang, and J. M. Lavine, Spectral output of semiconductor lasers, *J. Appl. Phys.*, **35,** 2581, 1964.

17. For example, S. Matsumoto, M. Fukuda, K. Sato, Y. Itaya, and M. Yamamoto, Highly reliable 1.55 μm GaInAsP laser diodes buried with semi-insulating iron-doped InP, *Electron. Lett.*, **30,** 1305, 1994.

18. For example, G. H. B. Thompson, *Physics of Semiconductor Laser Devices*, John Wiley & Sons, New York, 1980.

19. R. D. Dupuis, P. D. Dapkus, R. Chin, N. Holonyak, Jr., and S. W. Kirchoefer, Continuous 300 K laser operation of single-quantum well $Al_x Ga_{1-x}As$-GaAs heterostructure diodes grown by metalorganic chemical vapor deposition, *Appl. Phys. Lett.*, **34,** 265, 1979.

20. A. R. Adams, Band-structure engineering for low-threshold high-efficiency semiconductor lasers, *Electron Lett.*, **22,** 249, 1986.

21. E. Yablonovitch and E. O. Kane, Reduction of lasing threshold current density by the lowering of valence band effective mass, *IEEE J. Lightwave Technol.*, **4,** 504, 1986.

22. Y. Itaya, M. Fukuda, Y. Noguchi, and Y. Nakano, Improvement in differential quantum efficiency of 1.55 μm distributed feedback lasers grown by MOVPE, *Japan. J. Appl. Phys.*, **26,** L1289, 1987.

23. J. Nagle, S. Hersee, M. Krakowski, T. Weil, and C. Weisbuch, Threshold current of single quantum well lasers: The role of the confining layers, *Appl. Phys. Lett.*, **49,** 1325, 1986.

24. P. J. A. Thijs, Progress in quantum well lasers: Application of strain, *Proc. 13th Inter. Semicond. Laser Conf.*, Takamatsu, Japan, Sept., 1992, p. 2.

25. E. P. O'Reilly, G. Jones, A. Ghiti, and A. R. Adams, Improved performance due to suppression of spontaneous emission in tensile-strain semiconductor lasers, *Electron. Lett.*, **27,** 1417, 1991.

26. H. Kogelnik and C. V. Shank, Coupled-wave theory of distributed feedback lasers, *J. Appl. Phys.*, **43,** 2327, 1972.

27. M. Nakamura, H. W. Yen, A. Yariv, E. Garmire, S. Somekh, and H. L. Girvin, Laser oscillation in epitaxial GaAs waveguides with corrugation feedback, *Appl. Phys. Lett.*, **23,** 224, 1973.

28. W. Streifer, R. D. Burnham, and D. R. Scigres, Effect of external reflectors on longitudinal modes of distributed feedback lasers, *IEEE J. Quantum Electron.*, **QE-11**, 154, 1975.

29. H. Kressel and J. K. Butler, *Semiconductor Lasers and Heterojunction LEDs*, Academic Press, New York, 1977.

30. W. Streifer, D. R. Scifres, and R. D. Burnham, Coupling coefficients for distributed feedback single- and double-heterostructure lasers, *IEEE J. Quantum Electron.*, **QE-11,** 867, 1975.

31. T. Matsuoka, H. Nagai, Y. Noguchi, Y. Suzuki, and K. Kawaguchi, Effect of the grating phase at the cleaved facet on DFB laser properties, *Japan. J. Appl. Phys.*, Part 2, **23,** L138, 1984.

32. G. Bjork and O. Nilsson, A new exact and efficient numerical matrix theory of complicated laser structures. Properties of asymmetric phase-shifted DFB lasers, *Tech. Dig. Int. Conf. on Integrated Optics and Optical Fiber Communication (IOOC)/Euro. Conf. Optical Comm. (ECOC)'85*, Venezia, Italy, 1985, vol. 1, p. 107.

33. See for example, M. Asada, A. R. Adams, K. E. Stubkjaer, Y. Suematsu, Y. Itaya, and S. Arai, The temperature dependence of the threshold current of GaInAsP/InP DH lasers, *IEEE J. Quantum Electron.*, **QE-17,** 611, 1981.

34. T. Matsuoka, Temperature range for DFB mode oscillation in 1.5 μm InGaAsP/InP DFB lasers, *Japan. J. Appl. Phys.*, **25,** 1206, 1986.

35. H. Soda, Y. Kotaki, H. Sudo, H. Ishikawa, S. Yamakoshi, and H. Imai, Stability in single longitudinal mode operation in GaInAsP/InP phase-adjusted DFB lasers, *IEEE J. Quantum Electron.*, **QE-23,** 804, 1987.

36. Y. Yoshikuni, K. Oe, G. Motosugi, and T. Matsuoka, Broad wavelength tuning under single-mode oscillation with a multielectrode distributed feedback laser, *Electron. Lett.*, **22,** 1153, 1986.

37. M. Fukuda, K. Sato, Y. Kondo, and M. Nakao, Continuously tunable thin active layer and multisection DFB laser with narrow linewidth and high power, *J. Lightwave Technol.*, **7,** 1504, 1989.

38. Y. Suematsu and A. R. Adams, Ed., *Handbook of Semiconductor Lasers and Photonic Integrated Circuits*, Ohmsha, Ltd, Tokyo and Chapman & Hall, London, 1994.

39. V. Jayaraman, L. A. Cohen, and L. A. Coldren, Extended tuning range in a distributed feedback lasers with sampled grating, *Proc. Conf. on Optical Fiber Comm. (OFC)*, San Jose, CA, 1992, p. 165, WL12.

40. Y. Tohmori, Y. Yoshikuni, H. Ishii, F. Kano, T. Tamamura, Y. Kondo, and M. Yamamoto, Broad-range wavelength tunable super structure grating (SSG) DBR lasers, *IEEE J. Quantum Electron.*, **29,** 1817, 1993.

41. I. Hayashi, M. B. Panish, and F. K. Reinhart, GaAs-Al$_x$Ga$_{1-x}$As double heterostructure injection lasers, *J. Appl. Phys.*, **42,** 1929, 1971.

42. Y. Zou, J. S. Osinski, P. Grodzinski, P. D. Dapkus, W. C. Rideout, W. F. Sharfin, J. Schlafer, and F. D. Crawford, Experimental study of Auger recombination, gain, and temperature sensitivity of 1.5-μm compressively strained semiconductor lasers, *IEEE J. Quantum Electron.*, **29,** 1565, 1993.

43. A. R. Goodwin, J. R. Peters, M. Pion, G. H. B. Thompson, and J. E. A. Whiteway, Threshold temperature characteristics of double heterostructure Ga$_x$Al$_{1-x}$As lasers, *J. Appl. Phys.*, **46,** 3126, 1975.

44. M. Ettenberg, C. J. Nuese, and H. Kressel, The temperature dependence of threshold current for double heterojunction lasers, *J. Appl. Phys.*, **50,** 2949, 1979.

45. M. Yano, H. Nishi, and M. Takusagawa, Temperature characteristic of threshold current in InGaAsP/InP double heterostructure lasers, *J. Appl. Phys.*, **51,** 4022, 1980; M. Yano, H. Imai, and M. Takusagawa, Analysis of electrical, threshold, and temperature characteristics of InGaAsP/InP double-heterojunction lasers, *IEEE J. Quantum. Electron.*, **QE-17,** 1754, 1981.

46. G. H. B. Thompson and G. D. Henshall, Nonradiative carrier loss and temperature

sensitivity of threshold in 1.27 μm (GaIn)(AsP) D. H. lasers, *Electron. Lett.*, **16,** 42, 1980.

47. A. Sugimura, Band-to-band Auger recombination effect on InGaAsP laser threshold, *IEEE J. Quantum Electron.*, **QE-17,** 627, 1981.

48. T. Uji, K. Iwamoto, and R. Lang, Nonradiative recombination in InGaAsP/InP light sources causing light emitting diode output saturation and strong laser-threshold-current temperature sensitivity, *Appl. Phys. Lett.*, **38,** 193, 1981.

49. B. Etienne, J. Shah, R. F. Leheny, and R. E. Nahory, Influence of hot carriers on the temperature dependence of threshold in 1.3-μm InGaAsP lasers, *Appl. Phys. Lett.*, **41,** 1018, 1982.

50. M. Asada and Y. Suematsu, The effect of loss and nonradiative recombination on the temperature dependence of threshold current in 1.5–1.6 μm GaInAsP/InP lasers, *IEEE J. Quantum Electron.*, **QE-19,** 917, 1983.

51. J. O'Gorman, A. F. J. Levi, S. Schmitt-Rink, T. Tanbun-Ek, D. L. Coblentz, and R. A. Logan, On the temperature sensitivity of semiconductor lasers, *Appl. Phys. Lett.*, **60,** 157, 1992.

52. H. Oohashi, S. Seki, T. Hirono, H. Sugiura, T. Amano, M. Ueki, J. Nakano, M. Yamamoto, Y. Tohmori, M. Fukuda, and K. Yokoyama, High-power and high-efficiency 1.3 μm InAsP compressively-strained MQW lasers at high temperatures, *Electron. Lett.*, **31,** 556, 1995.

53. C. E. Zah, M. C. Wang, R. Bhat, T. P. Lee, S. L. Chuang, Z. Wang, D. Darby, D. Flanders, and J. J. Hsieh, High temperature modulation dynamics of 1.3 μm $Al_xGa_yIn_{1-x-y}$As-InP Compressive-strained multiple-quantum-well lasers, *Proc. 14th IEEE Int. Semiconduct. Laser Conf., Maui, HI,* 1994, p. 215.

54. S. Seki, H. Oohashi, H. Sugiura, T. Hirono, and K. Yokoyama, Study on the dominant mechanisms for the temperature sensitivity of threshold current in 1.3-μm InP-based strained-layer quantum-well lasers, *IEEE J. Quantum Electron.*, **32,** 1478, 1996.

55. W. E. Lamb, Jr., Theory of optical maser, *Phys. Rev.*, **134,** A1429, 1964.

56. H. Haug, Quantum-mechanical rate equations for semiconductor lasers, *Phys. Rev.*, **184,** 338, 1969.

57. K. Konnerth and C. Lanza, Turn-on characteristics of a GaAs injection lasers, *Appl. Phys. Lett.*, **4,** 120, 1964.

58. T. Ikegami and Y. Suematsu, Carrier lifetime measurement of a junction laser using direct modulation, *IEEE J. Quantum Electron.*, **QE-4,** 148, 1968.

59. T. L. Paoli, Changes in the optical properties of cw (AlGa)As junction lasers during accelerated aging, *IEEE J. Quantum Electron.*, **QE-13,** 351, 1977.

60. M. Fukuda, K. Wakita, and G. Iwane, Self-sustained pulsation appearance in InGaAsP/InP DH lasers during accelerated operaton, *Japan. J. Appl. Phys.*, **20,** L153, 1981.

61. J. P. Van der Ziel, J. L. Merz, and T. L. Paoli, Study of intensity pulsations in proton-bombarded stripe-geometry double-heterostructure Al_xGa_{1-x}As lasers, *J. Appl. Phys.*, **50,** 4620, 1979.

62. K. Lau, N. Bar-Chaim, and I. Ury, Direct amplitude modulation of short-cavity GaAs lasers up to x-band frequencies, *Appl. Phys. Lett.*, **43,** p. 1, 1983.

63. K. Lau and A. Yariv, Ultra-high speed semiconductor lasers, *IEEE J. Quantum Electron.*, **QE-21,** 121, 1985.

64. C. B. Su and V. Lanzisera, Effect of doping level on the gain constant and modulation bandwidth of InGaAsP semiconductor lasers, *Appl. Phys., Lett.*, **45,** 1302, 1984.

65. K. Uomi, T. Mishima, and N. Chinone, Ultrahigh relaxation oscillation frequency (up to 30 GHz) of highly p-doped GaAs/GaAlAs multiple quantum well lasers, *Appl. Phys., Lett.*, **51,** 78, 1987.

66. Y. Arakawa and T. Takahashi, Effect of nonlinear gain on modulation dynamics in quantum well lasers, *Electron. Lett.*, **25,** 169, 1989.

67. A. Ghiti and E. P. O'Reilly, Nonlinear gain effects in strained-layer lasers, *Electron. Lett.*, **26,** 1978, 1990.

68. S. Seki, T. Yamanaka, and K. Yokoyama, Theoretical analysis of gain saturation coefficients in InP based strained-layer quantum-well lasers, *J. Appl. Phys.*, **74,** 2971, 1993.

69. G. Agrawal, Gain nonlinearities in semiconductor lasers: Theory and application to distributed feedback lasers, *IEEE J. Quantum Electron.*, **QE-23,** 860, 1987.

70. K. Uomi, T. Tsuchiya, M. Aoki, and N. Chinone, Oscillation wavelength and laser structure dependence of nonlinear damping effect in semiconductor lasers, *Appl. Phys. Lett.*, **58,** 675, 1991.

71. W. Rideout, W. F. Sharfin, E. S. Koteles, M. O. Vassell, and B. Elman, Well-barrier hole burning in quantum well lasers, *IEEE Photon. Technol. Lett.*, **3,** 784, 1991.

72. See for example, M. Maeda, K. Nagano, M. Tanaka, and K. Chiba, Buried-heterostructure laser packaging for wideband optical transmission systems, *IEEE J. Trans. Comu.*, **COM-26,** 1076, 1978.

73. S. Matsumoto, M. Fukuda, K. Sato, Y. Itaya, and M. Yamamoto, Highly reliable 1.55 μm GaInAsP laser diodes buried with semi-insulating iron-doped InP, *Electron. Lett.*, **30,** 1305, 1994.

74. See for example, R. Nagarajan, T. Fukushima, S. W. Corzine, and J. E. Bowers, Effect of carrier transport on high-speed quantum well lasers, *Appl. Phys. Lett.*, **59,** 1835, 1991.

75. C. H. Henry, Theory of the linewidth of semiconductor lasers, *IEEE J. Quantum Electron.*, **QE-18,** 259, 1982.

76. K. Kondo, K. Sato, M. Nakao, M. Fukuda, and K. Oe, Extremely narrow linewidth ($\approx$ 1 MHz) and high-power DFB lasers grown by MOVPE, *Electron. Lett.*, **25,** 175, 1989.

77. K. Kikuti, Effect of $1/f$-type FM noise on semiconductor laser linewidth residual in high-power limit, *IEEE J. Quantum Electron.*, **QE-25,** p. 684, 1989.

78. M. Fukuda, T. Hirono, T. Kurosali, and F. Kano, $1/f$ noise behavior in semiconductor laser degradation, *Photon. Technol. Lett.*, **5,** 1165, 1993.

79. B. Tromborg, H. Olsen, and X. Pan, Theory of linewidth for multielectrode laser diodes with spatially distributed noise source, *IEEE J. Quantum Electron.*, **27,** 178, 1991.

80. X. Pan, B. Yronborg, and H. Olesen, Linewidth rebroadening in DFB lasers due to weak side modes, *IEEE Photon. Technol. Lett.*, **3,** 112, 1991.

81. H. Yasaka, M. Fukuda, and T. Ikegami, Current tailoring for lowering linewidth floor, *Electron. Lett.*, **24**, 760, 1988.

82. F. Kano, Y. Tohmori, Y. Kondo, M. Nakao, M. Fukuda, and K. Oe, Spectral linewidth reduction (580 kHz) in structure-optimized 1.5 μm butt-jointed DBR lasers, *Electron. Lett.*, **25**, 709, 1989.

83. M. C. Amann and R. Schimpe, Excess linewidth broadening in wavelength-tunable laser diodes, *Electron. Lett.*, **26**, 279, 1990.

84. H. Mawatari, R. Iga, H. Sugiura, Y. Tohmori, Y. Yoshikuni, Modulation-doped GaInAs/GaInAsP strained multi-quantum-well lasers grown by chemical beam epitaxy, *Appl. Phys. Lett.*, **65**, 277, 1994.

85. T. L. Koch and R. A. Linke, Effect on nonlinear gain reduction on semiconductor laser wavelength chirping, *Appl. Phys. Lett.*, **48**, 613, 1986.

86. D. E. McCumber, Intensity fluctuations in the output of cw oscillations, *Phys. Rev.*, **141**, 306, 1966.

87. D. J. Morgan and M. J. Adams, Quantum noise in semiconductor lasers, *Phys. Stat. Sol. (A)*, **11**, 243, 1972.

88. Y. Yamamoto, AM and FM quantum noise in semiconductor lasers—Part I: Theoretical analysis, *IEEE J. Quantum Electron.*, **QE-19**, 34, 1983; Y. Yamamoto, AM and FM quantum noise in semiconductor lasers—Part II: Comparison of theoretical and experimental results for AlGaAs lasers, *IEEE J. Quantum Electron.*, **QE-19**, 47, 1983.

89. C. H. Henry, Theory of the phase noise and power spectrum of a single mode injection laser, *IEEE J. Quantum Electron.*, **QE-19**, 1391, 1983.

90. R. Lang, K. Minemura, and K. Kobayashi, Low-frequency intensity noise in c. w. (GaAl)As DH lasers with stripe geometry, *Electron. Lett.*, **13**, 228, 1977.

91. T. Ito, S. Machida, K. Nawata, and T. Ikegami, Intensity fluctuation in each longitudinal mode of a multimode AlGaAs laser, *IEEE J. Quantum Electron.*, **QE-13**, 574, 1977.

92. D. Welford and A. Mooradian, Observation of linewidth broadening in (GaAl)As diode lasers due to electron number fluctuation, *Appl. Phys. Lett.*, **40**, 560, 1982.

93. K. Vahala and A. Yariv, Semiclassical theory of noise in semiconductor lasers—Part II, *IEEE J. Quantum Electron.*, **QE-19**, 1102, 1983.

94. R. Lang and K. Kobayashi, External optical feedback effects on semiconductor injection laser properties, *IEEE J. Quantum Electron.*, **QE-16**, 347, 1980.

95. A. P. Bogatov, P. G. Eliseev, and B. N. Sverdlov, Anomalous interaction of spectral modes in semiconductor laser, *IEEE J. Quantum Electron.*, **QE-11**, 510, 1975.

96. T. Kurosaki, T. Hirono, and M. Fukuda, Suppression of external cavity modes in DFB lasers with a high endurance against optical feedback, *IEEE Photon. Technol. Lett.*, **6**, 900, 1994.

97. D. Lenstra, B. H. Verbeek, and A. J. den. Boef, Coherence collapse in single-mode semiconductor lasers due to optical feedback, *IEEE J. Quantum. Electron.*, **QE-21**, 674, 1985.

98. T. Hirono, T. Kurosaki, and M. Fukuda, Transition from the lowest linewidth mode operation to coherence collapse in a semiconductor laser with feedback from a distant reflector, *IEEE J. Quantum Electron.*, **32**, 829, 1996.

99. J. Helms and K. Petermann, A simple analytic expression for the stable operation range of laser diodes with optical feedback, *IEEE J. Quantum Electron.*, **QE-26,** 833, 1990.

100. R. W. Tkach and A. R. Chraplyvy, Regimes of feedback effects in 1.5-μm distributed feedback lasers, *J. Lightwave Technol.*, **LT-4,** 1655, 1986.

101. S. O'Brien, D. F. Welch, R. A. Pavke, D. Mehuys, K. Dzurko, R. J. Lang, R. Waarts, and D. Scifres, Operating characteristics of a high-power monolithically integrated flared amplifier master oscillator power amplifier, *J. Quantum. Electron.*, **29,** 2052, 1993.

102. See for example, J. G. Endriz, M. Vakili, G. S. Browder, M. Devito, J. M. Haden, G. L. Harnagel, W. E. Plano, M. Sakamoto, D. F. Welch, S. Willing, D. P. Worland, and H. C. Yao, High power diode laser arrays, *J. Quantum Electron.*, **28,** 952, 1992.

103. See for example, D. Botez, L. J. Mawst, G. Peterson, and T. J. Roth, Resonant optical transmission and coupling in phase-locked diode laser arrays of antiguides: The resonant optical waveguide array, *Appl. Phys. Lett.*, **54,** 2183, 1989.

104. A. Chailertvanitkul, K. Iga, and K. Moriki, GaInAsP/InP surface emitting laser (λ = 1.4 μm, 77 K) with heteromultilayer Bragg reflector, *Electron. Lett.*, **21,** 303, 1985.

105. T. Sakaguchi, F. Koyama, and K. Iga, Vertical cavity surface-emitting laser with an AlGaAs/AlAs Bragg reflector, *Electron. Lett.*, **24,** 928, 1986.

106. F. Koyama, S. Kinoshita, and K. Iga, Room-temperature continuous wave lasing characteristics of a GaAs vertical cavity surface-emitting laser, *Appl. Phys. Lett.*, **55,** 221, 1989.

107. H. Deng, D. L. Huffaker, J. Shin, and D. G. Deppe, Gain-switching in a vertical-cavity laser with high-contrast mirrors, *Electron. Lett.*, **31,** 278, 1994.

108. Y. Hayashi, T. Mukaihara, N. Hatori, N. Ohnoki, A. Matsutani, F. Koyama, and K. Iga, Record low-threshold index-guided InGaAs/GaAlAs vertical cavity surface emitting laser with a native oxide confinement structure, *Electron. Lett.*, **31,** 560, 1995.

109. G. M. Yang, M. H. MacDougal, and P. D. Dapkus, Ultra low threshold current vertical-cavity surface-emitting lasers obtained with selective oxidation, *Electron. Lett.*, **31,** 886, 1995.

CHAPTER FOUR

Photodiodes

Photodiodes are typical optoelectronic *pn*-junction devices used under a reverse bias. Their operating mechanisms, like those of LEDs and laser diodes, are based on the electrical and optical properties of the *pn*-junction and semiconductor material. Their functions are, however, quite different from those of LEDs and laser diodes, and in the photodiodes the optical absorption processes are used. The basics of optical absorption in semiconductors are described in Section 1.1.3.1. The light absorption is the opposite of the stimulated and spontaneous emission process, and the operating wavelength therefore essentially depends on the band-gap energy of the material. The materials widely used in photodiodes are elementary semiconductors such as Si and Ge, and III-V and II-VI compound semiconductors such as GaAs, InP, and CdTe. Photodiodes composed of Si are mainly used in comsumer electronics, while those composed of Ge and InGaAs(P)/InP are used in optical fiber communication systems. In the longer-wavelength range, InAs and InSb are also used as photodiode materials.

4.1 BASICS OF PHOTODIODES

4.1.1 Photovoltaic Effect and Current-Voltage Characteristics

When a semiconductor is illuminated by light having an energy greater than its band-gap energy, the light is absorbed in the semiconductor and electron-hole pairs are generated (see Fig. 1.30). Those photoinduced electrons and holes recombine radiatively (photoluminescence) or nonradiatively. If an electrical field is applied to the semiconductor, some of the induced carriers take part in electric conduction and this leads to a decrease in electrical resistance of the semiconductor. This is called photoconduction. If there is a *pn*-junction in the illuminated area, the electrons and the holes are separated by the electrical field at the *pn*-junction without any electric bias, and an electromotive force between the *p*- and *n*-side semiconductors is generated. This is called the photovoltaic effect, and with regard to the photovoltaic effect there is basically no different between a *pn*-homojunction and a *pn*-heterojunction. A similar photovoltaic effect is observed at the Schottky barrier between a semiconductor and a metal. Based on the phenomena described above, light power can be converted into electrical power in photodiodes.

The conventional photodiodes convert light power into electrical power by using the photovoltaic effect. A bias circuit for photodiodes and current-voltage characteristics under light illumination are shown schematically in Fig. 4.1 (see Fig. 1.42). The dashed line indicates the current-voltage characteristic under no light illumination. The current under reverse bias is ideally equal to the saturation current given by Eq. (1.154) but is actually determined by the generation-recombination current and surface-recombination current as well, as discussed in Section 1.2.3.2 [see Eq. (1.161)]. This reverse current under conditions without illuminating light is called the dark current. When light is absorbed, the current-voltage characteristics change as shown in Fig. 4.1(b), and the reverse current increases because of the generation of photocurrent [see Fig. 4.1(a)]. Depending on the reverse voltage biased to the *pn*-junction, the operation behaviors of photodiode can be divided into solar cell, photodiode, and avalanche photodiode modes. Some suitable structures have been developed for each mode.

4.1.2 *pn*-Junction Under Reverse Bias and Reverse Current (Dark Current)

The basics of the *pn*-junction are described in Section 1.2. Ideally, the diffusion current appears under forward and reverse bias, and the diffusion current under reverse bias is called the saturation current. Several kinds of current, however, flow through the *pn*-junction in the forward and reverse directions. Although the exact causes of the additional current are unknown, its main components are surface leakage current, I_{ds}, generation-recombination current, I_{dg-r}, and tunneling current, I_{dt} (see Fig. 4.2). The reverse current is not constant at the saturation current but gradually increases because of those additional currents. The total dark current, I_d, is given approximately by

$$I_d = I_{dd} + I_{ds} + I_{dg-r} + I_{dt}, \qquad (4.1)$$

where I_{dd} is the diffusion current in the reverse direction and ideally corresponds to the saturation current. These currents result in the dark current and noise and strongly influence device reliability (see Section 7.2.2).

4.1.2.1 *Diffusion Current* The diffusion current under reverse bias (saturation current) is expressed with J_{s0} in Eqs. (1.152a) and (1.152b) and is determined by the band-gap energy of the material. The saturation current is almost independent of the reverse voltage and is constant over a wide range of reverse bias. In addition, the diffusion current changes exponentially with temperature as indicated by Eq. (1.152b).

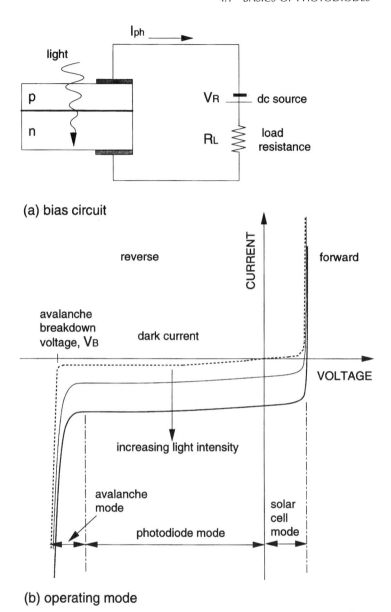

(a) bias circuit

(b) operating mode

FIGURE 4.1 Photodiode (a) bias circuit and (b) operation modes.

4.1.2.2 Surface Leakage Current The behavior of a *pn*-junction in the low-current range under reverse bias as well as forward bias is generally determined by the properties of the surface because the surface is more defective than the inner region (see Section 1.1.2.7). At the semiconductor surface, surface states are certainly formed. Interface states are also formed at the interface between

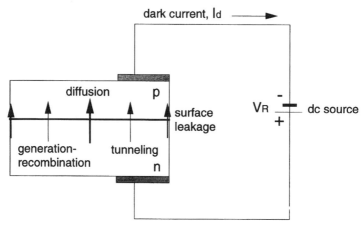

FIGURE 4.2   Main components of dark current in a photodiode under reverse bias.

the semiconductor and a dielectric passivation film when the surface is passivated with a dielectric film. The density of the surface and the interface states depends very much on device process and the passivation method (see Sections 6.2.2.1 and 6.2.3). The surface and interface states easily trap carriers as generation-recombination centers. If the trap level is near the mid-gap, and electron and hole capture cross sections are nearly equal, the generation current is increased by the following surface leakage current:[1,2]

$$I_{ds} = (q W_{sur}/2)n_i \sigma v_{th} N_{st},\qquad(4.2)$$

where W_{sur} is the surface (interface) area of the pn-junction depletion layer and σ and N_{st} are the carrier capture cross section and the trap density at the surface (interface) (see Fig. 4.3). Consequently, the surface leakage current flows via a pn-junction formed at such a defective surface or interface and often appears in the current-voltage characteristics of photodiodes under reverse bias. The defect density at the surface (interface) needs to be kept low during the fabrication process in order to reduce the dark current.

4.1.2.3   Generation-Recombination Current

The generation-recombination current is basically expressed with Eq. (1.155). Under forward bias it is mainly due to the recombination of injected carriers, whereas under reverse bias in which no carriers are injected it is due to the generation process. By multiplying Eq. (1.155) by the area of the pn-junction, A_{pn}, the generation-recombination current can be expressed by the following equation:[1,2]

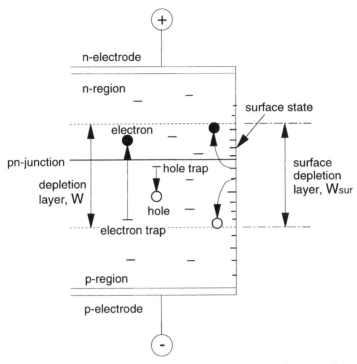

FIGURE 4.3 Generation of surface leakage current and generation-recombination current.

$$I_{\text{dg-r}} = A_{pn}J_{\text{NR}} = A_{pn}qn_i W/\tau_{\text{g}-\text{r}}$$
$$= A_{pn}qn_i W\sigma v_{\text{th}}N_t$$
$$= A_{pn}Wq\sigma v_{\text{th}}N_t(N_cN_v)^{1/2}\exp(-E_{\text{g}}/2k_{\text{B}}T), \qquad (4.3)$$

where Eqs. (1.25), (1.62), and (1.63) have been taken into account (see Fig. 4.3). The generation-recombination current exponentially increases as the temperature increases and the band-gap energy of the material decreases. It also increases when the reverse bias increases, because the depletion layer widens [see Eqs. (1.136) and (1.140)]. The current also depends on the defect density within the depletion layer, and the defect density needs to be kept low during the crystal growth and fabrication processes.

4.1.2.4 Tunneling Current Tunneling current originates from quantum-mechanical tunneling through the *pn*-junction as shown in Fig. 4.4. Because energy and the momentum are conserved during tunneling, the tunneling occurs easily in direct band-gap semiconductors, such as GaAs, InP, and GaAs. In indirect band-gap semiconductors, such as Si and Ge, in contrast, tunnel-

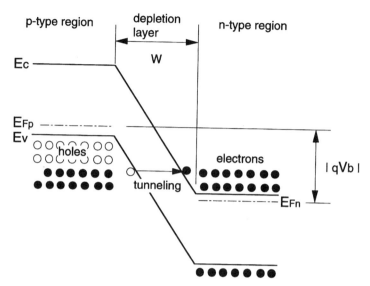

FIGURE 4.4   Schematic diagram of tunneling current under reverse bias.

ing probability is quite low because the transitions there are phonon-assisted types. This tunneling current is therefore mainly a problem in photodiodes composed of direct band-gap semiconductors and operated under a high reverse bias. The tunneling probability also increases with decreasing thickness of the potential barrier and this situation is generated at the band-gap energy under high reverse bias range as shown in Fig. 4.4. The tunneling current is given by the following equation:[3]

$$I_{dt} = A_{pn}[2^{1/2}q^3 F_{max}|V_b|/\pi h^2](m_e/E_g)^{1/2}$$
$$\times \exp(-\pi^2 m_e^{1/2} E_g^{2/3}/2^{1/2}hqF_{max}), \qquad (4.4)$$

where F_{max} is the maximum electric field in the *pn*-junction (see Fig. 1.36), m_e is the effective electron mass, and h is Plank's constant. As can be seen in Eq. (4.4), the tunneling current may be high in photodiodes composed of highly doped direct and narrow band-gap semiconductors having small effective electron masses under high reverse bias, V_b. This tunnel current finally leads to Zener breakdown as the reverse bias increases (see Section 1.2.3.2).

4.1.2.5 Dark Current and Temperature   When temperature increases, the dark current increases. The quantum-mechanical tunneling is itself independent of the temperature, but it increases as a result of the temperature dependence of the band-gap energy [see Eq. (4.4)]. The dominant current also depends on the operating mode. In the photodiode mode the generation-recombination current and diffusion current are dominant and mainly determine the temperature dependence of the dark current.[3] In the avalanche photodiode mode the tunneling current appears and strongly influences the temperature characteristics of the dark current. The temperature dependence of the dark current depends on the band-gap energy of the material, the structure, crystal quality, and so forth. The dark current in Si and Ge photodiodes, for example, increases by an order of magnitude if the temperature changes by about 30°C. It also tends to be larger in Ge photodiodes than in Si photodiodes because of the smaller band-gap energy of Ge. The dark current influences the noise behaviors of photodiodes, increasing the noise at higher temperatures (see Section 4.3.5).

4.1.3   Optical Absorption and Quantum Efficiency

Under light incidence, the photo-induced carriers are created by the intrinsic and extrinsic processes as shown in Fig. 4.5. In the extrinsic process electrons are excited from a deep level to the conduction band and then contribute to photocurrent. The light having an energy lower than the band-gap energy can be converted into electrical power. Typical examples are photoconductors composed of CdS, PbS, Zn-doped Ge, Ga, Sb, and so forth.[1] The photoconductor detects light by monitoring the change in resistivity of the semiconductor (photoconductor). Those photoconductors are used in the wavelength range over 5 μm but are not used widely because their responsivity and operating speed are low. The commonly used photodetector is an intrinsic type and detects light at wavelengths close to and shorter than that corresponding to the band-gap energy of the semiconductor.

The quantum efficiency, η_{ph}, of a photodiode is defined by

η_{ph} = (number of electron-hole pairs contributing to photo-induced current)/

(number of incident photons)

$$= [(I_{photo}/q)/(P_{inc}/h\nu)] \times 100 \qquad [\%], \qquad (4.5)$$

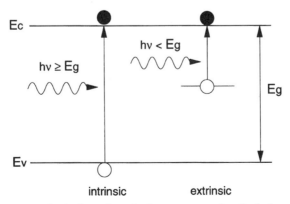

FIGURE 4.5   Intrinsic and extrinsic processes of optical absorption.

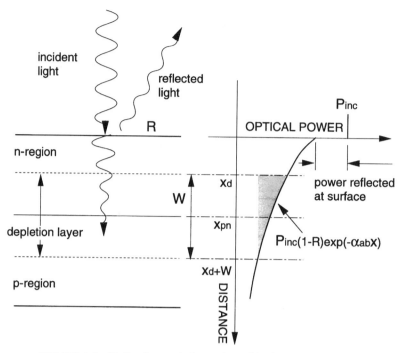

FIGURE 4.6   Reflection and absorption of incident light power.

where the I_{photo} and the P_{inc} are the photo-induced current and the incident optical power. The responsivity, S [A/W], is often used instead of the quantum efficiency and is given by

$$S = \text{(photo-induced current)/(incident optical power)}$$
$$= I_{photo}/P_{inc} = q\eta_{ph}/h\nu$$
$$= \eta_{ph}\lambda(\mu m)/1.24. \tag{4.6}$$

The quantum efficiency is influenced by the reflection of incident light at the surface of the photodiode, the recombination of photo-induced carriers at the surface and in the depletion layer, and the optical absorption outside of the depletion layer. About 30% of the incident light is reflected at the surface because the refractive index of the semiconductor material is about 3.5 (see Fig. 4.6). An antireflective dielectric film (see Section 6.2.2.3) is usually coated onto the surface to suppress the reflection, and the reflectivity at the surface can be reduced to less than 1%. The incident light power decreases by a factor of $(1 - R)$, where R is the power reflectivity at the surface. The problem of recombination of photo-induced carriers at the surface or in the depletion layer has already been discussed in Sections 1.1.2.7 and 4.1.2. To suppress the recombination and to increase the quantum efficiency, the introduction of defects

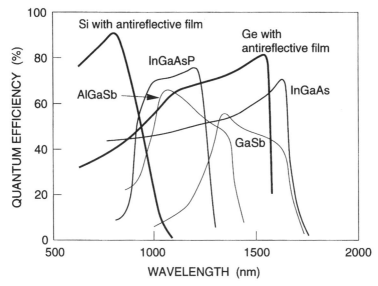

FIGURE 4.7   Quantum efficiency as a function of wavelength for various pin-photo-diodes. The Si and Ge photodiodes are coated with antireflective films, and the III-V compound semiconductor photodiodes are not coated with the dielectric films. From Y. Matsushima et al.[4] ©1982 John Wiley & Sons, Inc. Reprinted by permission of the authors and John Wiley & Sons, Inc.

during crystal growth and device fabrication processes has to be eliminated. The recombination at the surface is often reduced by covering the light absorption layer with another kind of semiconductor having a larger band-gap energy. A typical example of such a structure is the lattice-matched InP ($E_g \approx 1.35$ eV at 300 K)/InGaAs ($E_g \approx 0.75$ eV at 300 K) structure. The InGaAs layer is the absorbing layer and the InP layer is the cover (or window) layer. As shown in Fig. 4.6, the optical absorption can be analyzed by using Lambert's law (see Section 1.1.3.1). The light power reaching the active region decreases by $\exp(-\alpha_{ab}x_d)$, where α_{ab} is the absorption coefficient and x_d is the distance between the surface and the edge of the active region on the light-incident side (see Fig. 4.6). The actual quantum efficiency therefore deviates from the ideal value and the reduction is proportional to $(1 - R)\exp(-\alpha_{ab}x_d)$, where an influence from the recombination problem at the surface and in the active region is ignored. These phenomena result in the (external) quantum efficiency given by Eq. (4.5). The quantrum efficiencies of some typical photodiodes are shown in Fig. 4.7 as a function of wavelength of the incident light.[4] The gradual decrease in quantum efficiency at shorter wavelengths is mainly due to the increase in optical absorption between the illuminated surface and the active region and to the increased carrier loss caused by the surface (or the interface) recombination. The rapid decrease with increasing wavelength is determined by the band-gap energy of the active region.

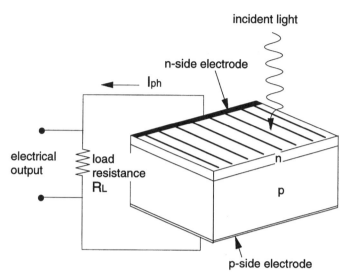

FIGURE 4.8   Schematic diagram of a solar cell.

4.2   SOLAR CELLS

Solar cells are used in equipment ranging from power plants to consumer electronics, such as handy calculators. Their most important application, of course, is in generating electricity from sunlight. The spectrum of the solar rays ranges roughly from 200 to 3000 nm, so most semiconductor materials can be used in solar cells. Popular materials for solar cells are polysilicon, Ge, GaAs, and CdTe. A solar cell is a simple photodiode, as shown in Fig. 4.8, and is used without any electrical bias (see Fig. 4.1). Comb-shaped ohmic contacts are usually formed as an *n*-side electrode. This shape suppresses the generation of inhomogeneous electrical potential at the *pn*-junction, which inhomogeneity is caused by the transverse current flow and the electrical resistance in the surface layer.

4.2.1   Basic Characteristics: Current-Voltage Characteristics

The energy band diagrams of a solar cell (or simple *pn*-homojunction photodiode) are shown in Fig. 4.9. As described in Section 1.2, under thermal equilibrium the *pn*-junction is the same as that of a normal diode. If the incident light has an energy larger than the band-gap energy, electron-hole pairs are generated in the depletion layer and in the *p*-type and *n*-type regions. Electrons and holes generated in the depletion layer are separated by the electric field built up within the depletion layer and drift to the *n*-type and *p*-type region. Electrons generated in *p*-type region (minority carriers) diffuse in the *p*-type region and there recombine with holes, and holes generated in the *n*-type region diffuse in the *n*-type region and recombine with electrons. Some of the carriers generated

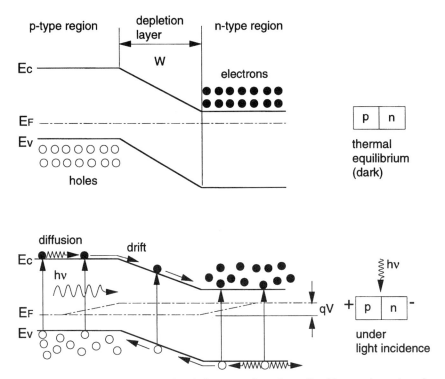

FIGURE 4.9   Schematic energy band diagram of a solar cell with a *pn*-homojunction.

within the carrier diffusion length from the edge of the depletion layer reach the edge of the depletion layer. Those electrons also drift within the depletion layer from the *p*-type region to the *n*-type region, and those holes drift within the depletion layer from the *n*-type region to the *p*-type region. These drifts of photo-induced carriers across the *pn*-junction result in photocurrent. The width of the depletion layer under no bias or under forward bias is less than the diffusion length. The photo-induced current, I_{photo}, is therefore approximated by the following equation:

$$I_{\text{photo}} \approx qA_{pn}G_{\text{ph}}(L_{ep} + L_{hn}), \tag{4.7}$$

where A_{pn} is the area of the *pn*-junction, G_{ph} is the rate of electron-hole pair generation (in $\text{cm}^{-3}/\text{sec}$), and L_{ep} and L_{hn} are the diffusion lengths of electrons in the *p*-type region and holes in the *n*-type region (see Section 1.2.3.1). The *pn*-junction is biased in the forward direction because of the photovoltaic effect as shown in Fig. 4.9. When the *p*-side and *n*-side are connected via an electrical resistance, photocurrent flows from the *p*-side to the *n*-side in the circuit, while the *pn*-junction is biased in the forward direction because of the voltage generated at the load resistance, R_{L} [see Fig. 4.10(a)]. The forward current is

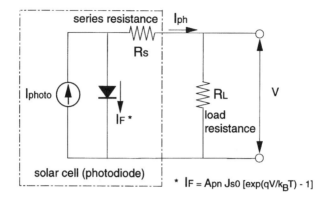

series resistance

R_s

I_{ph}

I_{photo}

I_F *

R_L

load resistance

V

solar cell (photodiode)

* $I_F = A_{pn} J_{s0} [\exp(qV/k_BT) - 1]$

(a) equivalent circuit

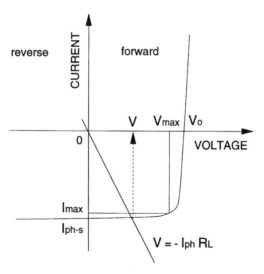

(b) current-voltage characteristics

FIGURE 4.10 (a) Equivalent circuit and (b) a current-voltage characteristics of a solar cell (*pn*-junction photodiode).

ideally the diffusion current [see Eq. (1.151)], and the total photocurrent under light incidence, I_{ph}, is given by a sum of the two opposite currents as

$$I_{ph} = -I_{photo} + A_{pn}J_{s0}[\exp(qV/k_BT) - 1], \tag{4.8}$$

where the second term on the right-hand side is the forward current.

Using Eq. (4.8), the equivalent circuit of a solar cell (simple *pn*-junction photodiode) can be drawn as shown in Fig. 4.10(a). If the load resistance is small or zero (short-circuit) or the intensity of the incident light is low, the voltage generated across the load resistance is small or zero. Consequently, the second term on the right-hand side of Eq. (4.8) is negligible and the short-circuit current, $I_{\text{ph-s}}$, is given by

$$I_{\text{ph-s}} = -I_{\text{photo}} = -qA_{pn}G_{\text{ph}}(L_{ep} + L_{hn}). \tag{4.9}$$

The photocurrent is proportional to the intensity of the light. When the load resistance is large or infinite (open-circuit), $I_{\text{ph}} = 0$ ($I_{\text{photo}} = I_F$) and the voltage generated in the open-circuit, V_O, is given by

$$\begin{aligned} V_O &= (k_B T/q) \ln[(I_{\text{photo}}/I_{s0}) + 1] \\ &\approx (k_B T/q) \ln(I_{\text{photo}}/I_{s0}), \end{aligned} \tag{4.10}$$

where $A_{pn}J_{s0}$ is replaced with I_{s0}. The voltage generated in the open circuit tends to increase as the saturation current, I_{s0}, decreases. As the intensity of incident light increases, the voltage generated across the load resistance increases logarithmically with the photo-induced current, that is, with the incident light intensity.

4.2.2 Output Power and Efficiency

Using Eq. (4.8), the output power of a solar cell (*pn*-junction photodiode), P_{photo}, is given as

$$P_{\text{photo}} = I_{\text{ph}} V = -I_{\text{photo}} V + I_{s0} V[\exp(q V/k_B T) - 1]. \tag{4.11}$$

The maximum output power can be obtained at $dP_{\text{photo}}/dV = 0$.[1] By differentiating Eq. (4.11) with respect to V,

$$I_{s0}[\exp(q V/k_B T) - 1] + I_{s0}(q V/k_B T) \exp(q V/k_B T) = I_{\text{photo}} \tag{4.12}$$

and then

$$\begin{aligned} V_{\text{max}} &= (k_B T/q) \ln\{[(I_{\text{photo}}/I_{s0}) + 1]/[(q V_{\text{max}}/k_B T) + 1]\} \\ &= V_0 - (k_B T/q) \ln[(q V_{\text{max}}/k_B T) + 1] \\ &\approx V_0, \end{aligned} \tag{4.13}$$

where the V_{max} is the generated voltage giving the maximum output power [see Fig. 4.10(b)]. From Eqs. (4.11), (4.12), and (4.13), the current giving the maximum output power, I_{max}, is given as

$$I_{max} = I_{s0}(q V_{max}/k_B T) \exp(q V_{max}/k_B T). \tag{4.14}$$

The maximum output power, $P_{photo\text{-}max}$, is therefore derived from Eqs. (4.13) and (4.14) as

$$P_{photo\text{-}max} = I_{max} V_{max}$$
$$\approx I_{s0}(q V_{max}^2/k_B T) \exp(q V_{max}/k_B T). \tag{4.15}$$

If the incident light power is denoted P_{inc}, the conversion efficiency (in %) of the solar cell, η_{cv}, is defined by

$$\eta_{cv} = \text{(electrical maximum output power/optical input power)}$$
$$= P_{photo\text{-}max}/P_{inc} = I_{max} V_{max}/P_{inc}$$
$$= [I_{s0}(q V_{max}^2/k_B T) \exp(q V_{max}/k_B T)/P_{inc}] \times 100 \tag{4.16}$$

The conversion efficiency is larger than 10% for some solar cells, such as silicon and GaAs diodes. Those experimental data are always lower than theoretical values because of the influence of defects, optical absorption outside of the depletion layer, and so forth (see Section 4.1.3). The fill factor which defines the power extraction efficiency is an important figure of merit in solar cell design and is given by the following equation:[5]

$$\text{Fill Factor} = I_{max} V_{max}/I_{ph\text{-}s} V_0. \tag{4.17}$$

The efficiency depends on the wavelength of incident light because optical absorption coefficient—that is, the penetration depth—changes with wavelength. The spectral response of solar cells is given by the variation of the short-circuit current, $I_{ph\text{-}s}$ ($= -I_{photo}$), as a function of the wavelength of the incident light.[1] This spectral response can be simply analyzed by using Lambert's law (see Fig. 4.6), taking into account the diffusion length of the carriers from the edge of the depletion layer, and can be expressed by the following equation:[1]

$$dI_{ph\text{-}s}(\lambda)/d\lambda \approx \alpha_{ab}\lambda(L_{ep} + L_{hn}) \exp(-\alpha_{ab}x_{pn}), \tag{4.18}$$

where L_{ep} and L_{hn} are the diffusion lengths of electrons in the p-type region and holes in the n-type region, and x_{pn} is the distance of the pn-junction from the surface (see Fig. 4.6).

4.2.3 *pn*-Heterojunction Solar Cell

For reasons of economy the pn-homojunction solar cells composed of polysilicon are widely used in many kinds of equipment and systems. Their perfor-

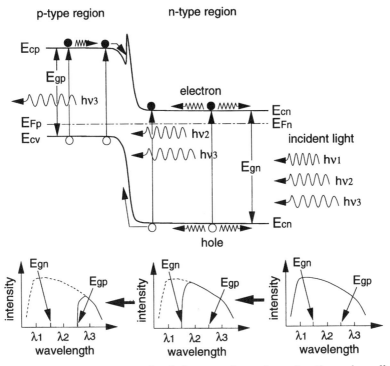

FIGURE 4.11 Schematic energy band diagram of a *pn*-heterojunction solar cell, and the variation of solar ray intensity within the cell.

mance is often limited by the structures. To enhance the response, especially at short wavelengths, the *pn*-junction needs to be close to the surface. This reduces the quantum efficiency, however, because of the surface (interface) recombination. In addition, the impurities doped into the *p*-type and *n*-type semiconductors also influence the efficiency. The optical absorption coefficient increases when the doping concentration is increased in order to reduce the sheet resistance of the layer, and therefore the optical absorption in the layer increases. This decreases the quantum efficiency. The reduction of the sheet resistance, however, is important in suppressing the generation of an inhomogeneous electrical potential at the *pn*-junction.

The *pn*-heterojunction solar cells[6] are released from those structural constraints as shown in Fig. 4.11. When light is incident from the side of the wide band-gap semiconductor, light having an energy higher than the wide band-gap energy, E_{gn}, is absorbed in the wide band-gap semiconductor, and then light having an energy between the wide and the narrow band-gap energies, E_{gn} and E_{gp}, is absorbed in the narrow band-gap semiconductor. Light having an energy lower than the narrow band-gap energy, E_{gp}, passes through the *pn*-junction. In the *pn*-heterojunction solar cell, the doping concentration and thickness of the wide band-gap semiconductor can be increased to reduce the sheet resistance of

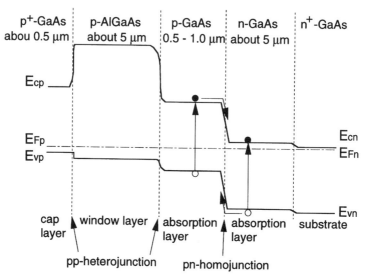

FIGURE 4.12 Schematic energy band diagram of an AlGaAs/GaAs heterojunction solar cell.

the layer without influencing the absorbing situation of the narrower band-gap semiconductor. Most of the photo-induced carriers contributing to photocurrent are released from the recombination at the surface. In addition, the saturation current is reduced and then V_0 in Eq. (4.10) increases. The notch at the heterointerface, however, is a disadvantage of the *pn*-heterojunction because it acts as a potential barrier to the photo-induced carriers and may greatly reduce the photocurrent. This disadvantage is avoided by combining an iso-heterojunction with a *pn*-homojunction (see Fig. 4.12). The wide band-gap layer (window layer) introduces the advantages described above and the disadvantage caused by the notch is eliminated. The photo-induced electrons and holes can flow without being obstructed by any potential barrier as shown in Fig. 4.12. The cap layer is required here in order to produce ohmic contact because the contact resistance between the semiconductor and metal electrode tends to decrease as the band-gap energy of the semiconductor decreases (see Section 6.2.1). The typical heterojunction solar cells are produced by using III-V compound semiconductors, such as AlGaAs/GaAs and InGaAs(P)/InP.

4.3 pin-PHOTODIODES

The pin-photodiodes are used in equipment ranging from optical fiber communication systems to consumer electronics (see Chapter 8). Their quantum efficiency is much higher than that of solar cells and their dark current and operation voltage are lower than those of the APDs described in Section 4.4. The solar cells use the photovoltaic effect without any electrical bias, whereas the pin-photodiodes use the photovoltaic effect under an electrical bias as shown in Fig. 4.1(b). The *pn*-junction is biased in the reverse direction by using a bias circuit [see Fig. 4.1(a)]. Popular pin-photodiode materials are Si for the wavelength range below the 800 nm band, and Ge, InGaAs, and InGaAsP for wavelengths longer than 1 μm.

Various kinds of structures have been developed and some of them are shown in Fig. 4.13. The *pn*-junction consists of a heavily doped region and an undoped (or lightly doped) region and are either diffused junctions formed by impurity diffusion or grown junctions formed during crystal growth (see Section 1.2). The undoped (lightly doped) region is ordinarily inserted between the *p*-type and *n*-type regions. This region is called *i*-layer ($n^- = \nu$-layer or $p^- = \pi$-layer), and the junction structure is the following: *p*-type region/*i*-region/*n*-type region (*p-i-n*). This type of photodiode is called the pin-photodiode. The light input region is ordinarily restricted to a diameter of 30-300 μm by a ring electrode and is coated with an antireflective dielectric film to reduce the amount of light reflected at the input surface. A guard ring is often formed by diffusion or ion-implanation around the diffused *pn*-junction region in order to eliminate the edge breakdown due to a high electric field generated at the edge of the diffused region. This structure is the same as that of APD (see Section 4.4). For high-frequency responsiveness a mesa structure is sometimes formed to reduce the junction capacitance. The performance of these pin-photodiodes is basically determined by the undoped (or lightly doped) layers.

4.3.1 *p-i-n* Structure and *i*-Layer

The characteristics of pin-photodiodes are strongly influenced by the thickness of and doping concentration in the *i*-layer. The *i*-layer is an undoped or very lightly doped layer (region) and is either an n^- or a p^--layer. The n^--layer and the p^--layer are often respectively called the ν-layer and the π-layer, and the carrier concentration in those layers is ordinarily between 10^{13} and 10^{16} cm^{-3}. The energy band diagram of an n^+-*i* (π)-p^+ structure is shown in Fig. 4.14. The carrier concentration of the n^+ and p^+ layers is usually set around or above 10^{18} cm^{-3}. Consequently, an abrupt step junction in which a depletion layer always expands to the *i*-layer (one-sided abrupt junction) is formed in the *p-i-n* structure [see Eq. (1.129)]. The thickness of the *i*-layer is about 3 μm in pin-photodiodes composed of III-V compound semiconductors and is more than a few tens of micrometers in Si and Ge pin-photodiodes (see Fig. 4.13). The applied reverse bias drops almost entirely across the *i*-layer, and the layer is fully depleted at a very low or zero reverse bias. A case of partial depletion within the *i*-layer under zero bias is also shown in Fig. 4.14. As a reverse bias, V_b, increases, the width of the depletion layer increases in proportion to the square root of the bias voltage, $|V_b|^{1/2}$ [see Eq. (1.132)] and the depletion layer reaches the p^+-layer. The situation in which the edge of the depletion layer reachs the p^+-layer is called the reach-through (punch-through) and the corresponding volt-

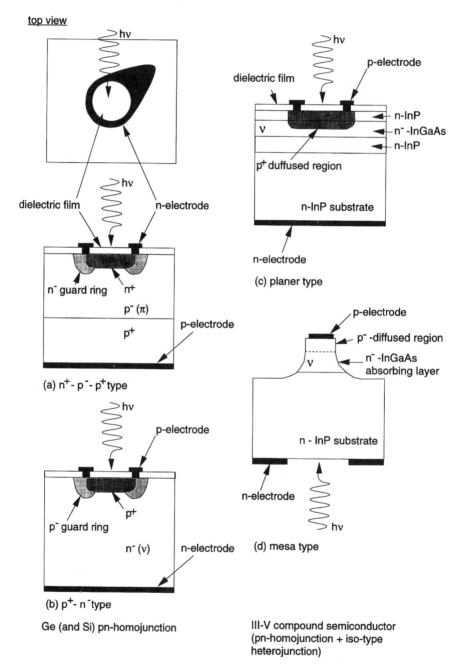

FIGURE 4.13 Schematic diagrams of some pin-photodiodes: (a) *n-p-p* type,[7] (b) *p-n* type,[8] (c) planar type,[9] (d) mesa type.[10]

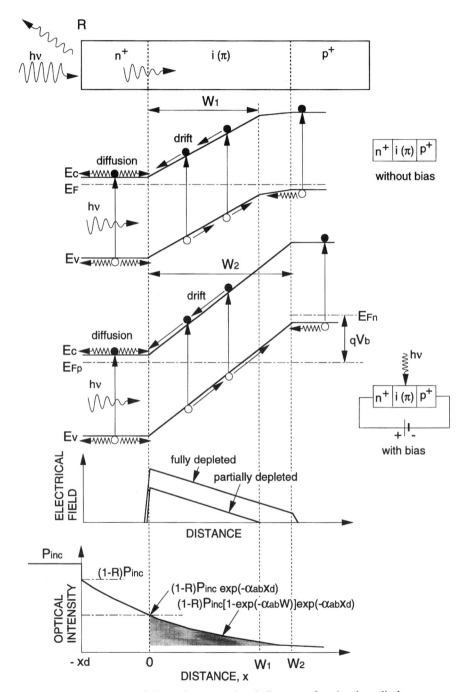

FIGURE 4.14 Schematic energy band diagram of a pin-photodiode.

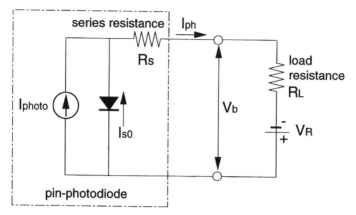

FIGURE 4.15   Equivalent circuit of a pin-photodiode.

age is called the reach-through voltage. The quantum efficiency also increases in proportion to the increase in the depletion region within the *i*-layer. Further increases in the reverse bias scarcely expand the edge of the depletion layer into the p^+-layer because of the heavy doping there, and the electric field within the *i*-layer is high as shown in Fig. 4.14. The thickness of the depletion layer is nearly equal to the thickness of the *i*-layer after reach-through.

4.3.2   Basic Characteristics: Current-Voltage Characteristics

4.3.2.1   Photocurrent and Dark Current   The equivalent circuit of the pin-photodiodes is basically the same as that of the solar cell (Fig. 4.10) and can be drawn as shown in Fig. 4.15. In photodiode-mode operation, however, the photodiode is reversely biased as shown in Fig. 4.15. From the bias circuit,

$$V_b = V_R - I_{ph}R_L. \tag{4.19}$$

The V_b is always negative in photodiode-mode operation, and the photocurrent, I_{ph}, is ideally given by

$$I_{ph} = I_{photo} + I_{s0}. \tag{4.20a}$$

Actually, it is given by using Eq. (4.1) and adding an effect due to background radiation in the ambience in which the photodiode is placed, I_{bg}:

$$I_{ph} = I_{photo} + I_d + I_{bg}. \tag{4.20b}$$

The photocurrent, I_{ph}, is proportional to the photo-induced current, I_{photo}, and

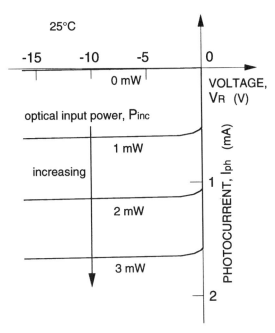

FIGURE 4.16 Photocurrent-voltage characteristics at various light input powers for a Si pin-photodiode with an input window 300 μm in diameter.

thus the incident light power, in the photodiode-mode operation. The change in photocurrent, I_{ph}, as a function of reverse voltage, V_R, is shown in Fig. 4.16 for a Si pin-photodiode with a light input window region 300 μm in diameter. The bias circuit is similar to that shown in Fig. 4.15, and the photocurrent is proportional to the incident input power.

4.3.2.2 Drift Current and Diffusion Current Under normal operating conditions, the photocurrent is several orders of magnitude larger than the dark current and the current caused by the background radiation. The photocurrent, I_{ph}, is therefore nearly equal to photo-induced current, I_{photo} in Eq. (4.20b). The photo-induced current comprises the drift current originated from the photo-induced carriers within the depletion region and the diffusion current due to the photo-induced minority carriers generated within their diffusion lengths from the edge of the depletion region (see Fig. 4.14). Consequently, by dividing each of these currents by the *pn*-junction area, A_{pn}, the photocurrent can be expressed as

$$J_{ph} = J_{ph\text{-}dr} + J_{ph\text{-}diff}, \tag{4.21}$$

where J_{ph}, $J_{ph\text{-}dr}$, and $J_{ph\text{-}diff}$ are the photocurrent density, the drift current density, and the diffusion current density. The electron-hole generation rate, $G_{ph}(x)$, can be expressed, according to Lambert's law and taking into consideration the incident input power, P_{inc}, and the reflectivity at the surface, R, as

$$G_{ph}(x) = (1 - R)(P_{inc}/A_{pn}h\nu)\alpha_{ab}\exp(-\alpha_{ab}x)], \quad (4.22)$$

where the incident light power $(1 - R)P_{inc}$ is changed into the photon flux (number of photons/sec/cm^2) by dividing it by the pn-junction area, A_{pn}, and the photon energy, $h\nu$. Integrating Eq. (4.22) from 0 to W gives the number of photo-induced electron-hole pairs within the depletion region. This can be easily understood from Fig. 4.14. The optical absorption within the depletion region is given by $(1 - R)P_{inc}[1 - \exp(-\alpha_{ab}W)]$, where the light absorption at the cap layer (n^+-layer in Fig. 4.14) is negligible because x_d is very small and $\exp(-\alpha_{ab}x_d) \approx 1$. If all of the light (photon) absorbed within the depletion region generates photo-induced carriers, the drift current is given by

$$J_{ph\text{-}dr} = q(1 - R)(P_{inc}/A_{ph}h\nu)[1 - \exp(-\alpha_{ab}W)]. \quad (4.23)$$

The diffusion current under normal operating conditions after reach-through is due to the photo-induced carriers generated in the region within the minority carrier diffusion length, L_{ep} [see Eq. (1.145a)], from the edge of the depletion region in the p^+-layer. The following one-dimensional diffusion equation holds for the electrons contributing to photocurrent:[1]

$$D_{ep}\partial n_p/\partial x^2 - (n_p - n_{p0})/\tau_{ep} + G_{ph}(x) = 0, \quad (4.24)$$

where D_{ep} and τ_{ep} are the diffusion constant for electron and the lifetime of the excess electrons in the p-type region, and n_p and n_{p0} are the excess electron density and the equilibrium electron density in the p-type region. By solving Eq. (4.24) under the boundary conditions (see Section 1.2.3.2), which are $n_p = n_{p0}$ for $x = \infty$ and $n_p = 0$ for $x = W$,

$$n_p = n_{p0} - [n_{p0} + C_0\exp(-\alpha_{ab}W)]\exp[(W - x)/L_{ep}] + C_0\exp(-\alpha_{ab}x), \quad (4.25a)$$

where

$$C_0 = (1 - R)(P_{inc}/A_{pn}h\nu)\alpha_{ab}L_{ep}^2/D_{ep}(1 - \alpha_{ab}^2L_{ep}^2)]. \quad (4.25b)$$

The diffusion current is therefore given by

$$\begin{aligned}
J_{diff} &= -qD_{ep}(\partial n_p/\partial x)|_{x=W} \\
&= q(1 - R)(P_{inc}/A_{pn}h\nu)[\alpha_{ab}L_{ep}/(1 + \alpha_{ab}L_{ep})]\exp(-\alpha_{ab}W) \\
&\quad + qn_{p0}D_{ep}/L_{ep}.
\end{aligned} \quad (4.26)$$

The photocurrent corresponding to the total current is then expressed by

$$J_{ph} = q(1 - R)(P_{inc}/A_{pn}h\nu)\{1 - [\exp(-\alpha_{ab}W)]/(1 + \alpha_{ab}L_{ep})\}$$
$$+ qn_{p0}D_{ep}/L_{ep}. \tag{4.27a}$$

The last term involving n_{p0} is small under normal operating conditions in pin-photodiodes, and thus

$$J_{ph} = q(1 - R)(P_{inc}/A_{pn}h\nu)\{1 - [\exp(-\alpha_{ab}W)]/(1 + \alpha_{ab}L_{ep})\}. \tag{4.27b}$$

The photocurrent is therefore proportional to the photon density flux, that is, to the incident light input power.

4.3.3   External Quantum Efficiency

Under normal operating conditions, the depletion region in pin-photodiodes is sufficiently thick. When the i-layer is thin as in InGaAs pin-photodiodes, the depletion layer after reach-through geometrically corresponds to the i-layer in pin-photodiodes. If all photo-induced carriers within the depletion layer and outside the layer but within the diffusion length contribute to photo-induced current, the quantum efficiency of a pin-photodiode under normal operating conditions is given by modifying Eq. (4.5) to

$$\eta_{ph\text{-}ex} = (J_{ph}/q)/(P_{inc}/A_{pn}h\nu)$$
$$= (1 - R)\{1 - [\exp(-\alpha_{ab}W)]/(1 + \alpha_{ab}L_{ep})\}. \tag{4.28}$$

All photo-induced carriers don't actually contribute to the photo-induced current, and this means that the internal quantum efficiency is not 100% (or 1). The internal quantum efficiency, generally defined as a percentage and given by

$$\eta_{ph\text{-}in} = [(\text{number of photo-induced electron-hole pairs})/$$
$$(\text{number of photons absorbed})] \times 100, \tag{4.29}$$

is usually high but not 100% because of defects at the surface and in the depletion region and so forth. Consequently, the external quantum efficiency is obtained from Eqs. (4.28) and (4.29) as

$$\eta_{ph\text{-}ex} = (J_{ph}/q)/(P_{inc}/A_{pn}h\nu)$$
$$= (1 - R)\{1 - [\exp(-\alpha_{ab}W)]/(1 + \alpha_{ab}L_{ep})\}\eta_{ph\text{-}in}. \tag{4.30}$$

The diffusion length, L_{ep}, is a few micrometers and $\alpha_{ab}L_{ep}$ is thus ordinarily negligible in Eq. (4.30). If some of the incident light reaches the deple-

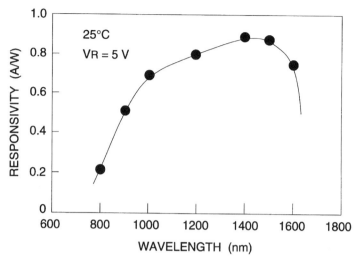

FIGURE 4.17 Responsivity of a planar type InGaAs/InP photodiode as a function of the wavelength of incident light.

tion region (i-layer) after passing through the cap layer (n^+-layer in Fig. 4.14), the light input power discussed above needs to be reduced by the factor $\exp(-\alpha_{ab}x_d)$.

The responsivity given by Eq. (4.6) holds for pin-photodiodes and can be related to the external quantum efficiency indicated in Eq. (4.30). The responsivity of a planar-type InGaAs/InP pin-photodiode with an input window 300 μm in diameter (see Fig. 4.13) is shown in Fig. 4.17. A high responsivity is obtained over the wide range of wavelengths. In pin-photodiodes composed of III-V compound semiconductors for communication system use, the responsivity is usually higher than 0.8 A/W in the 1300 nm and 1550 nm bands. These values correspond to a quantum efficiency of more than about 70% [see Eq. (4.6)]. The quantum efficiency of a pin-photodiode composed of Si or Ge is as high as that of a pin-photodiode composed of III-V compound semiconductors.

4.3.4 Frequency Response

The pin-photodiodes are often used in optical fiber communication systems and data links (see Chapter 8). In such applications, frequency response, in addition to high quantum efficiency, is one of the most important characteristics. The frequency response of a photodiode is mainly limited by the time needed for the photo-induced carriers to cross the depletion region (transit time) and by the CR-time constant, which is the product of capacitance and resistance in the bias circuit.

4.3.4.1 CR-Time Constant The limitation of frequency response due to the CR-time constant is a common problem in photodiodes, not just pin-photodiodes. Before reach-through, the *i*-layer is partially depleted. The depletion layer acts as a capacitance (junction capacitance), and the residual part of the *i*-layer acts as a high electrical resistance because of the low carrier concentration there. The frequency response of pin-photodiode is therefore low before reach-through. As the bias increases, the depletion layer widens and the junction capacitance decreases in inverse proportion to $|V_R|^{1/2}$ [see Eq. (1.137)], the highly resistive region narrows, and the electrical resistance decreases. The junction capacitance is given by Eq. (1.135), and the width of the depletion region after reach-through is nearly equal to the thickness of the *i*-layer. Consequently, the junction capacitance, C_j, can be expressed as

$$C_j = \epsilon_s \epsilon_0 / W_i, \tag{4.31}$$

where W_i is the thickness of the *i*-layer. The pin-photodiodes in equipment and systems requiring high-frequency responses are usually used after reach-through. By using the junction capacitance, the equivalent circuit of pin-photodiodes under a high bias is given by modifying Fig. 4.15 as shown in Fig. 4.18. The circuit is simplified by omitting an inductance of lead wire, a capacitance of package, and series resistance, such as, contact resistance of the diode (usually less than 10 Ω). From Fig. 4.18,

$$C_j dv_b(t)/dt + v_b(t)/R_{DL} = i_s(t), \tag{4.32}$$

where R_{DL} is equal to $1/(1R_D + 1/R_L)$. This equation can be used to analyze transient phenomena.

When input light is modulated by a sinusoidal signal, the generated photocurrent is given by $i_s(t) = i_0 \exp(i\omega t)$, and

$$v_b(t) = [i_0 R_{DL}/(1 + i\omega C_j R_{DL})] \exp(i\omega t). \tag{4.33}$$

The amplitude of output voltage is nearly $i_0 R_{DL}$ in the low-frequency range, so the ratio of the amplitude in the high-frequency range to that in the low-frequency range is $1/[(1 + (\omega C_j R_{DL})^2]^{1/2}$. Consequently, the frequency corresponding to the voltage amplitude decreasing by $2^{-1/2}$ (see Fig. 2.14) gives the cutoff frequency, $f_{c\text{-}CR}$:

$$f_{c\text{-}CR} = 1/2\pi C_j R_{DL}. \tag{4.34}$$

In the conventional InGaAs pin-photodiodes, the C_j and R_D are typically 100 fF and 100 MΩ, and the load resistance, R_L, is usually set at 50 Ω for matching with the circuit connected. The cutoff frequency for only the pin-photodiode is calculated to be about 30 GHz but the actual cutoff frequency is about an

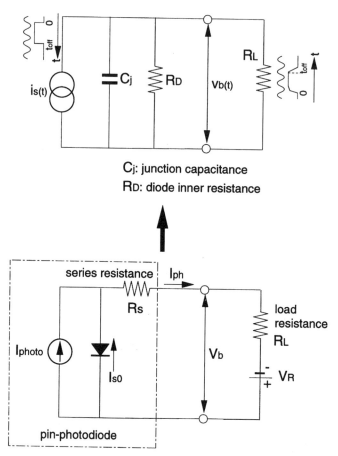

C_j: junction capacitance
R_D: diode inner resistance

FIGURE 4.18 Simplified equivalent circuit of a pin-photodiode under modulated light incidence at high reverse bias.

order of magnitude lower because of the resistances and capacitances due to the packaging and other parts of photodiode (series resistance and so forth). The parasitic capacitance, for example, is often over 1 pF after packaging and the cutoff frequency decreases to about 3 GHz when a parasitic capacitance of 1 pF is added in parallel to the circuit in Fig. 4.18.

For input light modulated by a rectangular pulse signal, current pulses are generated and then voltage pulses are generated at the load resistance, R_L. The step-pulse response can be analyzed by using the circuit in Fig. 4.18. The rise time and fall time are determined by the time constant, $C_j R_{DL}$, and are given by the following equations:[11]

$$v_{\text{b-rise}}(t) = i_0 R_{DL}[1 - \exp(-t/C_j R_{DL})] \qquad (4.35a)$$

and

$$v_{\text{b-fall}}(t) = i_0 R_{\text{DL}}[1 - \exp(-t_{\text{off}}/C_j R_{\text{DL}})]\exp[-(t - t_{\text{off}})/C_j R_{\text{DL}})], \quad (4.35b)$$

where t_{off} is the time at which the current turns off (see Fig. 4.18). When the modulation frequency is high, the output pulse is deformed because of the finite time constant, $C_j R_{\text{DL}}$. The rise and fall times of a rectangular pulse are usually specified in terms of the time interval between 10% and 90% of the plateau value (pulse peak). From Eqs. (4.34) and (4.35) and by using $C_j R_{\text{DL}} \gg t_{0.1}$, $t_{0.9}$, where $t_{0.1}$ and $t_{0.9}$ are the times corresponding to 10% and 90% of the pulse peak, the following relation can be deduced as

$$t_r = t_f = \ln9 / 2\pi f_{\text{c-CR}}, \quad (4.36)$$

where t_r and t_f are the rise time and fall time. For high-frequency responsiveness, pin-photodiodes are used at a high reverse bias after reach-through.

4.3.4.2 Transit Time Within the Depletion Layer

The transit time (the time required for photo-induced carriers to cross the depletion region) is determined by drift velocity of the photo-induced carriers. The drift velocity is given by Eq. (1.36) and is proportional to the mobility of the carrier and the intensity of the electrical field. As discussed in Section 1.1.2.6, the drift velocity never continues to increase with the increasing electrical field and is limited because carriers collide with the lattice (see Fig. 1.18). Some of the energy of the electrical field changes to Joule's heat (lattice vibration) during the collision. Finally, in sufficiently high fields which appear under sufficiently high reverse bias, the drift velocity of electrons and holes approaches a scattering-limited velocity and becomes constant. The constant drift velocity is called the saturation velocity ($v_{\text{ds}} \approx 10^7$ cm/sec for Ge, Si, and GaAs).[1] The following part of this section quantitatively analyzes the transit time in a depletion region with $W \gg 1/\alpha_{\text{ab}}$.[1] When the reach-through occurs and the drift velocity is the saturation velocity (this is the normal operating condition for pin-photodiodes) and the incident light intensity is modulated with $P_{\text{inc0}} \exp(i\omega t)$, the photocurrent density, $J_{\text{ph}}(t)$, can be expressed by the number of electrons reaching the edge of the i-layer ($x = 0$ in Fig. 4.14). If the electrons are generated at point, x, and at time, $(t - x/v_{\text{ds}})$, the photocurrent density is given by averaging the electrons within the depleted i-layer:

$$J_{\text{ph}}(t) = (1/W_i) \int_0^{W_i} (q P_{\text{inc0}}/A_{pn}h\nu) \exp i\omega(t - x/v_{\text{ds}})dx$$
$$= q P_{\text{inc0}}/A_{pn}h\nu \{[1 - \exp(-i\omega t_{\text{tr}})]/i\omega t_{\text{tr}}\} \exp(i\omega t), \quad (4.37)$$

where t_{tr} is the transit time of carriers and is equal to W_i/v_{ds}. The amplitude

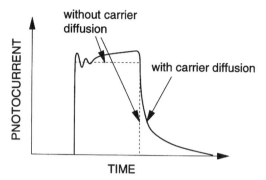

FIGURE 4.19   Illustration of a pulse response deformed by diffusion current.

of the photocurrent is nearly $qP_{\text{inc0}}/A_{pn}h\nu$ in the low-frequency range, and the ratio of the amplitude in the high-frequency range to that in the low-frequency range is $[1 - \exp(-i\omega t_{\text{tr}})]/i\omega t_{\text{tr}}$. Consequently, the frequency corresponding to the point at which the ratio decreases by $2^{-1/2}$ and giving the cutoff frequency, $f_{\text{c-tr}}$, is calculated as

$$f_{\text{c-tr}} \approx 2.8/2\pi t_{\text{tr}} \approx 0.4/t_{\text{tr}}. \tag{4.38}$$

Thus, the transit time of the photo-induced carriers within the depletion region determines the cutoff frequency in addition to the CR-time constant and strongly influences the frequency response of pin-photodiodes. If the thickness of the depletion region (*i*-layer) is 40 μm, the transit time is about 0.4 ns and the cutoff frequency is about 1 GHz. The cutoff frequency determined by the CR-time constant is ordinarily higher than 1 GHz, and this means that the frequency response of a pin-photodiode with such a thick *i*-layer is determined by the transit time. If the input light is modulated by a rectangular pulse signal, the photocurrent pulses generated are deformed if the transit time determines the frequency response as shown in Fig. 4.19. This is the reason why the diffusion current participates in the photocurrent (see Fig. 4.14). The diffusion current introduces the time delay determined by the transit time to the rise and fall of the output pulse current.

4.3.4.3   Relations Between Frequency Response, CR-Time Constant, and Transit Time   The frequency response of pin-photodiodes needs to be exactly analyzed using the equivalent circuit including the parasitic capacitances and inductances and taking into account the photocurrent due to the drift and diffusion of photo-induced carriers. The analysis has been performed in some papers.[5,12] The frequency response of a packaged Si pin $[n^+ - i(\pi) - p^+]$ photodiode with an optical input window 300 μm in diameter is shown, for various levels of reverse bias, in Fig. 4.20. The *i*-layer of the photodiode is fully depleted when the bias is more than 5 V. The bias circuit is shown in Fig. 4.15,

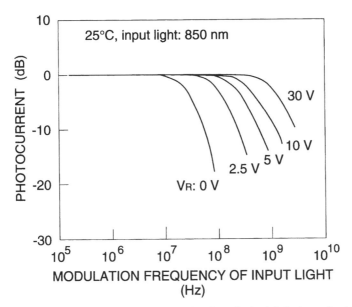

FIGURE 4.20 Frequency response of an Si pin-photodiode (n^+-i-p^+ type) with a light input window 300 μm in diameter.

where the load resistance, R_L, is 50 Ω. The frequency response is quite low at $V_R = 0$. This low-frequency response results from the large CR-time constant and the participation of the diffusion current which (as in solar cells) is due to the partial and narrow depletion region (see Section 4.3.4.1). As the reverse bias increases, the frequency response extends to higher frequencies because of the widening of the depletion layer (at $V_R = 2.5$ V). After reach-through (at $V_R > 5$ V), the response continues to extend to higher frequencies as the bias increases because the increase in electrical field intensity decreases the transit time within the i-layer.

For improving the frequency response, decreasing the thickness of the i-layer is effective from the viewpoint of the transit time but it increases the CR-time constant and decreases the (external) quantum efficiency. Those trade-off relations are therefore taken into consideration during device design. The CR-time constant can also be reduced by decreasing the area of the pn-junction. Most of the pin-photodiodes used in optical fiber communication systems have a pn-junction area 30 μm or less in diameter. Cutoff frequencies greater than a few gigahertz are easily obtained in III-V compound semiconductor pin-photodiodes by combining a small pn-junction with thin i-layers (see Section 4.3.1).

4.3.5 Noise and Signal-To-Noise Ratio For pin-Photodiodes

4.3.5.1 Quantum Noise and Thermal Noise   Noise is an important factor when pin-photodiodes are used for signal transmission in equipment and systems such as optical fiber communication systems and data links. The noise in (pin) photodiodes is quantum noise (or shot noise) and thermal noise (or Johnson noise or Nyquist noise) and often limits the performance of the equipment and systems. The shot noise is the essential noise in *pin*-photodiodes and, like that in laser diodes and LEDs, is induced by the behavior of the photo-induced carriers and incident photons as particles. The thermal noise is generated in the resistance connected to the photodiode and is caused by the random motion of carriers in the resistance.

The pin-photodiode generates photocurrent originated from the photo-induced carriers produced randomly and independently when the incident photons are absorbed. The motion of the generated carriers is also random, and these carriers are converted to photocurrent as indicated in Eq. (4.27). Consequently, a fluctuation due to the random and independent generation and motion of the carriers is introduced to the photocurrent as shot noise. A similar fluctuation is also introduced to the dark current and the current due to background radiation indicated in Eq. (4.20b). The mean-square shot noise current in the frequency range between f and $(f + \Delta f)$ is given by the following equation:[13]

$$\overline{i_{sn}^2} = 2q\bar{I}\Delta f, \tag{4.39}$$

where $\bar{I}$ is the steady-state current. The shot noise is almost independent of frequency (white noise). Using Eqs. (4.39) and (4.20b), the shot noise current in pin-photodiodes can be given as

$$\overline{i_{sn}^2} = 2q(I_{photo} + I_d + I_{bg})B_{bw}, \tag{4.40}$$

where Δf in Eq. (4.39) is replaced with a bandwidth, B_{bw}.

The Johnson noise originates from the random motion of carriers with average thermal energy $(3/2)k_BT$ in the electrical resistance at temperature T. This random motion produces a noise current in the load resistance, and the corresponding noise voltage across the load resistance introduces a noise current to the photodiode bias circuit. The mean-square Johnson noise current within the bandwidth, $B_{bw} (= \Delta f)$, is given by the following equation:[13]

$$\overline{i_{Jn}^2} = 4k_BTB_{bw}/R_{eq}, \tag{4.41}$$

where R_{eq} is the electrical resistance. The mean-square Johnson noise current, inversely proportional to the resistance, is also a kind of white noise.

By adding the noise sources indicated in Eqs. (4.40) and (4.41) to the equivalent circuit in Fig. 4.18, the noise equivalent circuit for pin-photodiodes can be expressed as shown in Fig. 4.21. Here, the electrical resistance, R_{eq}, in Eq. (4.41) is equal to $1/(1/R_D + 1/R_L)$. Both the shot noise and Johnson noise are

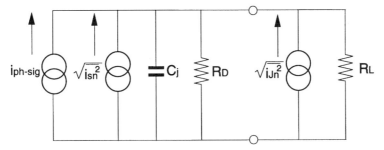

FIGURE 4.21   Noise equivalent circuit for pin-photodiodes.

white noises and are independent of the frequency. At frequencies lower than 1 kHz, $1/f$ noise (or flicker noise) increases and becomes dominant as frequency decreases.

4.3.5.2 Signal-to-Noise (Power) Ratio   The signal-to-noise power ratio (S/N) can be analyzed by using the equivalent circuit in Fig. 4.21. When the incident light is modulated as $P(\omega) = P_{inc0}\,[1 + M_i\,\sin(\omega t)]$ (analog modulation), where P_{inc0} and m_i are the average signal power and the modulation index, the steady-state average photo-induced signal current is given, according to Eq. (4.5), by

$$I_{ph\text{-}sig} = \eta_{ph}qP_{inc0}/h\nu, \qquad (4.42)$$

and the mean-square value of the signal current is

$$\overline{i^2_{ph\text{-}sig}} = (1/2)(m_i I_{ph\text{-}sig})^2 = (1/2)(m_i \eta_{ph} q P_{inc0}/h\nu)^2. \qquad (4.43)$$

The total signal power is given by $R_{eq}\,\overline{i^2_{ph\text{-}sig}}$ and the total noise power is R_{eq} $(\overline{i^2_{sn}} + \overline{i^2_{Jn}})$. The signal-to-noise power ratio is thus given by

$$\begin{aligned} S_{rms}/N_{rms} &= \overline{i^2_{ph\text{-}sig}}/(\overline{i^2_{sn}} + \overline{i^2_{Jn}}) \\ &= (1/2)(m_i I_{ph\text{-}sig})^2/[2q(I_{ph\text{-}sig} + I_d + I_{bg})B_{bw} + 4k_B TB_{bw}/R_{eq}], \quad (4.44) \end{aligned}$$

where the subscript, rms, means the root mean square. The resistance R_D ($\approx 10^{10}\,\Omega$) $\gg R_L$, and then $R_{eq} \approx R_L$, and the current due to background radiation, I_{bg}, is very small. The signal-to-noise ratio is influenced by the dark current, I_d, and the load resistance, R_L. If the effects of the parasitic capacitances of the package and the input resistance of the equipment appear in the characteristics, those elements are added to the circuit in Fig. 4.21 in parallel to the load resistance, and then the signal-to-noise power ratio can be calculated.

As a measure for the responsivity of photodiodes, the noise equivalent power

(NEP) is often used. The NEP indicates the input rms optical power per unit bandwidth at which the photocurrent is equal to the noise current and corresponds to the minimum input optical power detectable at $(S/N)_{rms} = 1$. Substituting $(S/N)_{rms} = 1$ into Eq. (4.44) and using Eq. (4.43),

$$\begin{aligned} \text{NEP} &= P_{inc0}/(1/2)^{1/2}|_{S/N=1} \\ &= (h\nu/m_i\eta_{ph}q)\{[2q(I_{ph\text{-}sig} + I_d + I_{bg})B_{bw} + 4k_BTB_{bw}/R_{eq}]/B_{bw}\}^{1/2} \\ &\qquad [W/Hz^{1/2}]. \end{aligned} \qquad (4.45)$$

The right-hand side of this equation is divided by bandwidth, B_{bw}, according to the definition of NEP, and the bandwidth is included in the equation because it cannot be directly eliminated from the measured values. The series resistance is omitted from the equivalent circuits in Fig. (4.21) and in Eq. (4.45) because it is small. The performance of the photodiode is inversely related to the NEP.

For digital modulation, the modulation index, m_i, is nearly equal to 1 in Eq. (4.43), and the signal current corresponds to the peak pulse level, P_{peak}. Consequently, the mean-square value of the signal current, $i_{ph\text{-}dig}$, is given by

$$\overline{i_{ph\text{-}dig}^2} = I_{ph\text{-}dig}^2 = (\eta_{ph}qP_{peak}/h\nu)^2. \qquad (4.46)$$

In a manner similar to that used in the case of analog modulation, the signal-to-noise power ratio can be expressed by

$$\begin{aligned} S_{p\text{-}p}/N_{rms} &= \overline{i_{ph\text{-}dig}^2}/(\overline{i_{sn}^2} + \overline{i_{Jn}^2}) \\ &= I_{ph\text{-}dig}^2/[2q(I_{ph\text{-}dig} + I_d + I_{bg})B + 4k_BTB/R_{eq}]. \end{aligned} \qquad (4.47)$$

By using Eq. (4.47), the minimum input optical power detectable for digital modulation can be calculated at a given $S_{p\text{-}p}/N_{rms}$.

4.3.6   Heterojunction and Waveguide-Type pin-Photodiodes

As described in Section 4.3.4.3, the quantum efficiency and the cutoff frequency are correlated. Decreasing the dark current is also important for improving the signal-to-noise ratio. As the band-gap of the semiconductor increases, the dark current decreases as indicated by Eqs. (4.2) and (4.3) and the breakdown voltage [see Fig. 4.1(b)] increases. The pn-heterojunction introduces such improvements. The high breakdown voltage leads to a high-frequency response because of the increase in the electrical field within the depletion ($i-$) layer. This improvement is especially effective in the pin-photodiodes used at long wave-

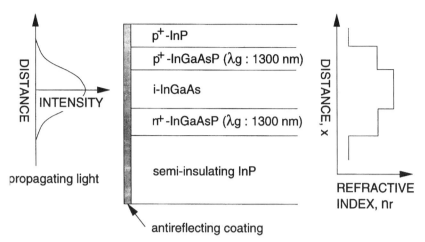

FIGURE 4.22   Schematic diagram of a waveguide type pin-photodiode with a *pn*-heterojunction.

lengths, such as the InGaAs/InP photodiodes. If a large notch due to a large band discontinuity is formed, however, it acts as a barrier to carrier flow and degrades the frequency response. The heterostructure needs to be formed carefully by selecting a suitable combination of materials as discussed in Section 4.2.3 (see Fig. 4.12).

A waveguide type photodiode is effective in increasing efficiency without degrading the frequency response because it is released from the requirement of the thickness for optical absorption. The input light incident to the pin-photodiode is in the direction parallel to the *pn*-junction plane. The structure is shown schematically in Fig. 4.22.[14] The *i*-layer and the adjacent layer forms a waveguide which the input light propagates along and is absorbed in. The electrical field can be set at a high value, and the barrier to carrier flow at the heterointerface is reduced by the multilayer heterostructure in which the band-gap of each layer gradually changes (see Fig. 4.22). A quantum efficiency of about 70% has been obtained in the waveguide-type pin-photodiode. The frequency response of the *pin*-photodide is shown in Fig. 4.23. The frequency response gradually increases as the bias increases because of the shortening transit time with increasing bias. The transit time is saturated in biases higher than 3 V, and the cutoff frequency exceeds 50 GHz without any deterioration of the quantum efficiency.[14]

4.4   AVALANCHE PHOTODIODES

Avalanche photodiodes (APDs) are also widely used in optical fiber communication systems. The APD is, however, not as frequently used in consumer

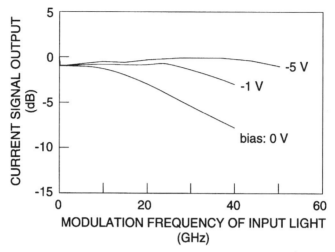

FIGURE 4.23   Frequency response of a waveguide type pin-photodiode, whose structure is shown in Fig. 4.22. (Courtesy of Dr. K. Kato).

electronics as the pin-photodiode is because its operating voltage is very high. The APD operates at nearly its breakdown voltage [see Fig. 4.1(b)] and can amplify the photo-induced current by avalanche multiplication. This process cannot occur in solar cells and pin-photodiodes, in which only photo-induced carriers contribute to the photocurrent. The operating mechanisms of APDs are more complicated than those of pin-photodiodes. The structures of APDs are similar to those of pin-photodiodes, but are slightly more complex. Popular materials for the APDs are Si in the 700 nm and 800 nm bands and Ge, InGaAs, and InGaAsP at wavelengths longer than 1 μm.

Some typical APDs are shown in Fig. 4.24. The active region consists of the light-absorbing region (layer) and the avalanche region (layer). The avalanche region is designed so that the highest electric field is generated under reverse bias. The light-absorbing region is adjacent to or coincides with the avalanche region. The structure in which the avalanche region is separated from the absorbing region is effective for increasing quantum efficiency and reducing noise, as will be discussed later. The window region for incident light is ordinarily restricted to a diameter of 30–300 μm by a ring electrode and is coated with an antireflective dielectric film to reduce the amount of light reflected at the input surface. A guard ring around the diffused region is formed by diffusion or ion-implantation in order to avoid edge breakdown because the strong electrical field is generated at the edge of the diffused region. Avalanche breakdown therefore always occurs in the central plane region. Most APDs have an $i(\pi$ or $\nu)$-layer [for an exception see Fig. 4.24(b)], and most APDs are used after reach-through in order to obtain the high-frequency response and the high quantum efficiency. As in pin-photodiodes, a mesa structure is sometimes used to reduce the junction capacitance for high-frequency response.

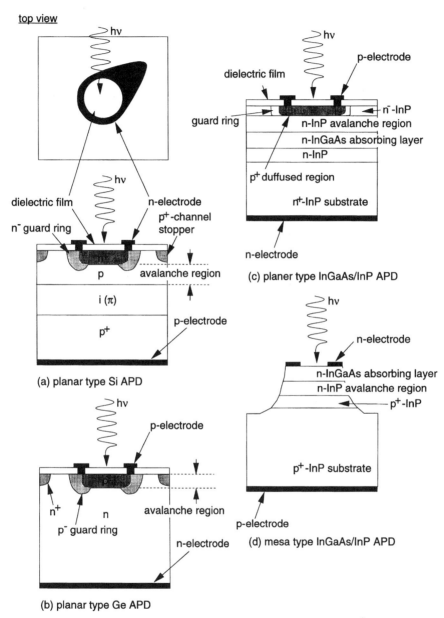

FIGURE 4.24 Schematic diagrams of some APDs: (a) planar type Si,[8] (b) planar type Ge,[15] (c) planar type InGaAs/InP,[16] (d) mesa type InGaAs/InP.[17]

4.4.1 Basic Characteristics

Although the characteristics of APDs are different from those of pin-photodiodes, APDs have operating mechanisms that are quite similar to those of pin-photodiodes except for the avalanche multiplication process. The energy band diagram of the avalanche region for a typical n^+-p-i (π)-p^+ structure (reach-through type) [see Fig. 4.24(a)] is shown in Fig. 4.25. The pn-junction is formed by the n^+-layer and the p-layer, and the depletion region mainly elongates into the p-layer from the abrupt step junction because of the low doping concentration [see Eq. (1.129)]. The light-absorbing region is separated from the avalanche region, so these APDs are called separated absorption and multiplication (SAM)-APDs. The p-layer is slightly doped, for example, to levels below 10^{16} cm^{-3} in Si and InGaAs APDs. The $i(\pi)$-layer in Si APDs, for example, is more than a few tens of micrometers thick and the doping (hole) concentration is on the order of magnitude of 10^{14} cm^{-3}. As in pin-photodiodes, under normal operating conditions the depletion region reachs the p^+-layer (reach-through). The input light is absorbed in the thick absorbing region [$i(\pi)$-layer], and electron-hole pairs are generated in the i-layer. Only the electrons generated drift toward the avalanche region; the holes drift to the p^+-layer in the i-layer. The electrons injected from the absorbing region to the avalanche region lead to the avalanche multiplication, and large numbers of electrons and holes reach the n^+-layer and the i-layer. As a result, the photo-induced current is amplified. Electron-hole pairs are also generated within the avalanche region, and the holes as well as electrons of these pairs participate in the avalanche multiplication.

In some APDs the light-absorbing region coincides with the avalanche region [see Fig. 4.24(b)], and in such devices electron-hole pair generation and avalanche multiplication occur in the same region. As discussed in Section 4.6.1, however, the structure in which the absorbing region is separated from the avalanche region is better from the viewpoint of noise.

4.4.1.1 Impact Ionization and Ionization Coefficients
As shown in Fig. 4.25, photo-induced carriers (electrons in Fig. 4.25) are injected into the avalanche region. The electrons are accelerated by the high electric field within the avalanche region and reach the saturation velocity because of their collisions with the lattice (see Section 4.3.4.2). The electrical field within the avalanche region is normally required to be between 10^4 and 10^5 V/cm. The saturation velocity is a mean value, and thus carriers having higher velocities also exist within the region. When a high-velocity carrier collides with the lattice, impact ionization (in which an electron is excited from valence band to conduction band) occurs. The impact ionization is a three-carrier collision process in which a high-energy primary carrier produces an electron-hole pair and is the inverse process of Auger recombination (see Section 1.1.2.7). Ionization requires at least the energy for excitation from the valence band to the conduction band, and for the impact ionization the energy of the primary carrier is ordinarily required to be equal to or greater than $(3/2)E_g$, where E_g is the band-gap energy

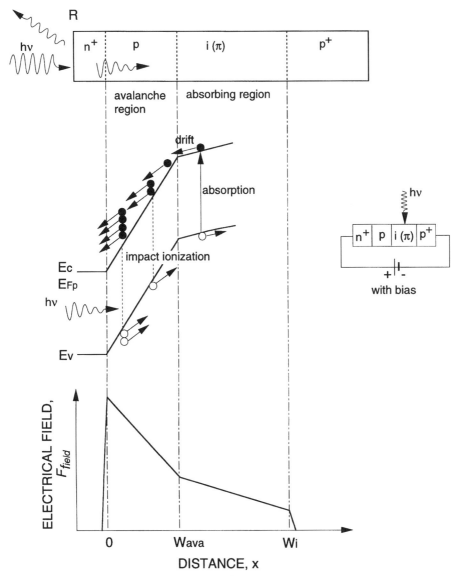

FIGURE 4.25   Simplified energy band diagram and electrical field of APDs—electron injection case.

of the semiconductor.[1] The carriers generated in the impact ionization process are also accelerated by the high electrical field within the avalanche region and themselves cause impact ionization. The number of ionizing collisions per unit length is called the ionization coefficient. As shown in Fig. 4.25, a large number of carriers are created as a result of such continuous impact ionization.

Though the impact ionization process is complicated, the process can be simply described with Baraff's theory.[18,19] Between the collisions (phonon scattering) or during the mean free time, an electron can move the mean free path, $L_{e\text{-mfp}}$ (see Section 1.1.2.6). The kinetic energy, $qF_{field}L_{e\text{-mfp}}$, is given to the electron from the electrical field, F_{field}, during the mean free time, and this energy is lost from the electron in the following collision. If the electron reaches the threshold ionization energy, $E_{e\text{-ith}}$, which is the minimum energy for impact ionization, the impact ionization occurs at the collision. The probability that the energy of the electron reaches the threshold ionization energy is given by the Boltzmann factor, $\exp(-E_{e\text{-ith}}/qF_{field}L_{e\text{-mfp}})$, and the electron moves by $E_{e\text{-ith}}/qF_{field}\ (=L_{e\text{-mfp}})$ during the acceleration process. Consequently, the ionization coefficient for electrons is given by

$$\alpha_{ion} = qF_{field}/E_{e\text{-ith}} \exp(-E_{e\text{-ith}}/qF_{field}L_{e\text{-mfp}}). \tag{4.48a}$$

Similarly the ionization coefficient for holes is given by

$$\beta_{ion} = qF_{field}/E_{h\text{-ith}} \exp(-E_{h\text{-ith}}/qF_{field}L_{h\text{-mfp}}). \tag{4.48b}$$

The ionization coefficient for electrons is not equal to that for holes. For Ge and InGaAsP,[20,21] the ionization coefficients are shown in Fig. 4.26. The ionization coefficient depends on the material and on the crystal orientation in their material. Each carrier has a few ionizing collisions within the avalanche region under the strong electrical field because the region is usually more than 1 μm thick [see Figs. 4.26(a) and 4.26(b)]. The ratio of the ionization coefficient for holes (electrons) to that for electrons (holes) is called the ionization coefficient ratio:

$$k_{ion} = \beta_{ion}/\alpha_{ion} \text{ or } \alpha_{ion}/\beta_{ion}. \tag{4.49}$$

The ionization coefficient ratio is determined by the material and is a measure of the excess noise in APDs.

4.4.1.2 Avalanche Breakdown (and Zener Breakdown) The continuous impact ionization leads to the increase in reverse current. The reverse current increases rapidly at the avalanche breakdown voltage (see Section 1.2.3.2) as indicated in Fig. 4.1(b). In the bias range before the breakdown voltage, the reverse amplified current can be controlled by the bias. The schematic illustration of the current distribution and energy band is shown in Fig. 4.27 for an avalanche process initiated by the electrons diffused from the p^+-layer. The total current density is kept constant within the avalanche region as well as in the whole region, and thus

$$J_{tot} = J_e(x) + J_h(x), \tag{4.50}$$

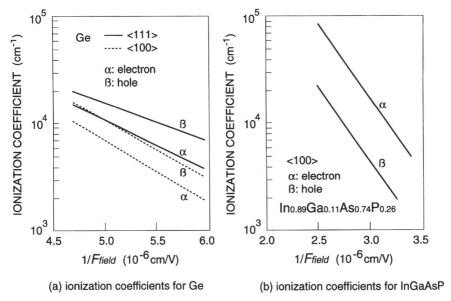

(a) ionization coefficients for Ge   (b) ionization coefficients for InGaAsP

FIGURE 4.26 Ionization coefficients for electrons and holes in (a) Ge and (b) InGaAsP. From T. Mikawa et al., for (a) Ge[20] © 1980 American Institute of Physics. Reprinted by permission of the authors and American Institute of Physics. From Y. Takanashi et al., for (b) InGaAsP[21] © 1979 *Japan. J. Appl. Phys.* Reprinted by permission of the authors and *Japan J. Appl. Phys.*

where $J_e(x)$ and $J_h(x)$ are the electron and hole drift currents, respectively. When electrons are injected into the avalanche region from the p^+-layer, the electron drift current is multiplied to $M_e J_{ei}$ at the end of avalanche region $(x = W)$. Consequently, the multiplication factor for electrons, M_e, can be expressed by

$$M_e = J_e(W)/J_{ei}. \qquad (4.51)$$

The increase in the number of electrons within the Δx at point x is equal to the sum of the electrons generated by electrons, $[J_e(x)/q]\alpha_{ion}\Delta x$, and the electrons generated by holes, $[J_h(x)/q]\beta_{ion}\Delta x$. The increase in current, ΔJ_e, is therefore given by

$$\Delta J_e = [J_e(x)/q]\alpha_{ion}\Delta x + [J_h(x)/q]\beta_{ion}\Delta x. \qquad (4.52)$$

By using Eq. (4.50),

$$dJ_e(x)/dx + (\beta_{ion} - \alpha_{ion})J_e(x) = \beta_{ion}J_{tot}. \qquad (4.53)$$

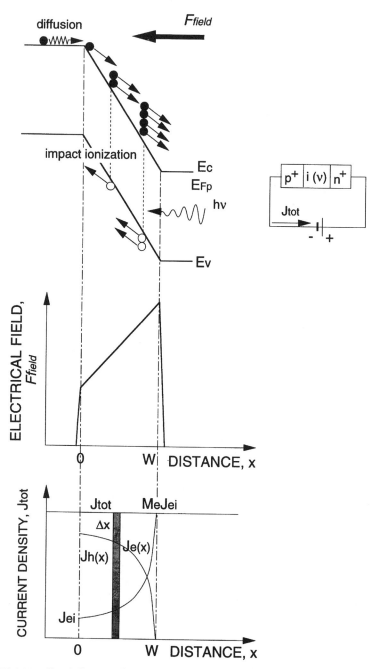

FIGURE 4.27 Band diagram for and illustration of electron and hole current densities in an APD.

Solving the differential equation for $J_e(x)$ in the structure shown in Fig. 4.27 and integrating $J_e(x)$ within the avalanche region (from $x = 0$ to $x = W$),

$$M_e = 1 / \left\{ 1 - \left\{ \int_0^W \alpha_{ion} \exp \left[\int_0^x (\beta_{ion} - \alpha_{ion}) dx' \right] dx \right\} \right\}. \qquad (4.54)$$

Similarly, when the avalanche process is initiated by holes diffused from the n^+-layer, the multiplication factor is given by

$$M_h = 1 / \left\{ 1 - \left\{ \int_0^W \beta_{ion} \exp \left[\int_x^W (\alpha_{ion} - \beta_{ion}) dx' \right] dx \right\} \right\}. \qquad (4.55)$$

Avalanche breakdown occurs when $M_e = \infty$ or $M_h = \infty$. Therefore, from Eqs. (4.54) and (4.55),

$$\int_0^W \alpha_{ion} \exp \left[\int_0^x (\beta_{ion} - \alpha_{ion}) dx' \right] dx = 1 \qquad (4.56)$$

and

$$\int_0^W \beta_{ion} \exp \left[\int_x^W (\alpha_{ion} - \beta_{ion}) dx' \right] dx = 1. \qquad (4.57)$$

Equations (4.56) and Eq. (4.57) are equivalent, and the avalanche breakdown voltage can be estimated from either of them. Avalanche breakdown tends to be suppressed by the carrier having a lower ionization coefficient because the impact ionization has to occur several times within the avalanche region to induce the avalanche breakdown. When the ionization coefficients for electrons and holes are the same ($\alpha_{ion} = \beta_{ion}$), Eqs. (4.56) and (4.57) are simplified to

$$\int_0^W \alpha_{ion} dx = 1. \qquad (4.58)$$

4.4.1.3 *Avalanche Multiplication and Photocurrent-Voltage Characteristics*
The multiplication factor is actually determined by the ratio of photocurrent amplified at a high bias in the avalanche photodiode mode, I_{ph-M}, to photocurrent at a low bias in the photodiode mode, I_{ph}, [see Fig. 4.1(b)]. Thus,

$$M = I_{ph-M} / I_{ph} = (I_{M-m} - I_{dM-m}) / (I_m - I_{dm}), \qquad (4.59)$$

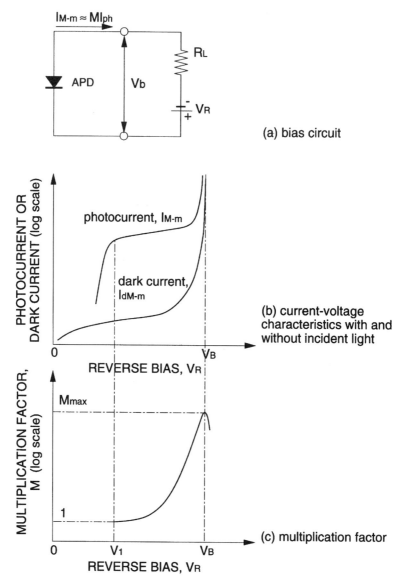

FIGURE 4.28 (a) Bias circuit, (b) current-voltage characteristics with and without incident light, and (c) multiplication factor as a function of reverse bias.

where $I_{\text{M-m}}$ and $I_{\text{dM-m}}$ are respectively the amplified current and the amplified dark current monitored experimentally in the avalanche photodiode mode (see Fig. 4.28) and I_{m} and I_{dm} are respectively the current and the dark current monitored experimentally in the photodiode mode. Equation (4.59) indicates the difference in photocurrent between pin-photodiode and APD.

The multiplication factor is given exactly by Eqs. (4.54) and (4.55), and those equations are simplified as follows:[22]

$$M = 1/(1 - |V_b/V_B|^{m_M})$$
$$= 1/\{1 - [(|V_R| - MI_{ph}R_{eq})/|V_B|]^{m_M}\}, \qquad (4.60)$$

where V_b is the bias voltage, V_B is the avalanche breakdown voltage (see Fig. 4.28), R_{eq} is the sum of the series resistance and the resistance in the i-layer or lightly doped layer, and m_M is a constant depending on the semiconductor material, doping profile, and wavelength. When the reverse bias approaches the breakdown voltage, V_B, Eq. (4.60) is simplified to $|V_B|/m_M M I_{ph} R_{eq}$.[23] The multiplication factor is the maximum at the breakdown voltage and can be expressed approximately by using Eq. (4.59) and the simplified Eq. (4.60):

$$M_{max} \approx (|V_B|/m_M I_{ph} R_{eq})^{1/2}, \qquad (4.61)$$

where it is assumed that $I_{Mm} \gg I_{d-M}$ and $I_m \gg I_d$. Equation (4.61) indicates that the maximum photomultiplication is inversely proportional to the square root of the photocurrent, I_{ph}. As shown in Fig. 4.28(c), the multiplication factor increases as the reverse bias increases gradually and reaches a maximum at the breakdown voltage. The maximum multiplication factor is over 50 in various kinds of APDs and is over 100 in Si APDs. The linearity between the amplified photocurrent and the input light power deteriorates under operation near the maximum multiplication factor. This results from the decrease in bias voltage to the avalanche region due to the voltage drop caused by the large photocurrent and the resistance R_{eq} (resistance in the i-layer and lightly doped layer, series resistance, and so forth). In the bias range higher than the breakdown voltage, the breakdown is generated by the dark current according to the current-voltage characteristics without light input and a large voltage drop is induced at each resistance. The multiplication factor is therefore again small as shown in Fig. 4.28(c).

When the input light power is very low or the dark current is higher than the photocurrent, the amplified dark current governs the avalanche breakdown process. This is sometimes the case in Ge APDs because of their high dark currents. The behavior of such APDs can be described in a manner similar to that describing the case discussed above, using the dark current instead of the photocurrent. The maximum multiplication factor is given by replacing photocurent with dark current in Eq. (4.61), yielding the following approximation:[24]

$$M_{max} \approx (|V_B|/m_M I_{dm} R_{eq})^{1/2}, \qquad (4.62)$$

where I_{dm} is the amplified dark current monitored [see Eq. (4.59)].

4.4.1.4 Device Structure and Avalanche Breakdown   The avalanche breakdown within the avalanche region is indispensable for normal operation of APDs. A strong electrical field within the avalanche region is required to generate the breakdown. Under such a strong field, however, a breakdown often occurs in other parts of APDs at a voltage lower than the normal avalanche breakdown voltage. Some structural improvements shown in Fig. 4.24 have therefore been used to prevent this extra breakdown.

Avalanche Breakdown and Zener (Tunnel) Breakdown   When the intensity of electrical field within the avalanche region is at or above 10^6 V/cm, the tunneling current given by Eq. (4.4) increases rapidly because of Zener breakdown (see Section 1.2.3.2). This breakdown tends to occur at the low-bias voltage, as the doping concentration of the depletion region is high and as the bandgap energy of material used in the depletion region is small. The multiplication of photocurrent disappears if Zener breakdown (tunneling breakdown) occurs. Consequently, the doping concentration to the layer corresponding to the depletion region needs to be maintained at a sufficiently low value: less than 10^{16} cm^{-3}. The avalanche breakdown voltages for Ge, Si, GaAs, and GaP are shown against impurity concentration for one-sided abrupt junctions in Fig. 4.29.[1] The dashed line indicates the maximum impurity concentration beyond which the tunneling mechanism will dominate the voltage breakdown characteristic. In APDs composed of the direct band-gap semiconductors, such as InGaAs and InGaAsP, the Zener breakdown needs to be paid particular attention because the tunneling probability is quite high in the direct band-gap semiconductors (see Section 4.1.2.4). A thick layer with a low doping concentration is needed in order to reduce the strength of the electrical field and avoid Zener breakdown.

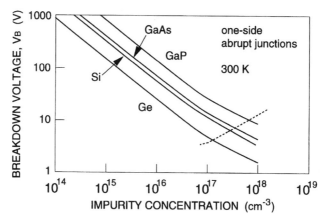

FIGURE 4.29   Avalanche breakdown voltage versus impurity concentration for one-sided abrupt junctions in Ge, Si, GaAs, and GaP. From S. M. Sze[1] © 1982 John Wiley & Sons, Inc. Reprinted by permission of John Wiley & Sons, Inc.

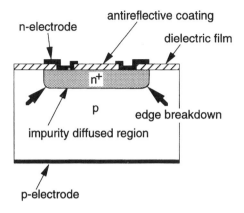

(a) region of edge breakdown

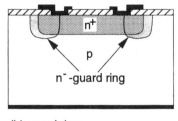

(b) guard ring

FIGURE 4.30   (a) A region of edge breakdown and (b) guard ring in a planar structure.

Edge Breakdown and Guard Rings   When the reverse bias gradually increases at the *pn*-junction shown in Fig. 4.30(a), the electrical field within the depletion layer also increases inhomogeneously. The electrical field tends to be higher at the edge (indicated with two arrows) than at the central plane region, and this situation can be analyzed exactly by solving Poisson's equation.[1] Breakdown at the edge of the *pn*-junction occurs at lower voltage than it does at the central part of the *pn*-junction region and is called the edge breakdown. If edge breakdown occurs, the avalanche breakdown at the central plane part, upon which most input light is received, hardly occurs under normal operating conditions. As a result, the diode shown in Fig. 4.30(a) cannot operate as an APD. As shown in Figs. 4.30(b) and 4.24, a guard ring is frequently used to increase the edge breakdown voltage and thus to avoid edge breakdown. The guard ring is formed by heating after impurity diffusion or ion implantation. The *pn*-junction for the guard ring is a linearly graded junction and the curvature of the ring edge is set at a large value. The junction is actually a combination of a linearly graded junction with a one-sided abrupt junction.[25] A similar improvement can also be carried out by forming a shallow *pn*-junction with a low concentration along the edge of the impurity-diffused region in Fig. 4.30(a).[24]

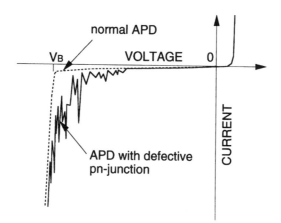

FIGURE 4.31 Current-voltage characteristics for a normal APD and a APD with defect at the *pn*-junction.

Breakdown and Defects If lattice defects, such as threading dislocations and stacking faults (see Section 7.2.1.1), or metallic precipitations, such as In and Ga precipitation in InGaAsP,[26,27] exist at the *pn*-junction, the current-voltage characteristic is often noisy and the breakdown occurs irregularly as shown in Fig. 4.31. This is due to the microplasma effect, which is localized breakdown at the high-field regions corresponding to the lattice imperfections or precipitations.[28] Current tends to concentrate in the small defective areas and plasma will form there under reverse bias. When such localized breakdowns occur, the current near the breakdown voltage fluctuates and noise is superimposed on the current-voltage characteristics. When the breakdown due to the microplasma occurs at a voltage lower than does the normal avalanche breakdown, the avalanche multiplication is impaired tremendously. The lattice imperfections and precipitations also increase the dark current and influence device life, that is, reliability (see Chapter 7). A similar breakdown also occurs at the surface if defects are introduced to the *pn*-junction perimeter during the fabrication process. Reducing the number of defects is quite important to obtain high performance and high reliability. The microplasma is sometimes observed as a bright spot under an optical microscope.

4.5 FREQUENCY RESPONSE

The frequency response is basically determined by the CR-time constant, transit time, and multiplication time. The limitation by the CR-time constant and the transit time is quite similar to that in the case of pin-photodiodes. In the worst case here the transit time is twice as large as that in pin-photodiodes because of the avalanche mechanism. One electron injected into an avalanche region, for example, causes repeated impact ionizations before reaching the opposite edge of the region. Holes generated by the impact ionization near the opposite edge drift toward the side on which the electrons were injected. These holes also contribute to the photocurrent, and thus the transit time becomes long.

Within the avalanche region the carriers induce the impact ionization during drifting. The impact ionization reduces the mean drift velocity of the carriers, and the mean time to drift a unit length is longer than the drift time at the saturation velocity (see Section 4.3.4.2). The time required for the avalanche multiplication is called the multiplication time, which corresponds to the rise time of the amplified current. The time dependence of the multiplication factor can be evaluated by using the following equation taking into account the multiplication process of current within the avalanche region:[29]

$$M(\omega) = M_0/[1 + (\omega M_0 \tau_{\text{d-ava}})^2]^{1/2}, \qquad (4.63)$$

where M_0 is the multiplication factor in the low-frequency range (or under dc) and $\tau_{\text{d-ava}}$ is the mean drift time within the avalanche region. The mean drift time depends on the impact ionization coefficient and is expressed by the following equation when electrons are injected to the avalanche region:

$$\tau_{\text{d-ava}} = N_{\text{ic}}(\beta_{\text{ion}}/\alpha_{\text{ion}})W/v_{\text{ds}}, \qquad (4.64)$$

where N_{ic} is a constant depending on the ionization coefficient ratio, W is the width of avalanche region, and v_{ds} is the saturation velocity (see Section 4.3.4.2). Here it is assumed that the saturation velocity for electrons and holes is equal. From Eq. (4.63), the 3 dB bandwidth can be calculated as the frequency at which $M(\omega)/M_0 = 1/2^{1/2}$ because the multiplication factor is for current amplitude and the 3 dB bandwidth is given by the frequency at which the power amplitude decreases to $\frac{1}{2}$ of the amplitude in the low-frequency range. Therefore, the gain-bandwidth (GB) product is given by the product of the 3 dB bandwidth and the multiplication factor in the low-frequency range or at dc:

$$GB = 1/2\pi\tau_{\text{d-ava}}$$
$$= (1/2\pi N_{\text{ic}})(\alpha_{\text{ion}}/\beta_{\text{ion}})(v_{\text{ds}}/W). \qquad (4.65)$$

The GB-product is inversely proportional to the mean drift time of carriers within the avalanche region and is proportional to the ionization coefficient ratio, $\beta_{\text{ion}}/\alpha_{\text{ion}}$, for the electron injection case. When the injected carriers are holes and avalanche breakdown is initiated by the holes, the GB-product is obtained by replacing $\alpha_{\text{ion}}/\beta_{\text{ion}}$ with $\beta_{\text{ion}}/\alpha_{\text{ion}}$ in Eq. (4.65). The correlation between bandwidth and multiplication factor is illustrated in Fig. 4.32. The bandwidth is determined by the transit time or the CR-time constant when the multiplication factor is small. As the multiplication factor increases, the band-

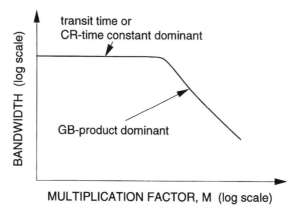

FIGURE 4.32 Bandwidth versus multiplication factor for APDs.

width gradually decreases because of the limitation of the gain-bandwidth product. The gain-bandwidth product often limits the performance of optical fiber communication systems operating at frequencies higher than 1 Gb/s.

4.6 NOISE AND SIGNAL-TO-NOISE RATIO FOR AVALANCHE PHOTODIODES

4.6.1 Quantum Noise, Thermal Noise, and Excess Noise

If an APD is used in the low-bias range without avalanche multiplication, the noise behavior is quite similar to that of a pin-photodiode and is mainly determined by shot (quantum) noise. Under the avalanche multiplication an additional noise is generated during the avalanche process in addition to the multiplication of the shot noise. That is called the excess noise and is introduced by the random fluctuation in successive ionizing collisions under the avalanche process. The excess noise is usually expressed by the excess noise factor multiplying the initial shot noise. When electrons are injected into the avalanche region, the excess noise factor for electrons, F_e, is given as[24]

$$F_e = M\{1 - (1 - k_{ion})[(M - 1)/M]^2\}. \tag{4.66}$$

For hole injection into avalanche region, the excess noise factor for holes, F_h, is given by

$$F_h = M\{1 + [(1 - k_{ion})/k_{ion}][(M - 1)/M]^2\}. \tag{4.67}$$

Here k_{ion} is the ionization coefficient ratio, β_{ion}/α_{ion}. The excess noise factor is sometimes simplified as

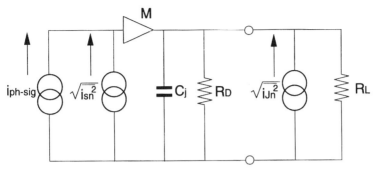

FIGURE 4.33 Simplified noise equivalent circuit for APDs.

$$F = M^x. \tag{4.68}$$

The excess noise factor increases according to the increase in multiplication factor as indicated in Eqs. (4.66), (4.67), and (4.68). It also depends on the ionization coefficient ratio, β_{ion}/α_{ion} and is smaller when $\alpha_{ion} \gg \beta_{ion}$ for the electron injection case and when $\beta_{ion} \gg \alpha_{ion}$ for the hole injection case. These relations mean that the excess noise in APDs can be reduced by injecting the carrier having the larger ionization coefficient.

4.6.2 Signal-to-Noise Ratio

The noise equivalent circuit for APDs can be given by adding an amplifier to that for pin-photodiodes because of the multiplication process. This is shown in Fig. 4.33, where the series resistance of the APD is omitted because it is small (see Figs. 4.10 and 4.21).

For analog modulation the mean-square shot noise current can be expressed with the steady-state current in a way similar to that in the case of pin-photodiodes [see Eq. (4.43)] but incorporating the multiplication factor, M:

$$\overline{i_{ph\text{-}sig}^2} = (1/2)(m_i M I_{ph\text{-}sig})^2 = (1/2)(m_i \eta_{ph} q M P_{inc0}/h\nu)^2. \tag{4.69}$$

Here the amplified photocurrent under steady-state conditions is given by

$$I_{ph\text{-}sig} = \eta_{ph} q M P_{inc0}/h\nu. \tag{4.70}$$

Also in a manner similar to that in the case of pin-photodiodes, the equivalent shot noise current in APDs can be expressed by

$$\overline{i_{sn}^2} = 2q(I_{photo} + I_{dM} + I_{bg})B_{bw}M^2F + 2qI_{ds}B_{bw}$$
$$\approx 2q(I_{photo} + I_{dM} + I_{bg})B_{bw}M^{2+x} + 2qI_{ds}B_{bw}, \tag{4.71}$$

where the current is multiplied by the multiplication factor M and the factor thus is squared. The dark current is the sum of diffusion current, generation-recombination current, surface leakage current, and tunneling current (see Section 4.1.2). The surface leakage current is not amplified because it scarcely passes through the avalanche region. The noise current term due to the surface recombination is added as $2qI_{ds}B_{bw}$. The mean-square Johnson noise current is given directly by Eq. (4.41) if the noise is independent of the avalanche multiplication. The signal-to-noise power ratio is therefore given in a way similar to that in the case of pin-photodiodes by the following equation:[30]

$$S_{rms}/N_{rms} = \overline{i_{ph\text{-}sig}^2}/(\overline{i_{sn}^2} + \overline{i_{Jn}^2})$$
$$= (1/2)(m_i M I_{ph\text{-}sig})^2/[2q(I_{ph\text{-}sig} + I_{dM} + I_{bg})M^{2+x}B_{bw}$$
$$+ 2qI_{ds}B_{bw} + 4k_BTB_{bw}/R_{eq}]. \tag{4.72}$$

Here the resistance R_D ($\approx 10^{10}$ Ω) $\gg R_L$, and thus $R_{eq} \approx R_L$, and the current due to the background radiation, I_{bg}, is very small. The signal-to-noise ratio is mainly determined by the amplified shot noise and the Johnson noise. The NEP can also be defined as the input optical power (in rms) at $(S/N)_{rms} = 1$ in Eq. (4.72) [see Eq. (4.45)].

The signal-to-noise ratio for digital modulation can also be analyzed as in the case of pin-photodiodes. The mean-square value of the signal current, $i_{ph\text{-}dig}$, is given by

$$\overline{i_{ph\text{-}dig}^2} = I_{ph\text{-}dig}^2 M^2 = (\eta_{ph}qMP_{peak}/h\nu)^2. \tag{4.73}$$

Consequently, the signal-to-noise power ratio can be expressed with the following equation:

$$S_{p\text{-}p}/N_{rms} = \overline{i_{ph\text{-}dig}^2}/(\overline{i_{sn}^2} + \overline{i_{Jn}^2})$$
$$= I_{ph\text{-}dig}^2 M^2/[2q(I_{ph\text{-}dig} + I_{dM} + I_{bg})M^{2+x}B_{bw} + 2qI_{ds}B_{bw}$$
$$+ 4k_BTB_{bw}/R_{eq}]. \tag{4.74}$$

This equation can also be used for calculating the minimum input optical power detectable in digital modulation at a given $S_{p\text{-}p}/N_{rms}$.

4.7 SUPERLATTICE APDs

As discussed in Sections 4.5 and 4.6, the frequency response and the signal-to-noise ratio depend very much on the ionization coefficient ratio. When this ratio is small or large, the GB-product increases and the excess noise decreases, and performance is best when the ratio = 0 or ∞. The impact ionization coefficients for electrons and holes are determined by the material property. Most materials used in APDs, except for Si, have an ionization coefficient ratio near 1, that is, $\alpha_{ion} \approx \beta_{ion}$. A material having a large difference between the ionization coefficients of electrons and holes further improves the performance, and the superlattice structure can introduce such a difference.

The band diagram of a conventional superlattice structure (see Section 1.3.1) is shown schematically in Fig. 4.34. The structures of GaAs (well)/AlGaAs (barrier) matching to GaAs substrate and of InGaAs (well)/InAlAs (barrier) matching to InP substrate are typical combinations for the superlattice.[31,32] In those combinations, the conduction band offset is larger than the valence band offset, respectively about 0.35 eV and 0.22 eV for $GaAs/Al_{0.43}Ga_{0.57}As$ and about 0.5 eV and 0.2 eV for $In_{0.53}Ga_{0.47}As/In_{0.52}Al_{0.48}As$. Electrons drifting

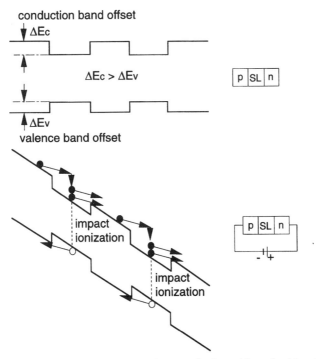

FIGURE 4.34   Schematic band diagram of a superlattice with and without bias (electrical field).

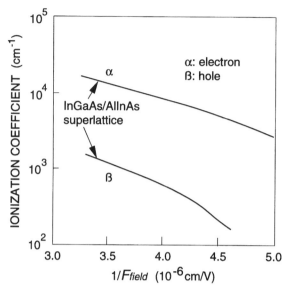

FIGURE 4.35 Ionization coefficient in an InGaAs/InAlAs superlattice. From T. Kagawa et al.[33] © 1989 IEEE. Reprinted by permission of the authors and IEEE.

within the superlattice adiabatically obtain the kinetic energy corresponding to the conduction band offset, and this process is repeated sequentially as shown in Fig. 4.34. The probability of the impact ionization for the electrons with high energies increases, and the ionization coefficient becomes high. Holes, on the other hand, cannot obtain such a large amount of energy and the ionization coefficient for holes is equal to or less than that in the bulk semiconductor. The relation between field strength and the ionization coefficients for electrons and holes in InGaAs/InAlAs superlattice is shown in Fig. 4.35.[33] There is a large difference between the ionization coefficients of electrons and holes. A super-lattice structure with 20 to 30 periods is inserted into the avalanche region in actual superlattice APDs, which show excellent characteristics. A GB-product of about 110 GHz and an excess noise factor of only 2 to 4, for example, is obtained in the InGaAsP/InAlAs superlattice APD.[34]

REFERENCES

1. S. M. Sze, *Physics of Semiconductor Devices*, 2nd ed., John Wiley & Sons, New York, 1981.

2. A. S. Grove, *Physics and Technology of Semiconductor Devices*, John Wiley & Sons, New York, 1967.

3. S. R. Forrest, R. F. Leheny, R. E. Nahory, and M. A. Pollack, $In_{0.53}Ga_{0.47}As$ photo-diodes with dark current limited by generation-recombination and tunneling, *Appl. Phys. Lett.*, **37**, 322, 1980.

4. Y. Matsushita and K. Sakai, Photodetector, in T. P. Pearsal, ed., *GaInAsP Alloy Semiconductors*, John Wiley & Sons, New York, 1982, Ch. 16.

5. See for example, P. Bhattacharya, *Semiconductor Optoelectronic Devices*, Prentice Hall, Englewood Cliffs, New Jersey 1994.

6. For example, A. G. Milnes and D. L. Feucht, *Heterojunctions and Metal-Semiconductor Junctions*, Academic Press, New York, 1972.

7. T. Mikawa, S. Kagawa, T. Kaneda, T. Sakurai, H. Ando, and O. Mikami, A low noise n^+np germanium avalanche photodiode, *J. Quantum Electron*, **QE-17**, 210, 1981.

8. O. Mikami, H. Ando, H. Kanbe, T. Mikawa, T. Kaneda, and Y. Toyama, Improved germanium avalanche photodiodes, *J. Quantum Electron*, **QE-16**, 1002, 1980.

9. T. Mikawa, S. Kagawa, and T. Kaneda, InP/InGaAs pin photodiodes in the 1 μm wavelength region, *Fujitsu Sci. Technol. J.*, **20**, 210, 1984.

10. T. P. Lee, C. A. Burrus, Jr., A. G. Dentai, InGaAs/InP p-i-n photodiodes for lightwave communications at the 0.95–1.65 μm wavelength, *J. Quantum Electron*, **QE-17**, 232, 1981.

11. H. Yonezu, *Hikari Tsuushin Sosi Kougaku* (Semiconductor Device Technology for Optical Fiber Communication Systems), Kougaku Tokyo, 1984 (Japanese).

12. J. E. Bower, and C. A. Burrus, InGaAs pin photodiodes with modulation response to millimeter wavelengths, *Electron. Lett.*, **21,** 812, 1985.

13. A. Yariv, *Introduction to Optical Electronics*, 3rd ed., Holt, Rinehart and Winston, New York, 1985.

14. K. Kato, S. Hata, K. Kawano, J. Yoshida, and A. Kozen, A high-efficiency 50 GHz InGaAs multimode waveguide photodetector, *IEEE J. Quantum Electron*, **QE-28**, 2728, 1992.

15. K. Nishida and K. Taguchi, N+-P-p-P+ Si avalanche photodiode optimized for optical communication use in 0.8–0.9 μm wavelength region, *NEC Research and Development*, **55**, 48, 1979.

16. T. Shirai, S. Yamazaki, K. Yasuda, T. Mikawa, K. Nakajima, and T. Kaneda, 1.0–1.6 μm planar avalanche photodiode by LPE grown InP/InGaAs/InP DH structure, *Electron. Lett.*, **18**, 575, 1982.

17. H. Kanbe, N. Susa, H. Nakagome, and H. Ando, InGaAs avalanche photodiode with InP *p-n* junction, *Electron. Lett.*, **16**, 163, 1980.

18. G. A. Baraff, Distribution functions and ionization rates for hot electrons in semiconductors, *Phys. Rev.*, **128**, 2507, 1962.

19. G. A. Baraff, Maximum anisotropy approximation for calculating electron distribution: Application to high field transport in semiconductors, *Phys. Rev.*, **133**, 1A, A26, 1964.

20. T. Mikawa, S. Kagawa, T. Kaneda, and Y. Toyama, Crystal orientation dependence on ionization rate in germanium, *Appl. Phys. Lett.*, **37**, 387, 1980.

21. Y. Takanashi and Y. Horikoshi, Ionization coefficient of InGaAsP/InP APD, *Japan J. Appl. Phys.*, **18**, 2173, 1979.

22. J. L. Moll, *Physics of Semiconductors*, McGraw-Hill, New York, 1964.

23. H. Melchior and W. T. Lynch, Signal and noise resonance of high-speed germanium avalanche photodiodes, *IEEE Trans. Electron Device*, **ED-13**, 829, 1966.

24. R. J. McIntyre, Multiplication noise in uniform avalanche diodes, *IEEE Trans. Electron Device*, **ED-13**, 164, 1966.

25. H. Yonezu and A. Kawaji, Computer-aided design of a Si avalanche photodiode, *IEEE Trans. Electron Device*, **ED-16**, 923, 1969.

26. S. Yamakoshi, M. Abe, O. Wada, S. Komiya, and T. Sakurai, Reliability of high radiance InGaAsP/InP LED's operating in the 1.2–1.3 μm wavelength, *IEEE J. Quantum Electron*, **QE-17**, 167, 1981.

27. M. Fukuda, K. Wakita, and G. Iwane, Dark defects in InGaAsP/InP double heterostructure lasers under accelerated aging, *J. Appl. Phys.*, **54**, 1246, 1983.

28. F. Capasso, P. M. Petroff, W. B. Bonner, and S. Sumski, Investigation of microplasma in InP avalanche photodiodes, *IEEE Electron Device Lett.*, **EDL-1**, 27, 1980.

29. R. B. Emmonds, Avalanche-photodiode frequency response, *J. Appl. Phys.*, **38**, 3705, 1967.

30. H. Melchier, Demodulation and photodetection technique, in F. T. Arecchi and E. O. Schulz-Dubois, eds., *Laser Handbook*, North-Holland, 1972.

31. F. Capasso, W. T. Tsang, A. L. Hutchinson, and G. F. Williams, Enhancement of electron impact ionization in a superlattice: A new avalanche photodiode with a large ionization ratio, *Appl. Phys. Lett.*, **40**, 38, 1982.

32. T. Kagawa, H. Iwamura, and O. Mikami, Dependence of the GaAs/AlGaAs superlattice ionization rate on Al content, *Appl. Phys. Lett.*, **54**, 33, 1989.

33. T. Kagawa, Y. Kawamura, H. Asahi, M. Naganuma, and O. Mikami, Impact ionization rates in an InGaAs/InAlAs superlattice, *Appl. Phys. Lett.*, **55**, 993, 1989.

34. T. Kagawa, Y. Kawamura, H. Iwamura, A wide-bandwidth low-noise InGaAsP-InAlAs superlattice avalanche photodiode with a flip-chip structure for wavelengths of 1.3 and 1.55 μm, *IEEE J. Quantum Electron.*, **QE-29**, 1387, 1993.

CHAPTER FIVE

Optical Modulators

Light-emitting *pn*-junction devices, such as LEDs and laser diodes, are easily modulated by superimposing signals onto the injected current. This is direct modulation, and laser diodes in high-bit-rate and long-span optical communication systems are frequently used under direct modulation. The direct modulation, however, results in chirping, which often limits transmission quality because of the dispersion in optical fibers (see Chapter 8). The modulation without chirping enables transmission distance to lengthen and bit-rate to increase. This chirpless modulation is important especially in systems employing optical fiber amplifiers (see Section 8.1.1.5) in which the transmission distance is nearly released from loss-limit. The chirping is due to the change in lasing situation caused by variation of carrier density and thus refractive index in lasing region under direct current modulation (see Section 3.5.2). An (external) optical modulator can modulate the light output power from laser diodes with little or no chirping.

There are two kinds of optical modulators: those composed of semiconductors and those composed of dielectric materials such as $LiNO_3$. The semiconductors used in modulators are ordinarily III-V compound semiconductors. These semiconductor optical modulators are *pn*-junction diodes and can be divided into two types. One is used under forward bias, as LEDs and laser diodes are, and the optical modulation is carried out by changing gain or loss within the modulators.[1] The other type is used, as photodiodes are, under reverse bias. Most high-performance semiconductor optical modulators are used under reverse bias which is needed to generate a strong electric field. Their operating behaviors are therefore somewhat similar to those of photodiodes. This chapter deals primarily with semiconductor optical modulators used under reverse bias.

5.1   PHYSICAL PHENOMENA USED IN OPTICAL MODULATION

Optical modulation is basically performed by modulating the refractive index or optical absorption coefficient of the modulators. These two physical parameters are not independent and related to each other by the Kramers–Kronig relations as noted in Section 1.1.3.3. Several physical phenomena are used to change the refractive index or absorption coefficient in optical modulators: mainly the electrooptic effect, the Franz–Keldysh effect, the (quantum-confined) Stark effect, and the Wannier–Stark effect.

5.1.1   Electrooptic Effect (Pockels Effect and Kerr Effect)

When a static electric field or an alternating electric field that has frequency much lower than the light frequency is applied to semiconductors and dielectric substances, their refractive indexes change. This phenomenon is known as the electrooptic effect,[2,3] and the field-dependent change in refractive index is described by

$$\Delta(1/n_r^2) = r_l F_{field} + s_q F_{field}^2 + \ldots, \tag{5.1}$$

where F_{field} is the strength of the electric field. If the first term is dominant, the electrooptic effect is called the linear electrooptic effect (the Pockels effect) and r_l is called the linear electrooptic (Pockels) coefficient. The linear electrooptic effect appears in optically noncentrosymmetric semiconductor crystals and dielectric substances such as $LiNbO_3$. The optical properties of some kinds of semiconductors, such as GaAs and InP, become anisotropic under an electric field and these materials then show the electrooptic effect. When the second term is dominant, the electrooptic effect depends on the square of electric field strength and is called the quadratic electrooptic effect (the Kerr effect) and s_q is called the quadratic electrooptic (Kerr) coefficient. The quadratic electrooptic effect is dominant in a centrosymmetric or amorphous material.

If light is incident to an electrooptic crystal in which an electric field is applied perpendicular to the incident light, the linear electrooptic effect in Eq. (5.1) can be described by

$$\Delta(1/n_r^2) = 2\Delta n_r/n_r^3 = r_l F_{field}, \tag{5.2a}$$

and thus the change in refractive index is given by

$$\Delta n_r = n_r^3 r_l F_{field}/2. \tag{5.2b}$$

The linear optoelectric coefficient here is, for example, 1.6×10^{-10} cm/V for GaAs and 3.08×10^{-9} cm/V for $LiNbO_3$.[3,4]

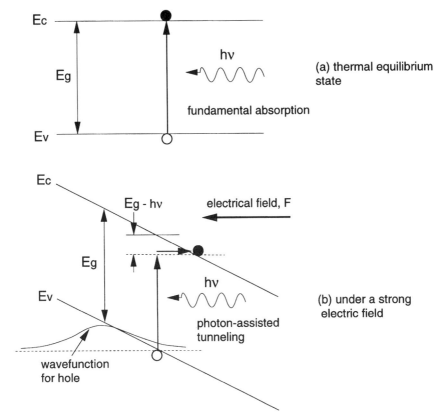

FIGURE 5.1 Schematic diagram of the tilting of energy bands under a strong electric field and of the Franz–Keldysh effect.

5.1.2 Franz–Keldysh Effect

A strong electric field applied to a semiconductor tilts the band edges, valence band maximum, and conduction band minimum, and the wavefunctions for electrons and holes penetrate into the band-gap (see Fig. 5.1). The photo-assisted tunneling therefore occurs at a wavelength longer than the wavelength corresponding to the band-gap energy. As a result, the fundamental absorption edge (see Section 1.1.3.1) shifts to a longer wavelength.[5,6] This phenomenon is called the Franz–Keldysh effect and it is remarkable when the field strength is more than 10^4 V/cm. The electric-field-dependent absorption coefficient becomes small as photon energy, $h\nu$, decreases from the band-gap energy, E_g, and it changes exponentially with $(E_g\text{-}h\nu)$. Consequently, light can be modulated by changing the absorption coefficient (intensity modulation) if the wavelength of the incident light is near the fundamental absorption edge. For light having a wavelength longer than that corresponding to the band-gap energy, however, the change in refractive index is dominant and a phase modulator is formed. These changes are proportional to the square of the intensity of the electric field.

5.1.3 Stark Effect

If an atom is placed in electric field, the orbits of its electrons are elliptically deformed and the atomic energy shifts by $\Delta E_{atom} = qdF_{field}$, where d is the eccentricity of the orbit. The emission or absorption wavelength, therefore, changes according to the electric field. This is the first-order (linear) Stark effect. There is also a second-order Stark effect related to deformation of the ground state orbits, in which the energy shift is proportional to the square of electric field intensity.

5.1.4 Quantum-Confined Stark Effect (QCSE)

The photon energy emitted or absorbed in a quantum well structure is given by Eq. (1.182). If the exciton absorption is considered, the transition in optical absorption occurs at a photon energy lower than the band-gap energy (see Section 1.1.3.1), and Eq. (1.182) can be rewritten by subtracting the binding energy of the exciton, E_{ex}:

$$h\nu = E_{gw} + E_{e1} + E_{hh1} - E_{ex}, \tag{5.3}$$

where E_{gw} is the band-gap energy of the well material, and where E_{e1} and E_{hh1} are the electron and hole subband energies [see Fig. 5.2(a)]. If the electric field applied to the quantum well structure is perpendicular to the layers, the ground-state wavefunctions of the electron and hole subbands are separated and pushed toward opposite sides of the well [see Fig. 5.2(b)]. Under the electric field, electron and hole subband energies shift downward and upward, respectively, and then the subband energies, E_e and E_h, are reduced simultaneously. Simultaneously, the binding energy of the exciton, E_{ex}, decreases and a very small change in the band-gap energy, E_{gw}, occurs (Stark effect). Consequently, the effective band-gap energy decreases with increasing electric field strength, and the refractive index also changes. This phenomenon is called the quantum confined Stark effect. In such phenomena, the change in effective band-gap energy related to exciton (exciton absorption) is mainly used for optical modulation. The exciton can be confined to a quantum well even under a strong electric field of about 10^5 V/cm at room temperature because the exciton in the well is surrounded by high potential energy barrier (see section 1.3.1.2). One example of the quantum-confined Stark effect for an InGaAs/InAlAs multiquantum well structure is shown in Fig. 5.3. As the applied voltage increases, the optical absorption edge shifts to longer wavelengths in proportion to the square of the electric field intensity. A shift by a few tens of millielectron volts, for example, occurs when an electric field on the order of 10^4 V/cm is applied to a GaAs/AlGaAs multiquantum well structure.

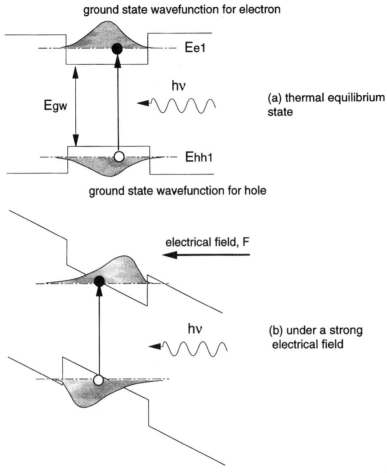

FIGURE 5.2   Schematic diagram of tilting of the energy bands under a strong electric field and of the optical absorption in a quantum well structure.

A similar phenomenon also occurs when a superlattice structure is subjected to a strong electric field: electron and hole wavefunctions and quantized energy levels change in a manner similar to that in which they change in the case of a quantum well structure. This is called the Wannier–Stark effect.[8]

5.1.5   Free Carrier Effect

The absorption coefficient and effective bnad-gap energy vary with the density of the free carrier because of the free carrier absorption and band-filling (or the Burstein–Moss shift) in semiconductors (see Section 1.1.3.1). The refractive index varies simultaneously, and these variations are correlated with each other by the Kramers–Kronig relation. The absorption coefficient due to free

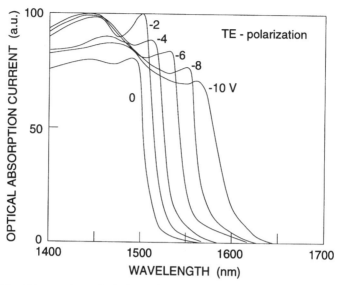

FIGURE 5.3 Changes in optical absorption edge with changes in the bias applied perpendicular to an InGaAs/InAlAs multiquantum well structure. From K. Wakita et al.[7] © 1989 IEEE. Reprinted by permission of the authors and IEEE.

carrier absorption is given by Eq. (1.84) and changes linearly with free carrier density. In the band-filing effect, the effective band-gap energy (that is, the absorption edge) changes with carrier density because electrons and holes fill the conduction and the valence bands from the lower energy level in regular order, respectively. The absorption edge is therefore a function of carrier density, and the change in the refractive index is expressed by the following equation:[9]

$$\Delta n_{\mathrm{r}} = B_{\mathrm{bf}}(\lambda)n, \tag{5.4}$$

where n is the carrier density and $B_{\mathrm{bf}}(\lambda)$ is the coefficient for the band-gap shift. The coefficient is a function of wavelength and increases dramatically as the wavelength approaches the band-gap energy. The absorption coefficient and the refractive index in a semiconductor can be modulated by changing the carrier density.

5.2 BASICS OF OPTICAL MODULATORS

Various kinds of optical modulators based on the physical phenomena described above have been developed as shown in Fig. 5.4. Most semiconductor optical modulators are *pn*-junction diodes because strong electric fields can be generated in a *pn*-junction. Schottky junction diodes are also used sometimes because

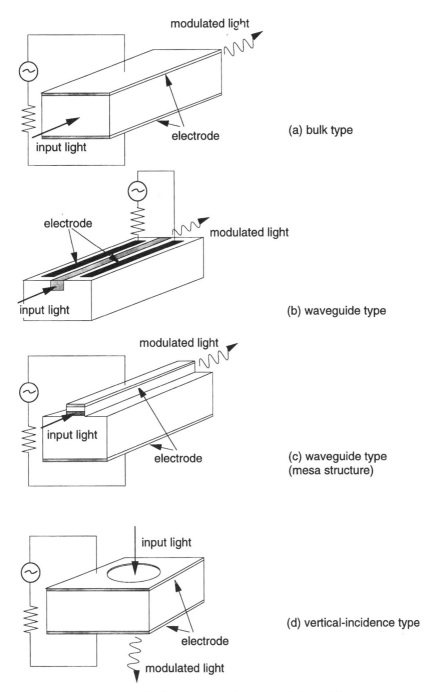

FIGURE 5.4   Basic structures of some optical modulators.

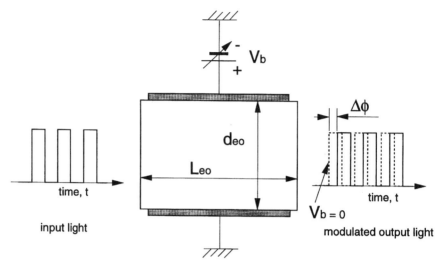

FIGURE 5.5 Schematic illustration of a waveguide type phase modulator.

a strong electric field can also be generated at the Schottky junction. These semiconductor modulators are divided into two types, phase modulation and intensity modulation (electroabsorption) types.

Semiconductor optical modulators frequently use the waveguide structure, in which incident light propagates along the waveguide. This leads to modulation efficiency better than that of bulk-type modulators, and this higher modulation efficiency results in a smaller size of modulator and lower operating voltages. The modulation voltage in the waveguide types is an order of magnitude or more lower than that in bulk types because of the smaller device size. Several kinds of waveguide structures have been fabricated: ridge-waveguide, mesa-structure, and so forth (see Fig. 5.4). The vertical incidence structure is also used in optical modulators, but the resultant modulation depth is not sufficient because the thickness of the modulating region is not so long as that in waveguide types.

5.2.1 Phase Modulators

5.2.1.1 Electrooptic Crystal Modulators A phase modulator ordinarily consists of an electrooptic crystal having a large electrooptic coefficient, such as $LiNbO_3$. The basic constitution of the phase modulator is shown in Fig. 5.5. This type of modulator has no *pn*-junction but is the basic modulator and understanding it helps us understand the semiconductor phase modulator. The refractive index of an electrooptic modulator changes under a changing electric field. If the bias voltage is V_b, the electric field is $F_{field} = V_b/d_{eo}$, where d_{eo} is the thickness of the modulator. Consequently, the change in refractive index is given by rewriting Eq. (5.2b) as

$$\Delta n_r = n_r^3 r_1 V_b / 2 d_{eo}, \tag{5.5}$$

and if the length of the modulator is L_{eo}, the phase shift, $\Delta \phi_{eo}$, is given by

$$\begin{aligned} \Delta \phi_{eo} &= (2\pi/\lambda_0) \Delta n_r L \\ &= \pi L n_r^3 r_1 V_b / \lambda_0 d_{eo}, \end{aligned} \tag{5.6a}$$

where λ_0 is the wavelength of incident light in a vacuum. The phase shift is proportional to the bias voltage. When the modulator is the waveguide type shown in Fig. 5.4(b), only light propagating along the waveguide is modulated. The optical confinement factor to the waveguide, Γ (see Section 3.2.4.1), therefore, has to be multiplied to Eq. (5.6a) as

$$\Delta \phi_{eo} = (2\pi/\lambda_0) \Gamma \Delta n_r L. \tag{5.6b}$$

The phase of the input light is therefore modulated by $\Delta \phi_{eo}$ according to Eqs. (5.6a) and (5.6b). The half-wavelength voltage, V_π, which is the voltage shifting the phase by π (half-wavelength), is frequently used as a performance measure for the phase modulator, and lower half-voltage is usually better. If the half-wavelength voltage is considered in Eq. (5.6a), the voltage is given as V_b at $\Delta \phi_{eo} = \pi$:

$$V_\pi = \lambda_0 d_{eo} / L n_r^3 r_1. \tag{5.7a}$$

The half-wavelength voltage is thus proportional to the wavelength of incident light modulated and inversely proportional to the length of the modulator. If the electric field is applied parallel to the incident light, the thickness of the modulator is equal to the length, $d_{eo} = L$, and

$$V_\pi = \lambda_0 / n_r^3 r_1. \tag{5.7b}$$

In this case, the half-wavelength voltage is almost independent of the scale of the modulator.

5.2.1.2 Semiconductor Modulators

The behaviors of modulators composed of semiconductors are more complicated than those of modulators composed of dielectric substances. The various effects described in Section 5.1 occur, and the simultaneous changes in refractive index and absorption coefficient under electric fields occur as shown in Fig. 5.6. As the electric bias increases, the optical absorption and the phase changes gradually become larger. The changes tend to be strongly enhanced when the wavelength of the light modulated approaches the absorption edge (in Fig. 5.6 this corresponds to shorter wavelengths). The

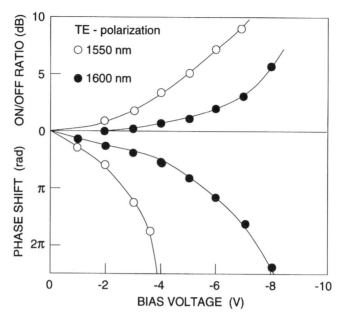

FIGURE 5.6 Intensity and phase changes as functions of bias voltage in an InGaAs/InAlAs multiquantum well modulator. From K. Wakita et al.[7] © 1989 IEEE. Reprinted by permission of the authors and IEEE.

basic mechanism of the phase modulation, however, is similar to that in the modulators composed of electrooptic crystals. Here the wavelength modulated needs to be a wavelength longer than that corresponding to the fundamental (and exciton) absorption edge in order to suppress the mixture of intensity modulation.

Bulk Type   A schematic diagram of a GaAs/AlGaAs phase modulator is shown in Fig. 5.7. The *i*-layer (see Section 4.3.1) is inserted between *p*- and *n*-GaAs layers in order to widen the region with a strong electric field under reverse bias. A similar structure without the *i*-layer is also frequently used. The GaAs layers form the waveguide, and the propagating light confined along the guide is modulated efficiently. Not only the propagating light is confined perpendicular to the *pn*-junction plane, but it is often confined in the direction parallel to the junction plane by a ridge waveguide structure or buried heterostructure. These waveguide structures are quite similar to those of laser diodes. They are usually grown on a (001) substrate, and the electric field is applied in the ⟨001⟩ direction. The refractive index of these semiconductor modulators can be expressed with the sum of the linear term and quadratic term of the electric field intensity as follows:[9]

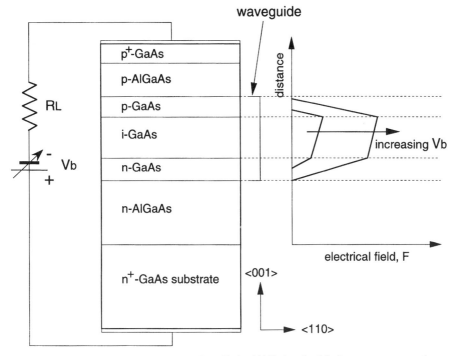

FIGURE 5.7   Schematic diagram of a GaAs/AlGaAs double-heterostructure phase modulator.

$$\Delta n_{\mathrm{r}} = (n_{\mathrm{r}}^3/2)\Gamma[\pm r_1 F_{\mathrm{field}} + s_{\mathrm{p\text{-}FK}}(\lambda)F_{\mathrm{field}}^2], \tag{5.8}$$

where the sign, $\pm$, depends on the direction of propagation in the active layer. The sign is $+$ in the $\langle 110 \rangle$ direction, which is normal direction in actual modulators.

The linear term is determined mainly by the Pockels effect and is dominant in bulk semiconductors. If there is a free carrier effect, the term indicated by Eq. (5.4) is added. The quadratic term here is mainly due to the Franz–Keldysh effect, and the coefficient, $s_{\mathrm{q\text{-}FK}}$, is a function of the wavelength of the light modulated and is very large near the wavelength corresponding to the bandgap energy. For GaAs the coefficients, $s_{\mathrm{q\text{-}FK}}$ (1.06 μm) $\approx 4.3 \times 10^{-16}$ cm^2/V^2 and $s_{\mathrm{q\text{-}FK}}$ (1.3 μm) $\approx 1.3 \times 10^{-16}$ cm^2/V^2, are estimated.[9] The phase shift due to the refractive index change is also a result of several factors but is roughly given by substituting Eq. (5.8) into Eq. (5.6b):

$$\Delta\phi_{\mathrm{eo}} = (2\pi/\lambda_0)\Gamma L(n_{\mathrm{r}}^3/2)\Gamma[\pm r_1 F_{\mathrm{field}} + s_{\mathrm{q\text{-}FK}}(\lambda)F_{\mathrm{field}}^2]. \tag{5.9}$$

Multiquantum Well Type   The structure of the quantum well modulators is similar to that of the bulk modulator shown in Fig. 5.7, in which the bulk layer
is replaced with a multiquantum well structure having a few tens of periods.
The electric field is applied perpendicular to the *pn*-junction plane and quantum
well layer, which is ordinarily the (001) plane. In a quantum well the dominant
electrooptic effect is the quantum-confined Stark effect. Consequently, the electrooptic effects are different from those in the bulk semiconductor modulators,
although effects similar to those in the bulk semiconductor modulators are also
generated. The quantum-confined Stark effect is proportional to the square of
electric field intensity (i.e., it is a quadratic effect). The change in refractive
index is given by

$$\Delta n_\mathrm{r} = n_\mathrm{r}^3 s_\text{q-QCSE} F_\text{field}^2 / 2, \tag{5.10}$$

where $s_\text{q-QCSE}$ is the quadratic electrooptic coefficient. For InGaAs/InAlAs
quantum well structures, for example, the coefficient is on the order of 10^{-14}
$\mathrm{cm}^2/\mathrm{V}^2$.[10] The phase change due to the QCSE is therefore given by

$$\Delta \phi_\text{eo} = (\pi/\lambda_0) \Gamma n_\mathrm{r}^3 s_\text{q-QCSE} F_\text{field}^2 L. \tag{5.11}$$

The refractive index change in strained quantum well structures (see Sections
1.3.2 and 3.3) is enhanced because the band structure varies with the strain. The
valence band edge corresponds to the heavy hole band in compressively strained
structures and corresponds to the light hole band in tensilely strained structures.
The coupling of the TE and TM modes depends on the valence band structure,
so the effect of the electric field in strained quantum well structures is slightly
different from that in lattice-matched quantum well structures. In InGaAs/InP
systems the change in refractive index in compressively strained quantum well
structures is greater than that in lattice-matched quantum well structures, while
that in the tensilely strained structures is less than that in the lattice-matched
structures.[11] In addition, the wavelength range of an optical modulator can be
widened by using the strained quantum well structures because the band-gap is
controlled to some extent in the strained structures.

5.2.2 Intensity Modulators

Intensity modulators can be formed by combining a phase modulator with
an interferometer or by using the electroabsorption effects (such as the
Franz–Keldysh effect and quantum-confined Stark effect). Both types of modulators have been widely used in optical fiber transmission systems. Intensity
modulation can also be accomplished by using optical switching devices, such
as directional couplers and *X*- or *Y*-branches, in which the change in refractive
index is used for switching.

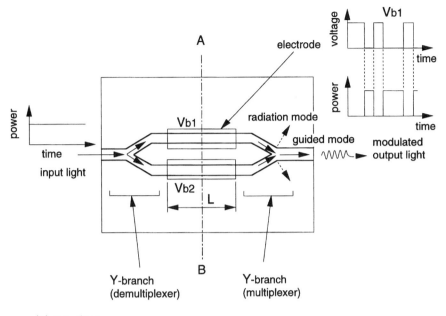

(a) top view

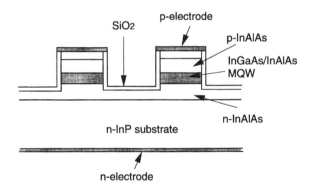

(b) cross-sectional view (A - B)

FIGURE 5.8   An example of Top and cross-sectional views of a Mach–Zehnder optical modulator.

5.2.2.1   *The Mach–Zehnder Type*

A schematic diagram of an InGaAs/InAlAs multiquantum well Mach–Zehnder type of optical modulator[11] is shown in Fig. 5.8. This interferometer consists of two Y-branches and two phase-modulation arms, all of which are ridge-waveguide structures. The input light is divided into two paths by the Y-branch (demultiplexer), at which the light is usually bisected. The light is guided to the two arms and the phase of the light in each path is modulated by applying voltage signals to electrodes. This part acts as a phase modulator. At the Y-branch (multiplexer), the phase modulation is converted into intensity modulation. When the phase of the light from one path is the same as that of the light from the other path, the light is added again and propagates along the waveguide toward the output port. If the phase difference between the light propagating each arm is π, they cancel each other and no optical power is emitted from the output port. Consequently, the half-wavelength voltage in Eq. (5.7a) gives the minimum output power from the modulator when only one arm is modulated. Here, the light never disappears at the multiplexer, and it changes from the guided mode to the radiation mode and is emitted to the outside of the waveguide. As described above, the phase modulation can be converted into intensity modulation in the Mach–Zehnder type modulator. If there is no optical absorption within the modulator, the ratio of output power to input power is given by the following equation:[12]

$$P_{out}/P_{in} = (1/2)\{1 + \cos[\Delta\phi(V_{b1}) - \Delta\phi(V_{b2}) + \phi_0]\}, \qquad (5.12)$$

where ϕ_0 is the phase difference between the two arms when no electric field is applied and $\Delta\phi(V_{b1})$ and $\Delta\phi(V_{b2})$ are the phase changes induced by the voltages V_{b1} and V_{b2} applied to each arm. The values of $\Delta\phi(V_{b1})$ and $\Delta\phi(V_{b2})$ are calculated using Eqs. (5.6b) and (5.10), and the ratio in Eq. (5.12) is a function of the refractive index change and the length of the modulating part and varies periodically with the applied voltage. If there is optical absorption within the arms, Eq. (5.12) is rewritten as

$$
\begin{aligned}
P_{out}/P_{in} = {}& (1/4)[\exp(-\Gamma\Delta\alpha_{ab1}L) + \exp(-\Gamma\Delta\alpha_{ab2}L)] \\
& + (1/2)\exp[-\Gamma(\Delta\alpha_{ab1} + \Delta\alpha_{ab2})L/2]\cos[\Delta\phi(V_{b1}) - \Delta\phi(V_{b2}) + \phi_0],
\end{aligned}
$$
$$(5.13)$$

where $\Delta\alpha_{ab1}$ and $\Delta\alpha_{ab1}$ are the changes in optical absorption coefficient and L is the length of the arms. The increase in absorption coefficient is ideally zero but actually is not because the mechanisms of the change in refractive index, except for the electrooptic effect, are associated with the effective absorption edge (the band-gap energy).

5.2.2.2 The Electroabsorption Type The electroabsorption type of modulator modulates light by changing the optical absorption coefficient under an electric field. The electrooptic effects used in the electroabsorption-type modulators are therefore mainly the Franz–Keldysh effect, the quantum-confined Stark effect, and the Wannier–Stark effect. Some basic structures of the electroabsorption type modulators are shown in Fig. 5.9. In the vertical type, ridge-waveguide type, and monolithically integrated type, the electric field applied to the optical absorbing layer is perpendicular to the *pn*-junction plane [(001) plane]. In the vertical type, the GaAs substrate is etched off to reduce the light absorption. If

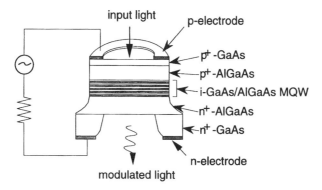

(a) mesa-etched vertical type

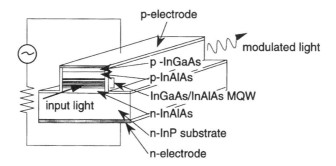

(b) ridge waveguide type

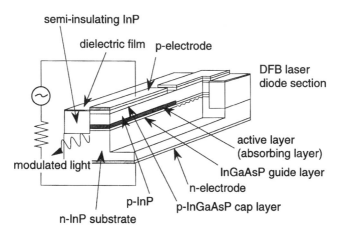

(c) monolithic integrated type

FIGURE 5.9 Schematic diagram of electroabsorption type modulators: (a) mesa-etched vertical type, (b) ridge waveguide type, and (c) monolithic integrated type.

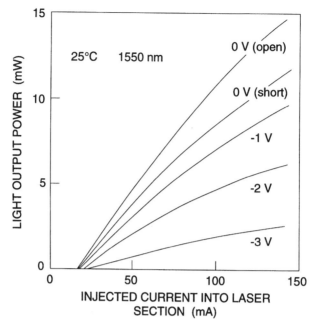

FIGURE 5.10 Current–light output power characteristics of a modulator monolithically integrated with DFB laser diode.

the substrate is transparent to the light, no etching is required. As in the phase modulators the impurity concentration in the absorbing layer is kept low, and the region with high electric field is widened at around the *pn*-junction. The wavelength of modulated light is set at around the absorption edge, and the absorption coefficient is changed according to the intensity of the electric field (see Fig. 5.3). The input light can be modulated by changing the absorption coefficient. The absorbed light is converted into photocurrent in the modulator and the photocurrent flows in the bias circuit in a manner similar to that in which it flows in photodiodes. Modulators employing the quantum well structures shown in Fig. 5.9(a) and 5.9(b) mainly use the quantum-confined Stark effect. The monolithically integrated modulator shown in Fig. 5.9(c), however, uses the Franz–Keldysh effect or the quantum confined Stark effect. The shift of the absorption edge is larger for the quantum-confined Stark effect than it is for the Franz–Keldysh effect. These modulation mechanisms, of course, depend on the structure of the light-absorbing layer.

The current-light output power characteristics of a monolithically integrated modulator in which a DFB laser diode is integrated are shown in Fig. 5.10 for various reverse voltages biased to the modulator. When the bias circuit is connected to the modulator ("short" in the figure), the light output power is less than that it is without the circuit ("open" in the figure). The lower power is due to the effect of the photo-induced carriers. The modulator without the bias circuit absorbs the light from the DFB laser diode and photo-induced carriers are separated by the *pn*-junction. The separated electrons and holes are initially stored in *n*-type and *p*-type regions, and then fill the conduction and the valence bands from the lower energy level in regular order. The carriers also bias the *pn*-junction in the forward direction. In such a situation (the Burstein–Moss shift), the optical absorption in the modulator section is suppressed. If the bias circuit is connected to the modulator but no electric bias is applied, the behavior of the modulator is quite similar to that of solar cells (see Section 4.2). The photo-induced carriers introduced by the optical absorption in the modulator flow in the bias circuit as the photocurrent. As the electric field intensity (reverse voltage) increases, the output power decreases because of the increase in the absorption coefficient, while the photocurrent increases. Similar phenomena are also characteristic of the modulator structures shown in Fig. 5.9(a) and 5.9(b).

5.2.3 Modulation Characteristics and Frequency Response

The previous sections have discussed the mechanisms of the phase and the intensity modulation under dc bias. The essential characteristics of a modulator, however, are of course the modulation characteristics and frequency response. These characteristics determine the performance of modulators in equipment and systems.

5.2.3.1 Extinction Ratio (Modulation Index) The extinction ratio is an important parameter in the intensity modulation performed by the phase modulator combined with an interferometer and by the electroabsorption type modulator. The extinction ratio is given by the on–off ratio which corresponds to the ratio of the intensity of high output power to that of low output power from the modulators. For the intensity modulators formed by combination of a phase modulator with a Mach–Zehnder interferometer, the extinction ratio in the ideal case is given by Eq. (5.12) and by Eq. (5.13). The small value in Eqs. (5.12) and (5.13) under electric bias gives the high extinction ratio. Consequently, the extinction ratio is influenced by the optical confinement factor, the length of the phase modulator section, and the absorption coefficient change due to the electric field. The change in the absorption coefficient of the arm is determined by the wavelength of input light. The wavelength is usually set so that the change in refractive index is large, where the change in refractive index becomes large as the wavelength approaches the value corresponding to the band-gap energy. At the wavelength range, however, the absorption coefficient is also large. Examples of the modulation characteristics of a multiquantum well Mach–Zehnder modulator are shown in Fig. 5.11.[10] The multiquantum well structure in this modulator consisted of 20 periods of undoped InGaAs/InAlAs, and the absorption edge due to the exciton absorption was 1.47 μm. The length of the modulator section was 1.45 mm and only one arm is biased. As the wavelength of light modulated approaches the absorption edge, the half-wavelength voltage decreases because the change in the refractive index increases. The damping of

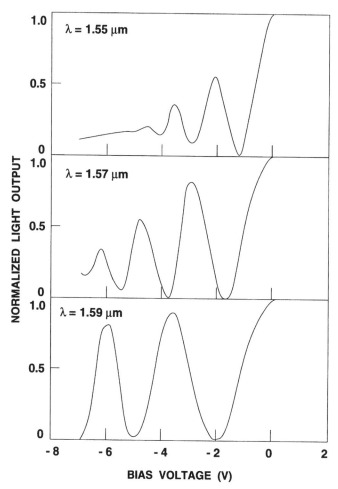

FIGURE 5.11 Examples of modulation characteristics (TE mode) for an InGaAs/InAlAs multiquantum well Mach–Zehnder type of intensity modulator. From S. Nishimura et al.[10] © 1992 IEEE. Reprinted by permission of the authors and IEEE.

the periodic change in output power, however, is large because of the increase in the absorption coefficient. An extinction ratio larger than 20 dB can be obtained in such an intensity modulator.

The extinction ratio for the electroabsorption type modulators is described by the following relation:

$$P_{out}/P_{in} \propto \exp(-\Gamma \Delta \alpha L), \tag{5.14}$$

where $\Delta \alpha$ is the change in the absorption coefficient under the applied electric field. The extinction ratio for electroabsorption type modulators is also influenced by the optical confinement factor and the length of the modulator. An extinction ratio of more than 20 dB is obtained in MQW modulators and an extinction ratio of more than 10 dB is obtained in monolithically integrated modulators.[13,14]

5.2.3.2 *Efficiency, Insertion Loss, and Chirping*

Low operation voltage and low insertion loss are favorable when the modulators are used in equipment and systems. In addition, suppressing a chirping of modulated light is required to keep signal transmission quality high.

The half-wavelength voltage is frequently used as a measure of the operation voltage for phase modulators, and a voltage giving a required extinction ratio is frequently used as a measure of the operation voltage for intensity modulators. For the operation voltage to be reduced, the structures must be arranged so that the physical effects causing modulation, such as the Franz–Keldysh effect and the quantum-confined Stark effect, are maximized. Waveguide structures are therefore indispensable. In semiconductors the change in absorption coefficient is larger than the refractive index change if the wavelength is near the one corresponding to the band-gap energy. The operation voltage, therefore, tends to be higher in phase modulators than intensity modulators.

The insertion loss, which is the loss between the input and output ports, is largely determined by the propagation loss, coupling loss, and reflection loss. The propagation loss corresponds to the absorption determined by the material properties described in Section 1.1.3.1. The coupling loss originates from the mode (size)-mismatching between the input or output port of the modulator and a fiber or waveguide. This problem is quite similar to the one in the case of laser diodes. The reflection loss is caused by the reflection of light at the input and output ports and is greatly reduced by antireflective coating films (see Section 6.2.2.3).

Chirping is related to the modulation mechanism and is an essential problem in semiconductor modulators using effects other than the electrooptic effect. It is induced by the simultaneous changes in refractive index and absorption coefficient under voltage biasing, and the wavelength of the modulated light fluctuates according to these changes. The degree of chirping is expressed by the linewidth enhancement factor, α, which is the ratio of the change in refractive index to the change in absorption coefficient (see Section 3.5.3.3). Chirping can be sufficiently suppressed in semiconductor modulators by setting the wavelength and absorption edge (band-gap energy) appropriately. This cannot be done in the case of laser diodes because they operate at the band-gap energy and the refractive index changes near the absorption edge are very large. The linewidth enhancement factor in optical semiconductor modulators can easily be reduced to less than 1.

5.2.3.3 *Frequency Response*

A few factors influence the frequency response of modulators. The response of change in refractive index and absorption coefficient is very high under modulation. The frequency response of the change in absorption coefficient is, for example, at around 100 GHz in III-V compound semiconductor intensity modulators using the electroabsorption effect. This response is determined by the plasma oscillation frequency. In most cases, the frequency response is not determined by the physical effects used for modulation but is limited by the structure (scale) of modulators. The scale factors limiting the frequency response are the CR-time constant and the lightwave transit time through the modulator.

Conventional Modulators: Lumped-Parameter Type   The modulators described in this chapter are lumped-parameter types similar to laser diodes and photodiodes. The equivalent circuit of the modulators, especially electroabsorption types, is similar to that of pin-photodiodes. A simplified bias circuit of modulators is shown in Fig. 5.12. The modulator is equivalently expressed with a capacitance, C_m. The impedance of the modulators gradually decreases as the modulation frequency increases and thus the intensity of the modulated electric field decreases. The 3-dB bandwidth defined by the frequency at which the voltage decreases by $2^{-1/2}$ from the value at low frequencies or under dc can be expressed as

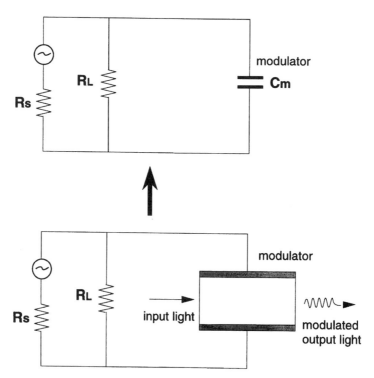

FIGURE 5.12   Simplified modulation circuit of optical semiconductor modulator.

$$f_{3\text{-dB/CR}} = 1/\pi R_L C_m, \tag{5.15}$$

where R_L is the load resistance. The CR-time constant of modulators can be reduced in a manner similar to that of laser diodes and photodiodes.

As the modulation frequency increases, the reciprocal of the frequency approaches the transit time for which the modulated light passes through the modulator. The modulating signal voltage begins to change before the modulated light has passed through the modulator, and the shape of the modulated output light is deformed by other voltage signals. The transit time for a light-wave, τ_m, is given by

$$\tau_m = n_r L_m / c_0, \tag{5.16}$$

where n_r and c_0, are the effective refractive index of the waveguide and the velocity of light in a vacuum. The 3-dB bandwidth is therefore given by the following equation:[2]

$$f_{3\text{-dB/tr}} \approx (1.4/\pi)/\tau_m = 1.4 c_0/\pi n_r L_m. \tag{5.17}$$

The refractive index of the semiconductor waveguide is about 3.2, and thus the bandwidth calculated for a modulator 500 μm long is about 84 GHz. The transit time of lightwave essentially determines the frequency response of modulators if they are used in the frequencies at around or more than 100 GHz. This essential limitation for the frequency response can be improved in traveling wave modulators.

Traveling Wave Modulators   The basic concept of the traveling wave modulator is shown in Fig. 5.13. The electrode is used as a transmission line, and

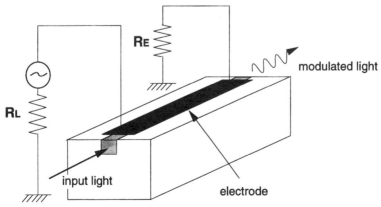

FIGURE 5.13   A simplified bias circuit for the traveling wave type of modulators.

the lightwave and modulating microwave interact in the distributed parameter circuit of the modulator. If the velocity of light traveling along the waveguide is the same as that of the modulating microwave traveling along the electrode, the lightwave is modulated without any deformation. The frequency response in such a modulator can be set at a very high range, and the 3-dB bandwidth can be given by the following equation:[15]

$$f_{3\text{-dB/t}} \approx 1.4 c_0 / \pi |n_r - n_{ele}| L_m, \tag{5.18}$$

where n_{ele} is the microwave effective refractive index of the modulator. The bandwidth is limited by the difference between the times of lightwave and microwave propagation within the modulator. The actual 3-dB bandwidth is also limited by the frequency response of the package, the electric connections, and so forth.

Other Factors   The space charge effect degrades the frequency response of electroabsorption modulators when the input light power is very high. The space charge effect appears in photodiodes under high input power. The photo-induced carriers distribute homogeneously within the absorbing layer under an electric field. Just after the modulation voltage signal is turned off, the electrons and the holes are separated by the electric field originating from the *pn*-junction. The Coulomb's force between them reduces the intensity of the electric field. As a result, the drift velocity of the photo-induced carriers in the central part of the absorbing layer decreases and a tail appears in the modulated waveform and the photocurrent. In addition, the photo-induced carriers, especially holes, tend to be stored at the heterointerfaces and also degrade the frequency response. These problems are solved by improving the heterostructures.[7,16]

Modulation Bandwidth and Modulation Power   The modulation bandwidth and the electric input power required for modulation are determined by several factors. Large bandwidth and low electric input power are essential characteristics of excellent modulators, but it is difficult to satisfy both simultaneously. The change in refractive index and absorption coefficient become large as the length of the modulator increases, but the operating voltage also increases. The bandwidth therefore tends to be inversely proportional to the electric input power.

5.2.3.4 *Temperature Characteristics*   Modulation characteristics are very sensitive to changes in temperature. From the application point of view, however, the temperature characteristics of semiconductor optical modulators are not actually important because the modulators can usually be used under temperature control, keeping the band-gap constant and stabilizing the modulation characteristics. If the temperature of the modulators is not controlled, the band-gap energy shifts when the temperature changes, and the wavelength spacing setting between modulated incident light and the fundamental absorption edge is not kept. The extinction ratio then changes under a constant signal voltage. For a 1550 nm-band electroabsorption type modulator monolithically integrated with a DFB laser diode, the fundamental absorption edge shifts by about 0.4 nm/°C while the shift of lasing wavelength is only about 0.1 nm/°C (see Section 3.4.1.3). This means that the wavelength spacing between the lasing wavelength and the absorption edge varies by 0.3 nm/°C. A large amount of light is absorbed in the modulator when the temperature is high, and then the extinction ratio decreases. Temperature control is therefore indispensable to the use of semiconductor optical modulators.

REFERENCES

1. See, for example, J. E. Zucker, J. L. Marshall, T. Y. Chang, N. J. Sauer, C. A. Burrus, and J. C. Centanni, 15 GHz bandwidth quantum well electron transfer intensity modulator at 1.55 μm, *Electron. Lett.*, **28**, 2206, 1992.

2. A. Yariv, *Quantum Electronics*, 2nd ed., John Wiley & Sons, New York, 1975.

3. P. Bhattacharya, *Semiconductor Optoelectronic Devices*, Prentice Hall, Englewood Cliffs, New Jersey, 1994.

4. I. P. Kaminov, Measurements of the electrooptic effect in CdS, ZnTe, and GaAs at 10.6 microns, *IEEE J. Quantum Electron.*, **QE-4**, 23, 1968.

5. T. E. Van Eck, L. M. Walpita, W. S. C. Chang, and H. H. Wieder, Franz–Keldysh electroreflection and electroabsorption in bulk InP and GaAs, *Appl. Phys. Lett.*, **48**, 451, 1986.

6. S. Aytac and A. Schlachetzki, Franz–Keldysh effect in InP, *J. Opt. Commun.*, **6**, 82, 1985.

7. K. Wakita, O. Mitomi, I. Kotaka, S. Nojima, and Y. Kawamura, High-speed electrooptic phase modulators using InGaAs/InAlAs multiple quantum well waveguides, *IEEE Photon. Technol. Lett.*, **1**, 441, 1989.

8. See for example, E. Bigan, M. Allovon, M. Carre, C. Braud, A. Carenco, and P. Voisin, Optimization of optical waveguide modulators based on Wannier–Stark localization: An experimental study, *J. Quantum Electron.*, **28**, 214, 1992.

9. S. S. Lee, R. V. Ramaswamy, and V. S. Sundaram, Analysis and design of high-speed high-efficiency GaAs-AlGaAs double-heterostructure waveguide phase modulator, *J. Quantum Electron.*, **27**, 726, 1991.

10. S. Nishimura, H. Inoue, H. Sano, and K. Ishida, Electrooptic effects in an InGaAs/InAlAs multiquantum well structure, *Photon. Technol. Lett.*, **4**, 1123, 1992.

11. J. Pamulapati, J. P. Loehr, J. Singh, and P. K. Bhattacharya, Refractive index and electro-optic effect in compressive and tensile strained quantum wells, *J. Appl. Phys.*, **69**, 4071, 1991.

12. R. A. Soref, D. L. McDaniel, Jr., and B. R. Bennett, Guide-wave intensity modulators using amplitude-and-phase perturbations, *J. Lightwave Technol.*, **6**, 437, 1988.

13. K. Sato, I. Kotaka, K. Wakita, Y. Kondo, and M. Yamamoto, Strained InGaAsP MQW electroabsorption modulator integrated DFB laser, *Electron. Lett.*, **29**, 1087, 1993.

14. H. Soda, M. Furutsu, K. Sato, N, Okazaki, S. Yamazaki, H. Nishimoto, and H. Ishikawa, High-power and high-speed semi-insulating BH structure monolithic electroabsorption modulator/DFB laser light source, *Electron. Lett.*, **26**, 9, 1990; M. Suzuki, Y. Noda, H. Tanaka, S. Akiba, Y. Kushiro, and H. Isshiki, Monolithic integration of InGaAs/InP distributed feedback laser and electroabsorption modulator by vapor phase epitaxy, *J. Lightwave Technol.*, **5**, 1277, 1987.

15. R. G. Walker, High-speed III-V semiconductor intensity modulators, *IEEE J. Quantum Electron.*, **27**, 654, 1991.

16. M. Suzuki, H. Tanaka, and Y. Matsushima, InGaAsP electroabsorption modulator for high-bit-rate EDFA system, *IEEE Photonic. Technol. Lett.*, **4**, 586, 1992.

CHAPTER SIX

Device Fabrication and Packaging

Chapters 2, 3, 4, and 5 discuss the basic characteristics and the device physics of LEDs, laser diodes, photodiodes, and semiconductor optical modulators. As shown in Fig. 6.1, these optoelectronic *pn*-junction devices are made of semiconductor materials, packaged, installed, and then used in equipment and systems. Various kinds of technologies are required from the material stage to the application stage, and the main technologies are those related to crystal growth, device processing, packaging, and installing. Each technology strongly influences the performance of the device chip itself, the module, and the equipment (or system). The device characteristics depend on the quality of semiconductor crystal, the surface and interface state density, parasitic capacitance of package, and so forth. The device fabrication technologies and packaging technologies are very important with regard to device performance. The characteristics of the products are ordinarily different even if the same materials are used for fabrication and if the fabricating and packaging conditions are similar ones. The appropriate conditions for fabrication and packaging are generally different for different production lines, but seeing the device fabrication and packaging technologies is quite important for understanding the devices, their performance, and their reliability. This chapter briefly describes some basic and important device fabrication and packaging technologies.

6.1 CRYSTAL GROWTH

Crystal growth is the first step in device fabrication. The active layers—which are the light-emitting layers in LEDs and laser diodes, the light-absorbing layers in photodiodes, and the light-modulating layers in modulators—are grown, with cladding and other layers, on the semiconductor substrate. Device performances

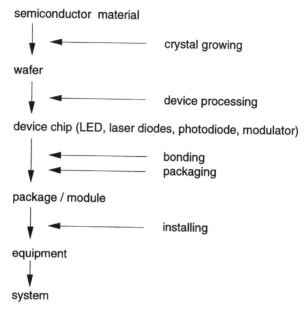

FIGURE 6.1   General flow from device fabrication to application.

are strongly influenced by the crystal quality, which is in turn determined by the growth technologies. Most of these devices consist of heterostructures as described in Chapters 2 through 5. When the heterostructures are formed, the lattice constant of each semiconductor material has to nearly coincide with that of the others. This requirement determines the suitable combinations of materials for the heterostructures (see Fig. 1.48). Metal-organic vapor-phase epitaxy (MOVPE or metal-organic chemical vapor deposition), liquid-phase epitaxy (LPE), and molecular beam epitaxy (MBE) are crystal growth methods frequently used in making optoelectronic *pn*-junction devices.[1–4]

6.1.1   Liquid-Phase Epitaxy

Liquid-phase epitaxy is used in making various semiconductor devices, and the crystal growth occurs near thermal equilibrium. When a supersaturated solution of semiconductor material comes into contact with a single-crystalline substrate, a lattice-matched epitaxal layer grows on the substrate. Graphite sliders are ordinarily used for forming multilayer structures, such as double heterostructures (see Fig. 6.2). The substrate set on the slide in a quartz reactor is successively slid under the solutions corresponding to each layer, and this procedure is performed automatically under computer control. A flow of H_2 prevents oxidation of the solutions and the substrate, and the growth is terminated by sliding the substrate out from under the solution. Temperature control is very important in LPE and is related to the growth rate, crystal quality, and so forth. The tem-

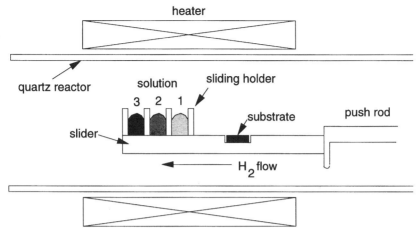

FIGURE 6.2 Basic setup for LPE.

perature over a length of about 30 cm in the reactor must be controlled with a precision of about ±0.1°C. During crystal growth, the temperature of the solution and substrate is set below the saturation temperature of the solutions by using the heater surrounding the reactor and is decreased at 0.1–1.0°C/min (see Fig. 6.3). This decrease in the temperature is required for the precipitation of each layer on the substrate. The temperature of the growth systems is the driving force of crystal growth in LPE, and the thickness of a grown layer is determined by the temperature and the length of time in which the substrate and solution are

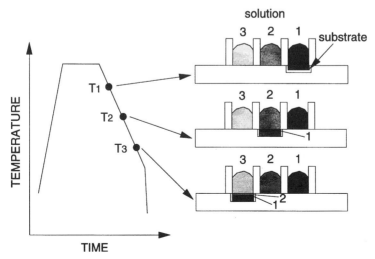

FIGURE 6.3 Formation of multilayer structure in LPE.

in contact. The temperature is, for example, set between 800°C and 850°C for AlGaAs material systems and between 600°C and 650°C for InGaAsP material systems.

The LPE can easily produce high-quality crystals and interfaces between the epitaxal layers, mainly because the crystal growth takes place under conditions close to thermal equilibrium. High-quality here simply means low-defect density and homogeneous composition (without any segregation). The LPE process is particularly suitable for regrowth on a chemically or physically damaged surface, such as one that has been etched, although crystal growth scarcely occurs during LPE if the surface is severely damaged or contaminated. The damaged layer on a slightly damaged surface tends to be wiped off by the solution, and then high-quality crystal grows on the cleaned surface. This results in a high-quality interface and is important for device structure fabrication. In addition, LPE is suitable for growing thick epitaxial layers quickly because the growth rate is high.[5]

6.1.2   Metal-Organic Vapor-Phase Epitaxy

Metal-organic vapor-phase epitaxy occurs far from the thermal equilibrium state. There are two types of MOVPE, atmospheric-pressure and low-pressure, and a basic setup for low-pressure MOVPE is shown in Fig. 6.4. In the low-pressure MOVPE, the reactor pressure is kept low by exhausting systems. The working pressure is set at 0.1–0.5 atm. The group-III metal-organic compounds,

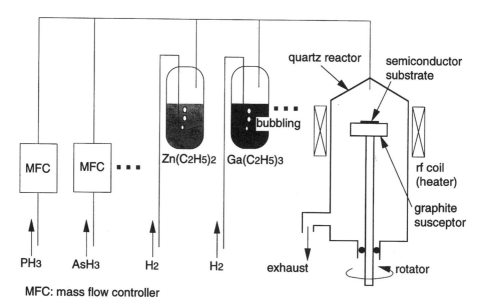

FIGURE 6.4   Basic setup for low-pressure MOVPE.

such as Ga(C$_2$H$_5$)$_3$, In(C$_2$H$_5$)$_3$, Al(C$_2$H$_5$)$_3$, and the group-V hydrides, such as AsH$_3$ and PH$_3$, are supplied to the quartz rector. In(C$_2$H$_5$)$_3$ and Ga(C$_2$H$_5$)$_3$ are liquids having appreciably high vapor pressures at room temperature; they are bubbled with H$_2$ and supplied as gases. Dopants, such as Zn and S, are also supplied as organic compounds or as hydrides, for example, as Zn(C$_2$H$_5$)$_2$ or H$_2$S. The substrate in the reactor is heated by the radio-frequency (rf) heater to about 700°C for AlGaAs material systems and to about 650°C for InGaAsP material systems, and the metal-organic compounds and hydrides introduced decompose on the hot substrate. The decompositions may produce the group-III and group-V elemental species, and then an epitaxial layer composed of those species will grow on the substrate. The composition of the epitaxial layer is changed by adjusting the gas flow.

MOVPE has some advantages over LPE. Large-scale epitaxial growth, on a 2-inch or 3-inch wafer, of an epitaxial layer of uniform composition and thickness can be accomplished easily. Consequently, MOVPE is suitable for the growth of thin epitaxial layers, such as those of (strained) quantum well structures, and for mass production. For increasing the controllability of the thickness and the sharpness of the heterointerface, low-pressure MOVPE (see Fig. 6.4) is frequently used. The interfaces between the epitaxial layers are excellent if the layers are grown sequentially, but regrowth on a chemically or physically damaged surface often results in a defective interface. The epitaxial layer grows on such defective surfaces because MOVPE occurs far from the thermal equilibrium state and this often causes problems, reducing the reliability of the buried heterostructure type of laser diodes (see Section 7.2.1.4).

6.1.3 Molecular Beam Epitaxy

Like MOVPE, MBE occurs far from the thermal equilibrium state. A basic constitution for MBE is illustrated in Fig. 6.5. Atoms or molecules in the effusion cells are evaporaed by heating and shot up to the substrate, where they are directly deposited on the heated substrate in a high vacuum. The group-III elements are adsorbed by themselves but the group-V elements are accompanied by the group-III elements. The high vacuum is therefore needed to lengthen the mean free path of the atoms or molecules and the base pressure is controlled at around or less than 10^{-7} Pa. The epitaxial layers composed of the atoms or molecules grow on the substrate, and the doping and the composition of the layers are changed by controlling the kinds of deposited atoms or molecules. The controlling is carried out by opening or closing the shutters near the effusion cells. MBE can also produce sharp interfaces and thin layers with a precision of atomic monolayer.

When compared with LPE, the advantages and disadvantages of MBE are similar to those of MOVPE. In addition, because MBE is carried out under high vacuum conditions, in-situ observations can be carried out using Auger electron spectroscopy (AES), electron probe microanalysis (EPMA), reflection high energy electron diffraction (RHEED), and so forth.

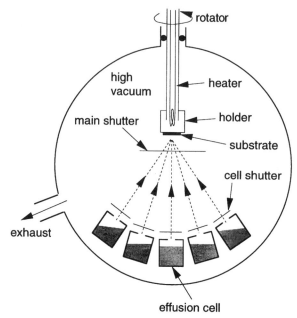

FIGURE 6.5   Basic setup for MBE.

6.2   DEVICE PROCESSES

After the active region for the device is produced by the epitaxial growth, fabrication processes start for producing device chips. This device processing strongly influences the performance, and especially the reliability, of optoelectronic pn-junction devices. The damage introduced during the processes (process damage) and the stability of materials used for the processes often determine device performance and reliability.

6.2.1   Electrode Formation

The ohmic contact is applied to most optoelectronic pn-junction devices as p-side and n-side electrodes. The electrical and thermal resistance of the ohmic contact must be low so that devices used under forward and reverse bias conditions require little electrical input power and have a high frequency response. In addition, the suppression of the heating in the active region also requies low electrical resistance ohmic contact so that there is little Joule's heating.

6.2.1.1   Materials for Electrodes   The potential barrier between the metal and the semiconductor is determined by difference between the work functions of the metal and semiconductor. That is, $\phi_B = (\phi_m - \phi_s)$ (see Section 1.2.4 and Fig. 6.6). If $\phi_s > \phi_m$, there is no potential barrier at the heterojunction for

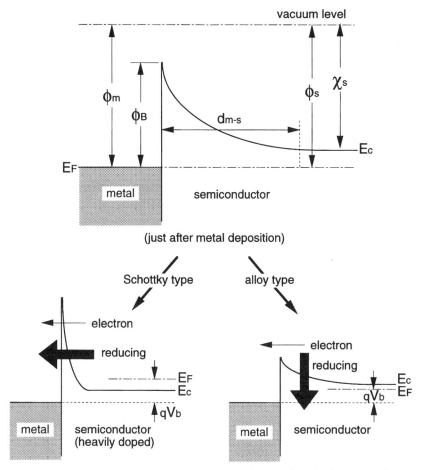

FIGURE 6.6 Basic concept of the energy band diagram for ohmic contact in n-type semiconductors ($\phi_m > \phi_s$): (a) alloy type and (b) Schottky type.

n-type semiconductors, and the ideal ohmic contact can be formed. For p-type semiconductors, $\phi_m > \phi_s$ is the condition for the ideal ohmic contact. Most semiconductors for optoelectronic devices, however, cannot satisfy the above conditions because of the surface states. A potential barrier between the metal and semiconductor is usually introduced, and some treatments are required for reducing the size of this barrier. There are two kinds of ohmic contacts used in the optoelectronic pn-junction devices: the alloy type and the nonalloy type (Schottky type).[6-8] Some typical metals are listed in Table 6.1. Elements in the group A column of the periodic table are used for Schottky-type ohmic contact and elements in the group B column are used for alloy-type ohmic contact.[9] Group A consists of the early transition metals—such as Ti, Cr, and Mo—which have relatively small electronegativities. Those metals form an inert interface

TABLE 6.1   Some Typical Metals Used for Optoelectronic *pn*-Junction Devices

Group A			Group B
. . . . . . Ti . . . . Cr	. . . . Ni	Cu	
. . . . Mo	. . . .	Ag	
. . . . (W)	. . . . Pt	Au	
Early transition metals	Near-noble metals (electronegative)	Noble metals (very electronegative)	
↓		↓	
Schottky type (inert interface)		Alloy type	

between the metal and the semiconductor and thus form Schottky-type electrodes. Group B consists of the noble metals—such as Cu, Ag, and Au—which are very electronegative. Those metals easily interact with semiconductors having band-gap energies less than about 2.5 eV at room temperature and thus form alloy-type electrodes.[10,11] Here, the interaction tends to occur easily as the band-gap energy of the semiconductor decreases. The use of a hard metal—such as W, which is frequently used in Si devices—is avoided in the devices composed of Ge or the III-V and II-VI compound semiconductors because most of the devices made of those materials are more brittle than Si devices.

6.2.1.2 Schottky-Type and Alloy-Type Electrodes   Just after the metal is deposited on the semiconductor, a high and wide potential barrier with a carrier depletion layer in the semiconductor side is usually formed at the interface between the electrode metal and the semiconductor (see Fig. 6.6).[12] This is the so-called Schottky contact. The width of the depletion layer, $d_{\text{m-s}}$, is proportional to the square root of the impurity concentration in the semiconductor:[13]

$$d_{\text{m-s}} \propto (\phi_B/N)^{1/2}, \qquad (6.1)$$

where ϕ_B is the potential barrier height and N is the impurity concentration in the semiconductor. If the barrier height is less than about 0.5 eV, the contact is nearly ohmic because electrons easily flow over the barrier under electrical bias. At the interface between most semiconductor materials used in the optoelectronic devices and the metals frequently used (such as Au, Ti, and Cr), relatively high barriers of 0.8 to 1.0 eV are ordinarily formed. At those contacts the main mechanism of the current flow is thermionic emission: the current is due to thermally activated electrons flowing over the barrier.[14] The contact resistance of such electrodes is quite high, and those electrodes cannot be used in optoelectronic *pn*-junction devices.

   If the metal is selected from group A in Table 6.1 and the semiconductor is heavily doped ($\geq 10^{19}$ cm^{-3}), an inert interface with a thin depletion layer [see Eq. (6.1)] is formed between the deposited metal and the semiconductor. As shown in Fig. 6.6, current can flow through a thin barrier under electrical bias by quantum-mechanical tunneling. This type of contact shows ohmic characteristics and is called the Schottky-type ohmic contact. The contact resistance, R_{c-S}, is closely related to the potential barrier as shown by the following relation:[14]

$$R_{c-S} \propto \exp(\phi_B/N^{1/2}). \qquad (6.2)$$

Typical Schottky-type electrodes are Ti/Pt/Au and Cr/Au, in which Pt is the barrier metal that prevents the diffusion of Au toward the semiconductor under operation and during bonding of the device chip on a heat sink (or stem) with solder.[7,8] The thicknesses of the Ti and Pt layers are, for example, 50 nm and 50 nm. The thicker Au layer is required for easy bonding of wire.

When a metal in group B in Table 6.1 is used and alloying the metal and the semiconductor or sintering (i.e., solid-state interdiffusion) is carried out under a relatively high temperature (ordinarily around 400°C), the width of the depletion layer decreases enough to be comparable to the extent of the diffuse conductor-semiconductor interface and the barrier height decreases. Consequently, when a bias is applied, current can easily flow over the reduced barrier by thermionic emission. This contact is also ohmic and is called the alloy-type ohmic contact, and its contact resistance is also closely related to the potential barrier:[14]

$$R_{c-a} \propto \exp(\phi_B/k_B T). \qquad (6.3)$$

Typical alloy-type electrodes for p-type and n-type semiconductors are AuZnNi and AuGeNi, respectively.

The reaction between the metal and the compound semiconductor tends to form the alloys of the metal and each element. For III-V compound semiconductors, alloys of metal-III elements and metal-V elements are formed separately.[9] In the case of an Au electrode on InP, for example,[15] Au-In alloy is initially formed during heat-treatment at temperatures between 320°C and 360°C. This is a result of the dissociation of the InP substrate by the deposited Au film (through a solid-state reaction) and diffusion of In and Au. With increasing treatment temperture or time, Au_3In and Au_2P_3 clusters are formed at the interface between Au-In and InP. Finally, the Au_2P_3 clusters and Au_3In grow inhomogeneously. The Au_2P_3 has a larger electrical resistance than does Au_3In. Appropriate conditions therefore need to be selected to form alloy-type electrodes with low resistance. If the deposited metal contains Zn, an alloy layer with low electrical resistance can be formed. These interface reactions are widely observed in Au-based electrodes.

The contact resistances for both types of electrodes depend very much on the potential barrier as indicated by Eqs. (6.2) and (6.3), and both types are widely used in optoelectronic *pn*-junction devices as well as semiconductor devices. Their contact resistances are reduced by a few ohms in actual devices as shown in Chapters 2 and 3 for LEDs and laser diodes.[8] Similar electrodes having low resistance are also formed in photodiodes and modulators. When high reliability is required, a Schottky-type electrode is usually used in the vicinity of the active region because of the inert interface (see Section 7.2.1.2), even though it is harder to form this type of an electrode than it is to form an alloy-type electrode. The alloy-type electrode is thus often used in the vicinity of the active region of devices for which stable operation over a long time is not required. The electrodes of the other side (on the substrate) are ordinarily formed as alloy-tyes because they are far from the active region.

6.2.2   Dielectric Film Deposition

Native oxide films of most compound semiconductors are unstable for using device processes and for passivating their surfaces, and the electrical resistances of the oxide films are too low to isolate semiconductor parts from electrodes. Consequently, dielectric films are required for those purposes. The dielectric films are widely used as masks for impurity diffusion and for chemical and physical etching during processing, as an electrical isolator between an electrode and a semiconductor, as passivation films for device surfaces, and as films for controlling reflectivity. The deposition of dielectric films on semiconductors is indispensable in device fabrication. Popular materials for dielectric films used for optoelectronic devices are SiO_2 and SiN_x. Al_2O_3 is also used for passivation films as well as antireflective and highly reflective films.

6.2.2.1   Adhesion, Damage, and Stability   Several kinds of methods, such as radio-frequency sputtering, chemical vapor deposition (CVD), and plasma CVD (PCVD), are used to deposit dielectric films.[16,17] The adhesion of the film and the damage introduced during the deposition depend on the depositing methods and conditions. The deposition of a dielectric film by radio-frequency sputtering is carried out as follows: (1) a cation such as Ar^+ is formed by glow discharge excited with radio-frequency waves in a vacuum chamber, (2) the ion collides with the target composed of the dielectric material and the target material is sputtered, (3) the sputtered material is deposited on the semiconductor (see Fig. 6.7). The use of radio-frequency waves is required in order to prevent the dielectric target from being positively charged. If the electric field were constant, the target would be charged and the cations would be reflected at the front of the target. Sputtering is a physical deposition process, whereas the dielectric film deposition with CVD is carried out by the chemical reaction of material gases in a vacuum chamber. The chemical reaction is enhanced by excitations such as heating, light incidence, and electrical discharges. The material gases for SiN_x films, for example, are SiH_4 and N_2.

The adhesion of a dielectric film is influenced by the cleanliness of the semiconductor surface and the depositing conditions, and the depositing method

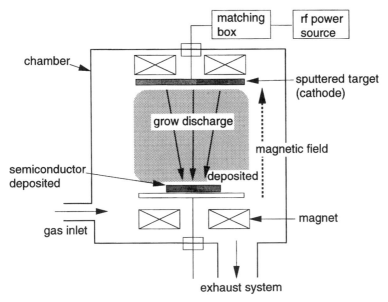

matching box — rf power source

chamber

sputtered target (cathode)

grow discharge

magnetic field

semiconductor deposited

deposited

gas inlet

magnet

exhaust system

FIGURE 6.7   Schematic diagram of radio-frequency (rf) sputtering apparatus.

strongly influences not only the adhesion of the film but also the damage of the deposited semiconductor surface (or interface). In the case of strong adhesion between the dielectric film and the semiconductor, a large amount of damage is ordinarily introduced to the semiconductor surface.

This happens in sputtering. The strong adhesion is generated by the physical deposition forming a thick interfacial layer composed of the dielectric film and the semiconductor (see Fig. 6.8). This figure shows that an interfacial layer of about 10 nm was formed between SiO_2 and GaAs. The interfacial layer, which is formed by crystal destruction at the semiconductor surface during the film deposition, may increase the adhesion of the film to the semiconductor. The photoluminescence intensity, for example, is therefore tremendously lower after dielectric film deposition by sputtering because of the formation of the interfacial state density around the interfacial layer. The case of severe damage for a planar type AlGaAs/GaAs laser diode is shown in Fig. 6.9. After SiO_2 deposition a large amount of leakage current is introduced by the damage of the pn-junction perimeter of the laser diode. The leakage current flows across the pn-junction perimeter through the damaged layer. Sputtering tends not to be used when making optoelectronic pn-junction devices that are operated under reverse bias because it often results in leakage current under reverse bias being introduced. The leakage current leads to the dark current in photodiodes, and the dark current degrades their performance (see Section 4.1.2). For those devices, CVD or PCVD are frequently used to deposit the dielectric film because, although the resultant adhesion is weaker than that achieved when sputtering is used, these deposition methods cause little damage.

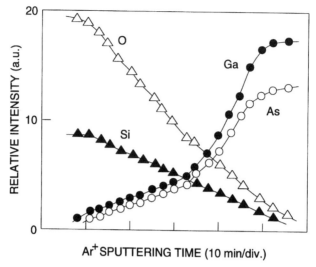

FIGURE 6.8 Auger depth profile at the interface between GaAs and a SiO_2 film deposited by rf-sputtering (100 W). The average rate of Ar sputtering in Auger microprobe is about 2 nm/10 min.

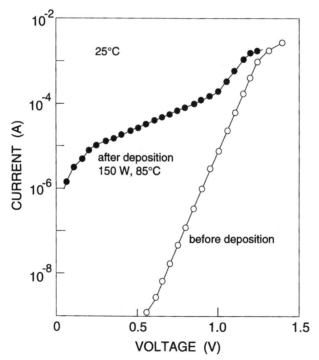

FIGURE 6.9 Leak current increase due to SiO_2 sputtering on the facet of a planar type AlGaAs/GaAs laser diode.

The leak path introduced during film deposition causes device degradation, and the long-term stability of a coated film-mirror facet interface is also often influenced by the material deposited. The aging characteristics of AlGaAs/GaAs laser diodes differ with different facet-coating materials.[18-20] At the SiO$_2$-coated facet of AlGaAs/GaAs laser diodes, the out-diffusion of Ga will occur and degrade the device characteristics.[20] Photodiodes also degrade at the interface between the dielectric film and the *pn*-junction perimeter (see Section 7.2.2.1). The material used for passivation or coating is therefore selected carefully.

Attention should also be paid to the mechanical stress due to the deposited dielectric film. When the mechanical stess is severe, slip dislocations or cracks are induced and degrade device performance (see Section 7.2.1.1). The photoelastic effect is also induced under severe mechanical stress, changing the refractive index of the waveguide and the polarization of light propagating in the waveguide.[21]

6.2.2.3   Antireflective and Highly Reflective Coatings

Single Dielectric Film   The reflectivity of the optical input and output ports is easily controlled by coating them with dielectric films. The reflectivity is ordinarily reduced by adjusting the thickness of the dielectric film. If there is no optical (absorption and scattering) loss in the dielectric film and the incident light is perpendicular to the film (see Fig. 6.10), the optical power reflectivity of the coated part, R_{coat}, can be expressed by the following equation:[22]

$$R_{coat} = (r_1^2 + r_2^2 + 2r_1 r_2 \cos 2\Phi)/(1 + r_1^2 r_2^2 + 2r_1 r_2 \cos 2\Phi), \qquad (6.4)$$

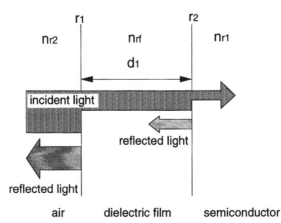

FIGURE 6.10   Reflection and transmission of light at the surface of a dielectric-film-coated semiconductor.

where

$$r_1 = (n_{r2} - n_{rf})/(n_{r2} + n_{rf}), \tag{6.5}$$

$$r_2 = (n_{rf} - n_{r1})/(n_{rf} + n_{r1}), \tag{6.6}$$

and

$$\Phi = 2\pi n_{rf} d_f/\lambda_0. \tag{6.7}$$

Here n_{r1} and n_{r2} are the refractive indexes of two media (semiconductor and air), n_{rf} is the refractive index of dielectric film, d_f is the thickness of the dielectric film, and λ_0 is the wavelength of the incident light in a vacuum. According to Eq. (6.4), the reflectivity periodically changes with Φ, that is, with the thickness of the dielectric film. The maximum value appears when $\cos 2\Phi = 1$—that is, $d_f = m\lambda_0/2n_{rf}$ (m = integer)—and is given for normal incidence by

$$R_{coat\text{-}max} = [(n_{r1} - n_{r2})/(n_{r1} + n_{r2})]^2. \tag{6.8}$$

The maximum reflectivity is independent of the refractive index of the dielectric film, n_{rf}, and coincides with the reflectivity before the dielectric film deposition. This type of film is often used for passivation of facets of AlGaAs/GaAs laser diodes and laser diodes emitting visible light. The minimum value of Eq. (6.4) appears at $\cos 2\Phi = -1$—that is, $d_f = m\lambda_0/4n_{rf}$ (m = odd number)—and is given for the normal incidence by

$$R_{coat\text{-}min} = [(n_{r1}n_{r2} - n_{rf}^2)/(n_{r1}n_{r2} + n_{rf}^2)]^2. \tag{6.9}$$

The minimum reflectivity gives the condition for the antireflective film coating. The minimum reflectivity would be strictly zero at

$$n_{rf} = (n_{r1}n_{r2})^{1/2}. \tag{6.10}$$

The dielectric material having a refractive index close to that given by Eq. (6.10) is therefore chosen for an antireflective film. For example, an Al_2O_3 film with thickness of $m\lambda_0/4n_{rf}$ and refractive index of about 1.7 (n_{rf}) is a very popular antireflective coating for AlGaAs and InGaAsP surfaces exposed to air ($n_{r2} = 1$) because each of these materials has a refractive index of about 3.5 (n_{r1}).

Periodically Stratified Films   Periodically stratified films are applied to some parts, such as interference filters and are highly reflective films. The highly reflective films are formed by multilayers of two kinds of dielectric materials (see Fig. 6.11) in which the thickness of each layer is usually set at the quarter-wavelength, $\lambda_0/4n_{rf}$, to increase the tolerance for thickness fluctuation. For normal incidence the reflectivity in such a case is given by[22]

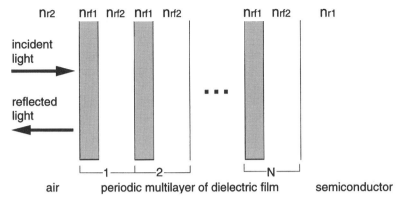

FIGURE 6.11   A periodic multilayer (highly reflective film).

$$R_{\text{coat-}2N} = \{[n_{r2} - n_{r1}(n_{rf1}/n_{rf2})^{2N}]/[n_{r2} + n_{r1}(n_{rf1}/n_{rf2})^{2N}]\}^2, \qquad (6.11)$$

where N is the number of periods of double layers. The reflectivity increases when the ratio n_{rf1}/n_{rf2} is increased or the number of periods is increased. The refractive indexes of the first and the second layers are respectively set at low and high values. Typical pairs for highly reflective films used in optoelectronic *pn*-junction devices are amorphous Si/SiO_2 (3.4/1.4) and TiO_2/SiO_2 (2.5/1.4). One or two pairs are usually coated onto the semiconductor surface such as facets of laser diodes, and the material with the lower refractivity is deposited first. In air ($n_{r2} = 1$) one pair of amorphous Si/SiO_2 layers and one pair of TiO_2/SiO_2 layers deposited on AlGaAs and InGaAsP ($n_{r1} \approx 3.5$) give reflectivities of about 83% and 70%, respectively.

6.2.3   Lithography and Chemical and Physical Etching

The semiconductor wafer processing for optoelectronic *pn*-junction devices consists mainly of dielectric or metal film deposition, photolithography, and chemical and physical etching. The photolithography is carried out as follows: formation of a photoresist film on the dielectric or metal film deposited on the semiconductor, exposure of the photoresist through a patterned mask by photoexposure or by electron-beam exposure without any mask (direct writing), developing of the photoresist, and rinsing of the patterned photoresist. The dielectric (or metal) film is then chemically or physically etched through the patterned photoresist, and the patterned dielectric film is used as a mask for the chemical or physical etching of the semiconductor. If a photoresist mask is directly formed on the semiconductor, the chemical or physical etching of the semiconductor is carried out through the photoresist pattern, though the photoresist does not have high stability during the etching processes.

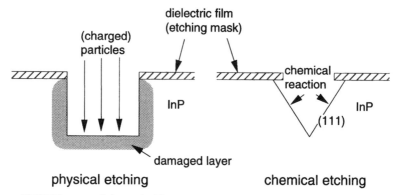

FIGURE 6.12 Schematic illustration of chemical and physical etching.

There are two types of etching: wet etching and dry etching. Various kinds of etchants are ordinarily used for wet etching; for example, hydrochloric acid, sulfuric acid, hydrobromic acid, and hydrofluoric acid are often used for etching GaAs and InP material systems. The etching rate and the shape of the etched pattern depends on combinations of semiconductor materials and etchants. Hydrochloric acid, for example, scarcely etches InGaAsP but rapidly etches InP. In addition, no physical damage is incurred during wet etching because it is a chemical process. There is little process damage and the defect density in the processed devices is low. The etched shapes and patterns are, however, determined by the combination and depend on the crystal orientation of the semiconductor. With wet etching it is therefore difficult to form fine patterns (see Fig. 6.12).

Physical etching can be used to form relatively fine patterns, and a typical kind of physical etching is dry etching, in which semiconductor materials are sputtered by collisions of particles, such as Ar atoms accelerated as ions by a strong electric field and then neutralized electrically. Fine patterns can be formed without the limitations imposed by a vacuum chamber, but a damaged layer is formed and the etched surface contains a large number of defects (see Fig. 6.12). This damage deteriorates device performance and reduces reliability. After dry etching, the etched surface is therefore lightly etched chemically to eliminate the damaged layer.

Most of the dry etching methods used in the processing of optoelectronic *pn*-junction devices combine physical and chemical etching. Typical methods are reactive ion etching (RIE) and reactive ion-beam etching (RIBE).[23–26] They are carried out in a vacuum chamber in which ions for etching are formed in plasma (for RIE) or are supplied by an ion source (for RIBE). In addition to the sputtering due to the ions, there are reactions between the semiconductor and radicals adsorbed on the semiconductor surface. The products of the reaction are easily sputtered by the incident ion. Here the radical is an electrically neutral molecule which has high internal energy and thus reacts easily with the semiconductor.

After device structures are formed in the semiconductor wafer, the electrodes are formed, and each device is separated from the wafer by using suitable techniques, such as cleavage. The cleaved surfaces serve as the mirror facets of laser diodes, the input and output ports of modulators, and the light-emitting facets for edge-emitting LEDs. They are coated with dielectric films for passivation or for controlling reflectivity. The device chips are then mounted on heat sinks or stems for packaging.

6.3 PACKAGING

Packaging is indispensable when optoelectronic devices are applied to equipment and systems. After packaging, optical input and output as well as electrical input and output can be performed via the optical ports and the electrical terminals. During packaging, bonding of the device chip on a heat sink or stem, for example, determines thermal conductance (or resistance), and the optical coupling between the device chip and the waveguide/fiber mainly determines the optical loss of modules. Consequently, the packaging greatly influences device performance and reliability. Cross-sectional views of some typical packages and modules for laser diodes are shown in Fig. 6.13. Various kinds of components and parts are used for fabricating them. The following sections describe some of the main technologies related to the packaging of optoelectronic devices.

6.3.1 Bonding

Device chip bonding (die bonding) on the heat sink or package stem is the first step in the assembly of the package, and the bonding configuration strongly influences the temperature characteristics of optoelectronic *pn*-junction devices, especially laser diodes and LEDs. Conductive plastics are sometimes used for bonding devices which operate under reverse bias and generate little heat. Some solders are usually used for bonding all kinds of optoelectronic devices. A schematic diagram of the baisc configuration is shown in Fig. 6.14. Here, the heat sink is a barrier of thermal conduction although it is called a heat sink. It is required to suppress the generation of mechanical stress between the device chip and the metal stem because of the difference in thermal expansion coefficient.

6.3.1.1 Solder For Bonding The most important factor for bonding by solder is wetting, which means the adhesion between solid (base metal) and liquid (solder). The strength of the adhesion is basically determined by the amount of the free energy reduction when the solid (base metal) surface changes into a solid-liquid (molten solder) interface. A capillary phenomenon, where the molten solder penetrates into the narrow gaps in the base metal, is also important in strengthening the adhesion. The formation of an alloy layer at the interface between the base metal and the leaching part also influences the strength of the

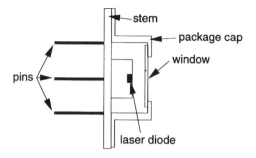

(a) coaxial type package

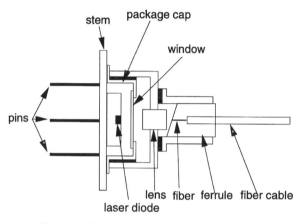

(b) coaxial type module

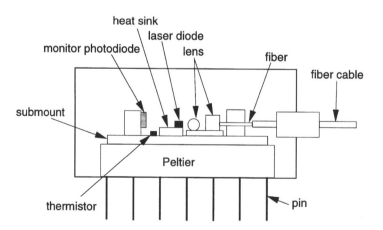

(c) rectangular type module

FIGURE 6.13   Cross-sectional view of hermetically sealed laser packages and modules.

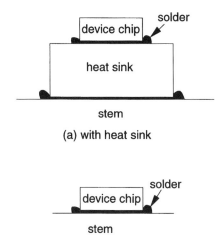

(a) with heat sink

(b) without heat sink

FIGURE 6.14   Schematic diagram of basic configuration for device chip bonding.

adhesion. Leaching here means the phenomenon that the base metal melts at a temperature lower than its melting point when it contacts with molten (solder) metal. The formation of a thick alloy layer should be avoided because such a layer is generally brittle and increases both electrical and thermal resistances. The base metal—electrode of the device chip and cover metal of the heat sink and stem in Fig. 6.14—is ordinarily Au in the optoelectronic *pn*-junction devices. The solder metals are selected taking into consideration those properties discussed above. Typical solder metals used in optoelectronic *pn*-junction devices are listed in Table 6.2.

TABLE 6.2   Typical Bonding Solder Metals Used for Optoelectronic *pn*-Junction Devices

(Eutectic) Material	Melting Point (°C)
Soft solder	
In	156
In-52 wt%Sn	117
Sn	232
Sn-60 wt%Pb	183
Sn-90 wt%Au	217
Hard solder	
Sn-20 wt%Au	280
Au-88 wt%Ge	356
Au-94 wt%Si	370

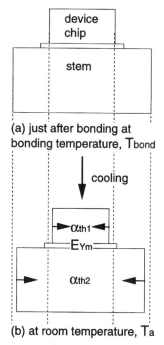

FIGURE 6.15   Conceptual illustration of the generation of mechanical stress during bonding.

Mechanical stress and increased electrical and thermal resistance are generally generated at the bonding part between device chip and heat sink and between heat sink and stem (see Fig. 6.15). The mechanical stress per unit length, S_{bond}, can be expressed by

$$S_{bond} = |\alpha_{th2} - \alpha_{th1}|(T_{bond} - T_a)E_{Ym}, \qquad (6.12)$$

where α_{th1} and α_{th2} are respectively the thermal expansion coefficients of the device chip and the stem, T_{bond} and T_a are the bonding temperature (near the melting point of solder) and the ambient temperature in which the device operates, and E_{Ym} is the Young's modulus of the device chip (nearly equal to that of the semiconductor). Consequently, the mechanical stress increases with the melting point of the solder. In the case of bonding GaAs to a Cu stem, for example, the compressive stress generated in the GaAs chip at 25°C is about 2×10^9 dyn/cm² when Au-rich AuSn solder ($T_{bond} \approx 290$°C) is used and is about 9×10^8 dyn/cm² when In solder ($T_{bond} \approx 130$°C) is used. (Here α_{th1} and α_{th2} are respectively about 6.6×10^{-6} and 1.7×10^{-6} K^{-1}, and the E_{Ym} of GaAs is about 8.6×10^{11} dyn/cm².) At the bonding part with a solder having a low melting point, the stress can be released to the solder layer because such solder is generally soft and plastic deformation is easily generated at the solder layer. Under high mechanical stress (2×10^9 dyn/cm^2), the device would be degraded because of the growth slip dislocations (see Section 7.2.1.1), and thus a heat sink is required.

6.3.1.2 Soft Solder and Hard Solder

The solder metals used in optoelectronic *pn*-junction devices can be divided into two groups: soft solders and hard solders (see Table 6.2). A soft solder has a low melting point, and a hard solder has a high melting point. These two types of solder metals show different properties as bonding materials. The mechanical stress is generally larger at the part bonded with hard solder because of the bonding temperature difference [see Eq. (6.12)]. In addition, the soft solders with low melting point absorb the mechanical stress generated at the interface between the device chip and heat sink or stem because they are deformed nonelastically under the stress (see Fig. 7.35). The soft solder can be used to directly mount a chip on a metal stem [see Fig. 6.14(b)]. The soft solders, however, become unstable during long-term operation because thermal fatigue occurs as a result of the mechanism by which the mechanical stress is absorbed. The stability of the solder metal itself generally tends to be low when the melting point is low. Consequently, several problems related to long-term stability are generated at the parts bonded with soft solder. Bonding with a hard solder, on the other hand, is very stable for the long term and is only slightly more difficult than bonding with a soft solder. Au-rich AuSn solder is used for bonding the device chip and heat sink/stem in most optoelectronic *pn*-junction devices.

6.3.1.3 Thermal Resistance and Heat Sinks

Thermal Resistance As discussed in the previous section, the heat sink in optoelectronic *pn*-junction devices is necessary for suppressing the generation of mechanical stress. The heat sink itself and the bonding parts are barriers to thermal conductance (thermal resistance) because the heat sink is between the device chip and stem [see Fig. 6.14(a)]. The thermal resistance of *pn*-junction devices is easily monitored by measuring the junction voltage change under forward bias. The forward junction voltage is determined by the band-gap of the semiconductor material and as shown in Fig. 6.16 is nearly linearly related to the junction temperature [see Eq. (2.28)]. As this figure shows, the forward junction voltage at 1 mA changes linearly with the ambient temperature. When the forward current is set at a value so small that the Joule's heat generated at the *pn*-junction is negligible, the slope gives the temperature coefficient of the junction voltage, K_j:

$$dV_j/dT = -K_j, \qquad (6.13)$$

where V_j is the forward junction voltage. This relation is used to monitor the junction temperature under operation. The basic concept of the method monitor-

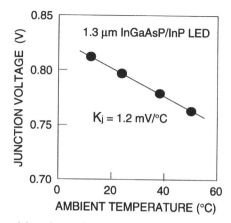

FIGURE 6.16 Forward junction voltage (at a current bias of 1 mA) as a function of ambient temperature for an InGaAsP/InP LED.

ing the junction temperature in *pn*-junction devices is shown in Fig. 6.17. The junction voltage change, ΔV_f, corresponds to the junction temperature change from the operating temperature, T_j, to ambient temperature, T_a. Consequently, ΔV_f is given by the next equation taking the monitoring time delay, t_d, into consideration:

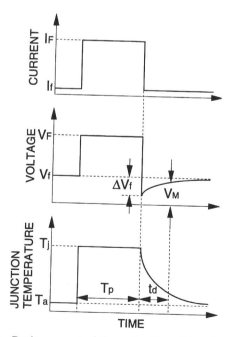

FIGURE 6.17 Basic concept of the estimation of junction temperature.

$$\Delta V_{\rm f} = \lim_{t_{\rm d} \to 0} [V_{\rm f} - V_{\rm M}(t_{\rm d})], \tag{6.14}$$

where $V_{\rm M}(t_{\rm d})$ is the junction voltage monitored at time delay, $t_{\rm d}$. Using Eqs. (6.13) and (6.14), the junction temperature change is given by $\Delta V_{\rm f}/K_{\rm j}$ ($=T_{\rm j}-T_{\rm a}$). Using the thermal resistance, $R_{\rm th}$, which indicates the temperature rise under operation (at a certain input power) and which is expressed in units of $°C/W$, the junction temperature under operation can be given as

$$\begin{aligned} T_{\rm j} &= T_{\rm a} + R_{\rm th}I_{\rm F}(V_{\rm b} - I_{\rm F}R_{\rm s}) \\ &= T_{\rm a} + R_{\rm th}I_{\rm F}V_{\rm j}, \end{aligned} \tag{6.15}$$

where $V_{\rm b}$ is the applied voltage. The thermal resistance is therefore given by

$$R_{\rm th} = \Delta V_{\rm f}/K_{\rm j}I_{\rm F}V_{\rm j}, \tag{6.16a}$$

and if the optical output power, $P_{\rm out}$, is emitted,

$$R_{\rm th} = \Delta V_{\rm f}/K_{\rm j}(I_{\rm F}V_{\rm j} - P_{\rm out}). \tag{6.16b}$$

The thermal resistance is a function of operating time—that is, the pulse width of the electrical input—because the heat generated at the *pn*-junction spreads from the chip to the heat sink, stem, and package frame. One example of the relation between the pulse width and the change in junction voltage is shown in Fig. 6.18. As the pulse width increases, the thermal resistance (measured here as the difference in junction voltage) increases, and steps appear at two points (corresponding to the bonding part) because of the low thermal conductance at

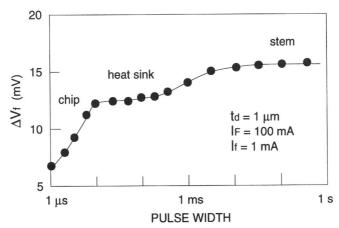

FIGURE 6.18   Change in junction voltage as a function of biased pulse width for a 1300 nm-band InGaAsP/InP LED mounted in the junction-down configuration.

the bonded parts. When the injected current is 100 mA and the junction voltage is 1.6 V, the thermal resistance (including that of the chip, the heat sink, and the stem) is calculated using Eq. (6.16a) and the data in Fig. 6.16 is about 83°C/W. The thermal resistance further increases as duration of the electrical bias lengthens. The thermal resistance strongly influences the current–light output power characteristics of LEDs and laser diodes and the lasing wavelength of laser diodes. The temperature rise due to reverse current also degrades device characteristics of photodiodes and modulators. Reducing the thermal resistance is therefore an important part of improving the device characteristics.

The temperature rise of the junction can also be monitored by other methods, for example, by measuring the shift of peak emission wavelength in LEDs [see Eq. (2.30)]. The difference between the current–light output power characteristics measured under pulsed and dc operation also corresponds to the junction temperature rise in LEDs and laser diodes.

Heat Sinks As discussed in the previous section, the thermal expansion coefficient of the heat sink material must be close to that of device chips and the thermal conductivity of the heat sink material must be high. High electrical resistivity and low dielectric constant are also important for high-frequency modulation. The heat sink materials frequently used in the optoelectronic *pn*-junction devices are summarized in Table 6.3. Most materials are electrical insulators and thus have high electrical resistances. Those materials, however, cannot be directly bonded to metal with solder. Metals such as Ti/Pt/Au and Cr/Au are therefore deposited on the surface of these materials when they are used as heat sinks (metalization). Another property required of heat sink materials is thus that the metallization can be performed easily. The mechanical stress generated at the bonded part is given by Eq. (6.12). From the data listed in Table 6.3, a generation of a large amount of mechanical stress can be recognized when a device chip is bonded to a metal heat sink, such as one made of Au or Cu, because of the large difference between the thermal expansion coefficients of the chip and the heat sink. The mechanical stress, in contrast, can be reduced by using a heat sink composed of diamond, Si, or a ceramic material. The mechanisms of heat conduction differ between the metals and other heat sink materials in Table 6.3. Heat in metals is carried mainly by free electrons, whereas heat in electrical insulators such as ceramics is carried mainly by phonons (lattice vibration). The insulators used as heat sinks are therefore materials with little phonon scattering (see Section 1.1.2.6). Those are materials that are composed of atoms with small masses and with high bond energy and that have a simple crystal structure without defects, impurity atoms, or precipitations.[27,28] The materials such as SiC, AlN, c-BN (cubic-BN) in Table 6.3 are typical ones satisfying those conditions.

6.3.2 Optical Coupling

Optical coupling in optoelectronic devices is as important as electrical connection. Alignment precision of micron order or higher is required for optical coupling because of the small optical beam sizes in optoelectronic devices. In

TABLE 6.3 Typical Heat Sink Materials Used in Optoelectronic *pn*-Junction Devices and Some Semiconductor Materials (at about 300 K)

Material	Thermal Conductivity ($W\ K^{-1}\ cm^{-1}$)	Resistivity (Ω cm)	Thermal Expansion Coefficient ($\times 10^{-6}\ °C^{-1}$)
Diamond	20	$>10^{14}$	2.3
AlN (ceramic)	0.26	$>10^5$	4.2
SiC (ceramic)	0.7	$>10^{13}$	3.7
Al_2O_3 (ceramic)	0.17	$>10^{14}$	6.5
BN (cubic)	6	$>10^{14}$	3.7
CuW	2		6
Si (single-crystal)	1.4	$>10^{-3}$–10^3	3.5–4.0
GaAs	$0.54^{a,b}$		6.6^c
$Al_{0.3}Ga_{0.7}As$	0.13^d		6.2^e
$Al_{0.03}Ga_{0.97}As$	0.29^d		6.6^e
InP	$0.7^{f,g}$		4.5^h
$In_{0.82}Ga_{0.18}As_{0.4}P_{0.6}$	0.04^i		5.1^j
$In_{0.76}Ga_{0.24}As_{0.55}P_{0.45}$	0.036^i		5.3^j
$In_{0.65}Ga_{0.35}As_{0.79}P_{0.21}$	0.037^i		5.6^j
$In_{0.47}Ga_{0.53}As$	0.053^i		5.9^j
Au	3.2		14.2
Cu	4		16.8

[a]R. O. Carlson et al., *J. Appl. Phys.*, **36**, 506, 1965.
[b]J. Blanc et al., *J. Phys. Chem. Solids*, **25**, 225, 1964.
[c]P. W. Sparks et al., *Phys. Rev.*, **163**, 779, 1967.
[d]M. A. Afromowitz, *J. Appl. Phys.*, **44**, 1292, 1973.
[e]H. Yonezu, *Hikaritsushin sosi-kougaku* (Japanese), Kougaku Tosyo Co., Tokyo, 1984.
[f]S. A. Aliev et al., *Sov. Phys. Solid State*, **7**, 1287, 1965.
[g]I. Kudman et al., *Phys. Rev.*, **133**, A1665, 1964.
[h]H. Welker et al., *Solid State Phys.*, F. Seitz and D. Turnbull, eds., Academic Press, New York, 1956, vol. 3, p. 51.
[i]W. Both et al., **21**, K85, 1986.
[j]N. A. Bert et al., *Sov. Phys. Tech.*, **26**, 610, 1981.

addition, as discussed in Section 3.2.4.2, diffraction occurs at the output port of the optoelectronic devices. The optical coupling for optoelectronic devices is therefore very difficult and complicated, especially for edge-emitting LEDs and laser diodes. As shown in Fig. 6.19, the incident angle of light into the fiber/waveguide has to be less than the angle at which the total reflection occurs at the interface between core and cladding layer in the optical fiber or waveguide. The light incident at an angle larger than the critical value, θ_{max}, leaks out to the cladding layer and is lost. The critical situation is usually expressed in terms of numerical aperture (NA):

$$NA = n_{r1} \sin\theta_c$$
$$\approx n_{r1}[2(n_{r1} - n_{r2})/n_{r1}]^{1/2}, \tag{6.17}$$

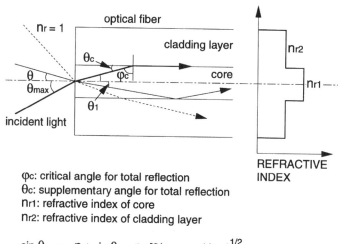

φc: critical angle for total reflection
θc: supplementary angle for total reflection
nr1: refractive index of core
nr2: refractive index of cladding layer

$$\sin \theta_{max} = n_{r1} \sin \theta_c \approx n_{r1}[2(n_{r1} - n_{r2})/n_{r1}]^{1/2}$$

(nr1 - nr2)/nr1: relative refractive-index difference

FIGURE 6.19 Correlation between light incident angle and optical coupling (for a step-index optical fiber).

where n_{r1} and n_{r2} are the refractive indexes of the core and the cladding layer, respectively, and θ_c is the supplementary angle of the critical angle for the total reflection in the fiber/waveguide. Here $(n_{r1} - n_{r2})/n_{r1}$ is called the relative refractive index difference. The critical angle can be given as a function of NA:

$$\theta_{max} = n_r \sin^{-1} NA, \tag{6.18}$$

where n_r is the refractive index of the medium outside of the fiber/waveguide and for air is equal to 1. Similar considerations apply in the case of a graded-index fiber/waveguide. The condition described above also holds for the light focused by lenses and by some other passive components. The use of parallel rays is effective in improving the optical coupling, and several combinations of passive components with fibers/waveguides have been tried in order to reduce the incident angle and thus decrease the optical coupling loss.

There are two kinds of alignment methods used in constructing the optical coupling systems: active and passive alignments. In active alignment the optical alignment between devices and optical fibers/waveguides, lenses, and other passive optical components is performed under operation of the optoelectronic device to be packaged, whereas in passive alignment there is no operation of the device. Most optical modules are fabricated using active alignment. After the device is mounted on the stem or submount (see Fig. 6.13), lenses and fibers are aligned and fixed one by one so that the monitored power through the lens or

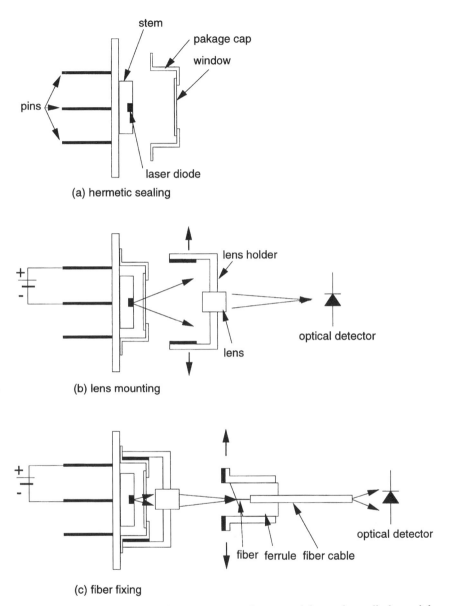

FIGURE 6.20 Alignment and fixing processes for a coaxial type laser diode module.

the optical fiber is the maximum for LEDs and laser diodes and so that the monitored photocurrent for optical input power through the lens or the optical fiber is the maximum for photodiodes and modulators. One example of alignment for a coaxial-type laser module is shown in Fig. 6.20. Surface-emitting LEDs and normal photodiodes can be fabricated relatively easily because the sizes of their active region (optical ports) are large. For passive alignment, bonding pads for devices and lenses and fiber guides/waveguides are patterned on the submount or heat sink, and the optoelectronic devices, passive components, and fibers or waveguides are mounted on the patterns.[29] Flip-chip bonding is also used for bonding device chips. The passive alignment is suitable for mass-production, although precise patterning is indispensable.

Plastics, soldering, and laser welding are used for bonding and fixing devices, passive components, and fibers. During long-term operation, however, plastics are degraded by humidity, ultraviolet rays, temperature changes, and so forth, and creep occurs at parts fixed with soft solder if mechanical stress is generated.[30] Creep is a phenomenon in which the solder deforms nonelastically with time when it is subjected to a mechanical stress stronger than the substance's elasticity limit. This degradation and creep gradually decrease the optical coupling. Laser welding and hard solder are therefore used for bonding and fixing in highly reliable packages and modules (see Section 7.4).

6.3.3 Temperature Control: Peltier Cooler and Thermistors

Optoelectronic *pn*-junction devices are inherently sensitive to the operating temperature because they are composed of semiconductors, and device characteristics directly related to the band-gap energy of the semiconductor material change a great deal when the operating temperature changes. The operating temperature is therefore controlled electrically when, for example, the wavelength of the input and output light needs to be kept constant. The laser diodes used as optical pumping sources for optical fiber amplifiers and as transmitters for coherent communication systems also require temperature control. When optical modulators are used with an input light whose wavelength nearly corresponds to the band-gap energy, temperature control is required to keep their performance stable. Temperature control is also often used to improve device characteristics (for exmaple, the output power of long-wavelength laser diodes) which are very sensitive to temperature. For high power semiconductor lasers (or laser arrays), water coolers are frequently used because they can generate as much heat as gas lasers and solid-state lasers do. For most optoelectronic *pn*-junction devices, however, the operating temperature is controlled using thermoelectric (Peltier) coolers and thermistors (see Fig. 6.13).

6.3.3.1 Peltier Cooler The Peltier cooler controls temperature by means of the Peltier effect. The Peltier effect is a phenomenon in which electrical power is converted into heat by electrons and holes.[12] A schematic diagram of the unit cell of a Peltier cooler is shown in Figs. 6.21 and 6.22. The *p*-type and *n*-type semiconductors are connected in series by electrodes such as Cu, and then the whole part is sandwiched with two ceramic boards. When an electrical bias is applied to the electrodes as shown in Fig. 6.21, electrons and holes with certain energies move in the same direction. As a result, the side from which electrons and holes leave cools down and the opposite side heats up. The amount of heat absorbed or generated, Q_{heat}, is given by

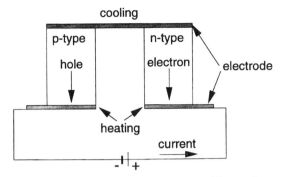

FIGURE 6.21   Basic structure of a Peltier cooler.

$$Q_{heat} = \pi_P I, \tag{6.19}$$

where π_P and I are the Peltier coefficient and the bias current. The Peltier coefficient can be given by

$$\pi_P = \eta_{te} T, \tag{6.20}$$

where η_{te} is called the thermoelectric power (or Seebeck coefficient). From Eqs. (6.19) and (6.20), a semiconductor with high Peltier coefficient, and thus a high Seebeck coefficient, is required for making an efficient Peltier cooler. Thermoelectric materials such as Bi_2Te_3, Bi_2Te_3-Sb_2Te_3, Pb (Te, Se), Ge-Si, and a solid solution of InAs-GaAs are frequently used for this purpose. One example of a Peltier cooler is shown in Fig. 6.22. For controlling temperature quickly, two or three Peltier coolers are stacked sometimes. For the same purpose, a water cooler is often used. The water cooler is set at the Peltier cooler

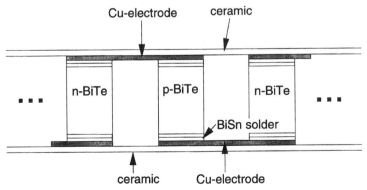

FIGURE 6.22   Cross-sectional view of one example of a Peltier cooler.

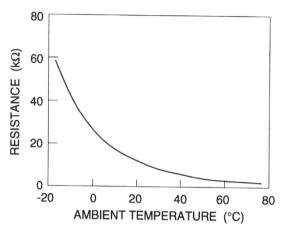

FIGURE 6.23 An example of thermistor resistance as a function of ambient temperature.

on the opposite side of the optoelectronic device and keeps the temperature of that side constant.

6.3.3.2 Thermistors To control operating temperature or to keep it constant, the temperature needs to be monitored and then the current biased to the cooler has to be changed. The temperature is frequently monitored with thermistors. The word "thermistor" is a shortened form of "thermally sensitive resistor." It is a kind of semiconductor and is formed from a solid solution or a sintered alloy of oxides or carbonates of Fe, Ni, Mn, Mo, and Cu. As shown in Fig. 6.23, a thermistor has a large negative temperature coefficient for electrical resistance. The temperature is monitored by measuring the change in the electrical resistance. As the thermistor is a kind of resistor, its size can easily be reduced and it can be easily mounted, with solder, in modules (see Fig. 6.13).

6.3.4 Hermetic Sealing and Plastic Packaging

Most of the modules used in optical fiber communication systems are sealed hermetically because of their high reliability. For reasons of economy, however, most of the optoelectronic *pn*-junction devices used in consumer electronics are molded in plastics, although hermetic sealing is also used for some applications in consumer electronics. The following sections describe some typical packages and modules.

6.3.4.1 Hermetic Sealing The optoelectronic *pn*-junction devices are mounted on stem or submounted with other passive components and electrical cooler and then hermetically sealed by bonding the cap or the metal cover in dry N_2 or inactive gases such as Ar. For such hermetic packaging, some parts of packages

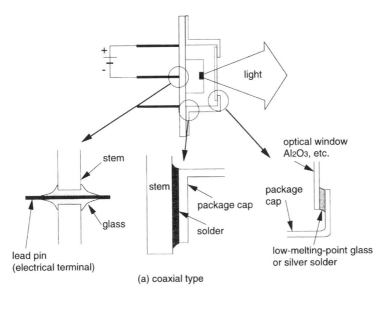

(a) coaxial type

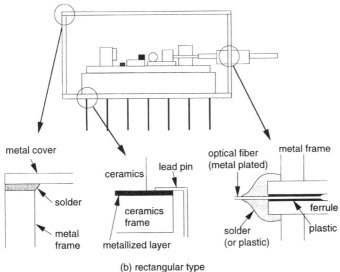

(b) rectangular type

FIGURE 6.24 Some examples of hermetic sealing for optoelectronic *pn*-junction devices.

and modules, which are mainly electrical and optical input and output terminals or ports, have to be also sealed hermetically. Those parts are shown in Fig. 6.24.

In the conventional coaxial type package, the lead pins (electrical terminals) are isolated from the package stem and sealed with glass. The thermal expansion coefficient of the stem and pin need to coincide with that of the glass. The material for the stem and the pin is usually kovar or Fe-Ni and is sometimes Fe. The sealing is carried out by melting glass powder packed between the pin and the stem. When the sealing is done, the surface of the connecting part of the stem and the pin is intentionally oxidized. The molten glass strongly combines with the stem and the pin through the metal oxide. The optical window is glass coated with an antireflective film to increase the light power passing through the window. In packages for communication uses, an Al_2O_3 glass is frequently used to ensure long-term stability. The glass window is combined with the package cap, which is ordinarily composed of kovar, by using a low-melting-point glass. The Al_2O_3 window is partially metallized along its edge, and the metallized part is joined to the cap by brazing with silver solder. After baking, the package stem with lead pins and the cap with the window are finally joined with solder under dry N_2 or an inert gas. The soldering is carried out either by using Joule's heating generated by applying an instantaneous and high electrical bias between the stem and the cap or by using the heating due to high-power laser beam irradiation.

Most of the rectangular-type packages are used in optical communication systems. Ceramics are often used as their package frames, and a metallized layer on the ceramics is used for joining. This layer is formed with a metal such as Ag, Ag-Pd, Cu, or Ni. At the electrical input and output terminals, the metallized layers on the ceramics connect the devices with the lead pins as shown in Fig. 6.24(b) and hermetically seal the interface between the ceramic parts. At the optical ports, fibers can be inserted into the ceramic ferrule and fixed with plastic, or fibers plated with metal can be soldered. A hard solder is needed here in order to suppress the creep at the joint. At the joint of the ceramic frame and the metal cover, the metallized layer is also used for soldering. After baking, the soldering is carried out under dry N_2 or an inert gas. If the package frame is metal, laser welding in addition to the soldering is frequently used for joining.

The packaging materials and fabrication technologies, especially for the modules used in communication systems, are chosen to provide long-term stability under various operating conditions. The hermetic sealing with high reliability, however, has led to complicated packaging processes and to increases in the packaging cost. There are of course some exceptions, such as laser diode packaging for optical disk systems, in which the packaging cost is greatly reduced by the use of standardized components for packaging and by the cost-effectiveness of mass production.

6.3.4.2 Plastic Packaging   Plastic packaging can be carried out quite easily and is much less costly than hermetic sealing. The long-term stability under various operating conditions is, however, much inferior to that of hermetically sealed packages. The plastic materials have been improved for application in the packaging of optoelectronic devices over the last two decades and today's plastic packaging is relatively reliable. Plastic packaging is currently used in many of the optoelectronic devices in consumer electronics, such as LEDs for

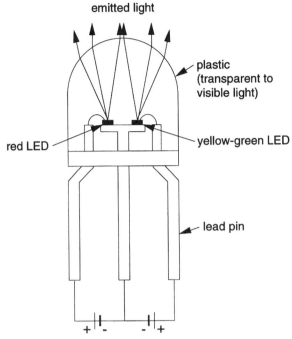

FIGURE 6.25 One example of cross-sectional view of a dichromatic LED lamp.

displays and indicators and LEDs and photodiodes for home electronics (see Chapter 8). A dichromatic LED lamp packaged in plastic is shown in Fig. 6.25. Lenses and domes can be easily formed with plastics, and a dome-shape molding is applied to the lamp in Fig. 6.25 in order to increase the amount of light emitted to the front.

The main requirements for the plastics used in the optoelectronic *pn*-junction devices are the following: low mechanical stress, high electrical resistivity, wet-resistance, light-resistance, and heat-resistance. Most of those requirements are related to reliability. The low mechanical stress is very important in plastic packaging because the device chips are directly covered with plastics. If the mechanical stress is large, the devices crack or easily degrade during operation (see Section 7.2.1.1). The mechanical stress generated at the interface between the device and the plastic can be basically expressed with Eq. (6.12) by replacing the stem with the plastic. The mechanical stress is due to the difference between the thermal expansion coefficients of the chip and the plastic. For pure plastic this coefficient is about an order of magnitude larger than that of semiconductors. When the coefficient needs to be decreased, fillers which are powders of some materials such as SiO_2 are added to the plastic. A wet-proof plastic packaging material is also essential because optoelectronic *pn*-junction devices are easily degraded by humidity. It is therefore important that the plastic and the packaging structure are wet-proof. Devices packaged in plastic are often coated with dielectric films to protect them from humidity. The light-resistance and heat-resistance are especially important for LEDs and laser diodes because of the intense light and large amount of heat generated at their active regions. The plastics used to package these devices, however, must of course be transparent to the input and output light. For a photodiode application, it is favorable that plastic is transparent only to the wavelength of input signal light and absorbs the other light for reducing background noise (dark current) (see Section 4.3.2.1).

Some kinds of epoxy resins are currently used for plastic packaging. They are basically transparent to input and output light and visible light. For photodiodes used in the infrared wavelength range, a kind of dye is added to the epoxy in order to block the visible light. These epoxy resins appear black or dark red. Plastic packages of various shapes are formed by curing at a relatively high temperature, after the devices are mounted on lead pins and dipped into liquid epoxy resin filling vessels of various shapes (lens-shaped or dome-shaped). Consequently, a large number of devices can be packaged in plastic simultaneously.

Some kinds of silicone resins without impurities are also used to cover optoelectronic *pn*-junction devices mounted in modules. Those are very soft plastics and very little mechanical stress is generated at the interface. These resins, however, are used less frequently than epoxy resins because they cannot protect the devices from humidity and external mechanical forces. They are often combined with an epoxy cover as shown in Fig. 6.26 or are used to cover devices mounted in ceramic and metal frames.[31,32] The epoxy cover and the ceramic and metal frames protect the devices from mechanical stress and humidity. The refractive indexes of the silicone resins are about 1.4 and nearly the same as

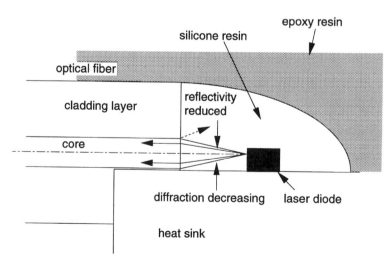

FIGURE 6.26   Simplified basic structure of plastic packaging with silicone resin and epoxy resin.

those of the cores of the silica fibers and silica-based waveguides. The reflection of incident light at the input port of the fiber/waveguide can be suppressed, and the diffraction of output light from the device or of input light from the fiber also decreases, if the silicone is filled up to the gap between the device and the fiber/waveguide. This increases the optical coupling.

The long-term stability of the plastic packages is determined by the properties of plastic materials. Various plastics for packaging optical devices have been developed and some have been used in the modules for data links. Plastic packaging will soon be another of the key technologies for optoelectronic *pn*-junction devices used in public optical fiber communication systems.

REFERENCES

1. H. M. Manasevit, Single-crystal gallium arsenide on insulating substrates, *Appl. Phys. Lett.*, **12**, 156, 1968.

2. J. P. Hirtz, M. Razeghi, M. Bonnet, and J. P. Duchemin, *GaInAsP Alloy Semiconductors*, John Wiley and Sons, New York, 1982.

3. H. C. Casey, Jr., and M. B. Panish, *Heterostructure Lasers*, Academic Press, New York, 1978.

4. W. T. Tsang, *Semiconductors and Semimetals*, Academic Press, New York, 1985, vol. 22, part A.

5. J. J. Hsieh, Thickness and surface morphology of GaAs LPE layers grown by supercooling, step-cooling, equilibrium-cooling and two-phase solution techniques, *J. Cryst. Growth*, **27**, 49, 1974.

6. See, for example, B. L. Sharma, Ohmic contacts to III-V compound semiconductors, in *Semiconductor and Semimetals*, eds. R. K. Willardson and A. C. Beer, Academic Press, New York, 1981, vol. 15, ch. 1; A. Piotorowska, A. Guivarch, and G. Pelous, Ohmic contacts to III-V compound semiconductors: A review of fabrication techniques, *Solid-State Electron.*, **26**, 179, 1983.

7. V. G. Keramidas, H. Temkin, and S. Mahajan, Ohmic contact to InP and InGaAsP, *Inst. Phys. Conf. Ser.*, **56**, ch. 5, p. 293, 1981 (Gallium Arsenide and Related Compounds, 1980).

8. See, for example, M. Ettenberg and I. Ladany, Metallization for diode lasers, *J. Vacuum Sci. Technol.*, **19**, 799, 1981; A. K. Shinha, T. E. Smith, M. H. Read, and J. M. Poate, *n*-GaAs Schottky diodes metallized with Ti and Pt/Ti, *Solid-State Electron.*, **19**, 489, 1976.

9. A. K. Sinha and J. M. Poate, Metal-compound semiconductor reactions, in J. M. Poate, K. N. Tu, and J. W. Mayer, eds., *Thin Films: Interdiffusion and Reactions*, John Wiley & Sons, New York, 1978, ch. 11.

10. A. Hiraki, S. Kim, W. Kammura, and M. Iwami, Dynamical observation of room temperature interfacial reaction in metal-semiconductor system by Auger electron spectroscopy, *Surface Sci.*, **86**, 706, 1979.

11. A. Hiraki, A model on the mechanism of room temperature interfacial intermixing

reaction in various metal-semiconductor couples: What triggers the reaction? *J. Electrochem. Soc.*, **127**, 2662, 1980.

12. S. A. G. Milnes and D. L. Feucht, *Heterojunctions and Metal-Semiconductor Junctions*, Academic Press, New York, 1972.

13. S. M. Sze, *Physics of Semiconductor Devices*, 2nd ed., John Wiley & Sons, New York, 1981.

14. A. Y. C. Yu, Electron tunneling and contact resistance of metal-silicon contact barriers, *Solid-State Electron.*, **13**, 239, 1970.

15. A. Piotrowska, P. Auvray, A. Guivarc'h, G. Pelous, and P. Henoc, On the formation of binary compounds in Au/InP system, *J. Appl. Phys.*, **52**, 5112, 1981.

16. G. S. Anderson, W. N. Mayer, and G. K. Wehner, Sputtering of dielectrics by high-frequency field, *J. Appl. Phys.*, **33**, 2991, 1962.

17. See for example, T. Miyazaki, N. Nakamura, A. Doi, and T. Tokuyama, Electrical properties of gallium arsenide-insulator interface, *Proc. 2nd Int. Conf. on Solid Surface*, Kyoto, 1974; *Japan. J. Appl. Phys.*, **13**, supplement 2-2, 441, 1974.

18. Y. Shima, N. Chinone, and R. Ito, Effects of facet coatings on the degradation characteristics of GaAs-$Ga_{1-x}Al_x$As DH lasers, *Appl. Phys. Lett.*, **31**, 625, 1977.

19. H. Imai, M. Morimoto, K. Hori, M. Takusagawa, and H. Saito, Long-lived high-power GaAlAs DH laser diodes, *IEEE J. Quantum Electron.*, **QE-16**, 248, 1980.

20. M. Fukuda and K. Wakita, The stability of coating film-mirror facet interfaces of AlGaAs/GaAs laser diodes, *Japan. J. Appl. Phys.*, **19**, 1969, 1980.

21. T. M. Benson, Strain-induced effects in GaAs directional coupler switches, *J. Appl. Phys.*, **54**, 6221, 1983.

22. M. Born and E. Wolf, *Principles of Optics*, 5th ed., Pergamon Press, New York, 1975.

23. L. A. Coldren, K. Iga, B. I. Miller, and J. A. Rentchler, GaInAsP/InP stripe-geometry laser with a reactive-ion-etched facet, *Appl. Phys. Lett.*, **37**, 681, 1980.

24. M. Uchida, S. Ishikawa, N. Takado, and K. Asakawa, An AlGaAs laser with high-quality dry etched mirrors fabricated using an ultrahigh vacuum in situ dry etching and deposition processing system, *IEEE J. Quantum Electron.*, **QE-24**, 2170, 1988.

25. S. J. Fonash, An overview of dry etching damage and contamination effects, *J. Electrochem. Soc.*, **137**, 3885, 1990.

26. C. Constantine and D. Johnson, Plasma etching of III-V semiconductors in CH4/H2/Ar electron cyclotron resonance discharges, *J. Vac. Sci. Technol.*, **B8**, 596, 1990.

27. C. L. Julian, Theory of heat conduction in rare-gas crystals, *Phys. Rev. 1A*, **137**, A128, 1965.

28. G. A. Slack, Nonmetalic crystals with high thermal conductivity, *J. Phys. and Chem. Solids*, **34**, 321, 1973.

29. See, for example, Y. Yamada, S. Suzuki, K. Moriwaki, Y. Nibino, Y. Tohmori, Y. Akatsu, Y. Nagasuga, T. Hashimoto, H. Terui, M. Yanagisawa, Y. Inoue, Y. Akahori, and R. Nagase, Application of planar lightwave circuit platform to hybrid integrated optical WDM transmitter/receiver module, *Electron. Lett.*, **31**, 1366, 1995; Special Feature (2): Technologies for economical optical access system components, *NTT Review*, **9** (6), 1997.

30. O. Mitomi, T. Nozawa, and K. Kawano, Effects of solder creep on optical component reliability, *IEEE Trans. Comp. Hybrids, and Manufacturing Technol.*, **CHMT-9**, 265, 1986.

31. M. Fukuda, F. Ichikawa, Y. Yamada, Y. Inoue, K. Kato, H. Sato, T. Sugie, H. Toba, and J. Yoshida, Highly reliable plastic packaging for laser diode and photodiode modules used for access network, *Electron. Lett.*, **33**, p. 2158, 1997.

32. See, for example, J. W. Osenbach and T. L. Evanosky, Temperature-humidity-bias-behavior and acceleration model of InP planar pin photodiodes, *J. Lightwave Technol.*, **14**, 1865, 1996.

CHAPTER SEVEN

Reliability

Optoelectronic *pn*-junction devices, such as lasers, LEDs, photodiodes, and optical modulators, have been described in the previous chapters from the viewpoint of their basic design, performance, and fabrication technology. These are the main devices in various kinds of optoelectronic equipment and systems, so their reliability must be assured. Reliability is the key factor for determining whether the devices can be put on the market.

A reliable optoelectronic *pn*-junction device is one that will operate in a specified environment for a required period of time. Reliability is therefore expressed as the probability that a device will operate in the normal situation under the required conditions for the required time period without failure. Failure is defined at the point when the device loses this required ability. This chapter summarizes the basic and main failure modes from the physical point of view and gives some examples of actual reliability assessments for optoelectronic devices.

7.1 BASICS OF RELIABILITY

7.1.1 Reliability Function and Failure Rate

Reliability is generally expressed as the probability of operating for a certain time and is given by a reliability function. The reliability function, $R(t)$, gives the percentage of devices operating until a certain time. The value of $R(t)$ is equal to 1 at $t = 0$ and is 0 at infinite time. An unreliability function denoted $F(t)$ is defined by

$$R(t) + F(t) = 1, \tag{7.1}$$

and indicates the probability of failure by operating time t. It is called the failure distribution function. These reliability and failure distribution functions corre-

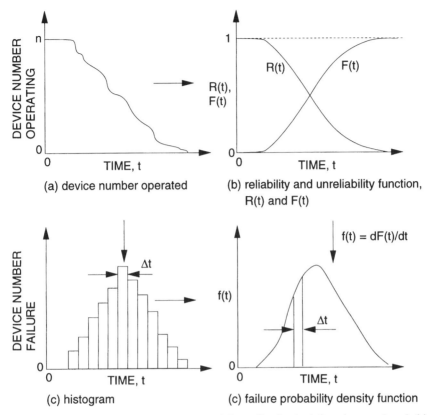

(a) device number operated

(b) reliability and unreliability function, R(t) and F(t)

(c) histogram

(c) failure probability density function

FIGURE 7.1   Reliability and unreliability (failure distribution) functions and probability density function.

late with actual test data as shown in Fig. 7.1(a). The number of devices operating decreases monotonically as a result of device failure, and a histogram corresponding to the failed devices can be obtained as shown in Fig. 7.1(c). When the sampling interval, Δt, is infinitely small, the shape of the entire histogram corresponds to that of the failure probability density function $f(t)$. The failure probability density function indicates the velocity of the increase in $F(t)$ and is given by

$$f(t) = dF(t)/dt = -dR(t)/dt, \qquad (7.2a)$$

where

$$F(t) = \int_0^t f(t)\, dt \qquad (7.2b)$$

and

$$R(t) = 1 - F(t) = \int_t^\infty f(t) \, dt. \tag{7.2c}$$

The probability of failure, P_F, for a time between t and $t + dt$ is given by the product of $R(t)$ and the failure rate at t, $\lambda_f(t)dt$ as

$$P_F = R(t)\lambda_f(t)dt = f(t)dt. \tag{7.3}$$

Thus

$$\lambda_f(t) = f(t)/R(t) = [-dR(t)/dt]/R(t). \tag{7.4}$$

The failure rate, $\lambda_f(t)$, is also called the hazard rate. Equation (7.4) can be rewritten to express the reliability by an exponential function:

$$R(t) = \exp\left[-\int_0^t \lambda_f(t)dt \right]. \tag{7.5}$$

Optoelectronic *pn*-junction devices are generally nonrepairable items and cannot be reused after failure. The failure rates for those nonrepairable devices changes with operating time as shown in Fig. 7.2. The curve is called the bathtub curve and consists of an early (infant) failure period, a random (chance) failure period, and a wearout failure period. The failure in the early failure period is ordinarily eliminated by the implementation of screening tests, and the devices passing the screening tests are subject to random and wearout failures.

7.1.2 Expression of Reliability for Nonrepairable Devices

The lifetime of a nonrepairable device is defined by the time at which the device fails after starting to operate. This time is called the time to failure (TTF), and the mean value of TTF is called the mean time to failure (MTTF). The device life is characterized by using the MTTF, $E(t)$, and the variance, $V(t)$, indicating a degree of deviation from the mean value:

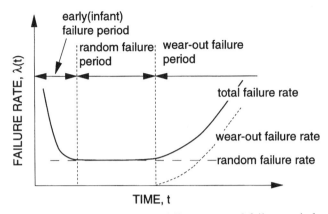

FIGURE 7.2   Bathtub curve, failure rate, and failure period.

$$E(t) = \int_0^\infty t f(t) \, dt$$

$$= \int_0^\infty R(t) \, dt \tag{7.6}$$

$$V(t) = \int_0^\infty [t - E(t)]^2 f(t) \, dt$$

$$= \int_0^\infty t^2 f(t) \, dt - E(t)^2. \tag{7.7}$$

This $E(t)$ corresponds to the area bounded by the curve $R(t)$ and horizontal axis in Fig. 7.1(b). The MTTF and variance can be estimated by the failure probability density function and then the failure rate as indicated in Eqs. (7.6) and (7.7).

7.1.3   Reliability Estimation for Nonrepairable Devices

As described above, laser diodes, LEDs, photodiodes, and modulators are non-repairable devices, and their lives and reliability are defined by their TTFs. They are used under various operating conditions, and their reliability under actual operating conditions therefore needs to be assessed. LEDs usually operate at a constant current and laser diodes are generally used under conditions in which output power is constant, although they are also sometimes used under those in which current is constant. A decrease in optical output is observed during degradation in LEDs as shown in Fig. 7.3. For laser diodes, an increase in operating current is generated during degradation, as a result of which the operating

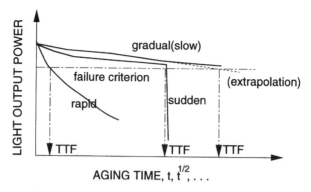

(a) LEDs and laser diodes under a constant current operation

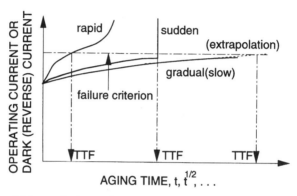

(b) laser diodes under a constant power, photodiodes, and modulators

FIGURE 7.3   Typical degradation pattern and time-to-failure for LEDs, laser diodes, photodiodes, and optical modulators.

current automatically increases in order to maintain the constant output power as shown in Fig. 7.3. When laser diodes operate at a constant current, their output power decreases like that of LEDs. Photodiodes ordinarily operate under a constant voltage bias or without bias. Consequently, a reverse leakage current (dark current) at a certain constant reverse voltage is usually used as a measure of their degradation. For modulators used under reverse bias, the dark current or extinction ratio is used as a measure of degradation. The relations between degradation modes and TTFs are summarized for these devices in Fig. 7.3. The reliability or the probability density function of failure is estimated from those aging data (TTFs). As shown in Fig. 7.3, the failure of the devices scarcely corresponds to the point at which LEDs stop emitting light or laser diodes stop lasing. The situation for photodiodes and modulators is similar to that for LEDs and laser diodes. Before the devices fail completely, they fail to provide the

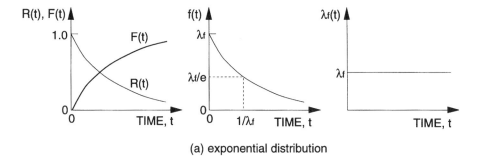

(a) exponential distribution

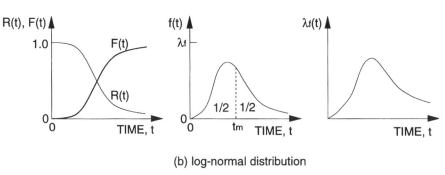

(b) log-normal distribution

FIGURE 7.4 Exponential and log-normal distribution functions: Cumulative failure distribution function $F(t)$ [and reliability function $R(t)$], failure probability density function $f(t)$, and failure rate $\lambda_f(t)$.

performance required by the equipment or systems. The device lives are therefore determined by taking into consideration equipment or system requirements and device performance. A typical failure criterion for LEDs, for example, is a 1-dB decrease in optical output power from the initial value, and a typical failure criterion for laser diodes is an increase in operating current to 1.5 times the initial value. For photodiodes, one criterion is a tenfold increase in dark current. These criteria are used to estimate the TTF for each aging device, and then the probability density function can be deduced from the cumulative failure distribution for the devices tested as indicated in Eq. (7.2). Then the device lives and failure rates are calculated. The log-normal distribution and exponential distribution (see Fig. 7.4) are often used as the failure distribution functions for LEDs, laser diodes, photodiodes, and modulators. The device life and failure rate in wearout failure modes are often estimated by assuming a log-normal distribution, while those in random failure modes are generally obtained by using an exponential distribution. Actually, the probability paper is used for analyses of wearout failure as shown in Fig. 7.5. If the TTF is linearly plotted on the probability paper, the failure data fit the failure distribution of the probability paper and can be treated with the distribution function. The median time

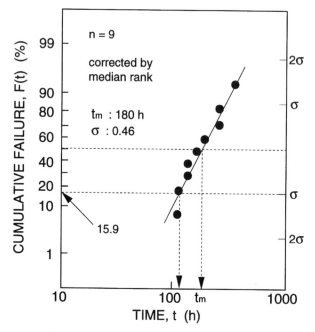

FIGURE 7.5 Use of log-normal distribution probability paper in reliability analysis.

to failure (MTF or median life) at which the 50% of devices tested fail and the standard deviation which is determined by the slope of the linear plot are obtained from the plots. The failure rate for wearout failure, $\lambda_w(t)$, is then estimated as shown in Fig. 7.6 (this is a professional procedure, See Refs. 1–3). The standard deviation, σ, of the log-normal distribution is given as

$$\sigma = \ln(t_m/t_1), \tag{7.8}$$

where t_m and t_1 are the MTF and the time corresponding to a cumulative failure of 15.9%. The failure rate for random failure is easily estimated by using the exponential distribution function and is given by

$$\lambda_r(t) = n_{sample}/t_{comp} = 1/MTTF, \tag{7.9}$$

where n_{sample} and t_{comp} are the number of samples tested and the component hour (or device hour) given by the sum of aging time of each device. After the failure rates, $\lambda_w(t)$ and $\lambda_r(t)$, are estimated, their certainty is assessed. This assessment is omitted here but is described in some other books.[1–3] The total failure rate is given by the sum of the two kinds of failure rates (see Fig. 7.2):

$$\lambda_f(t) = \lambda_w(t) + \lambda_r(t). \tag{7.10}$$

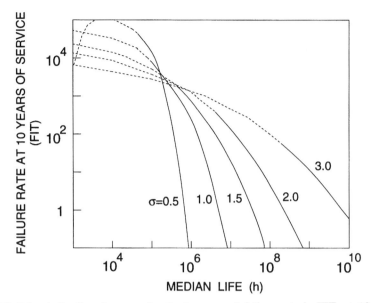

FIGURE 7.6 A family of curves for the log-normal failure rate in FITs at 10 years' service. The area indicated with dashed line has no meaning because the maximum failure rate appears before the service term (10 years).

A quantity designated FIT is usually used as the unit of the failure rate and is defined as

$$\text{FIT} = [n_{\text{failed}}/t_{\text{comp}}] \times 10^9, \tag{7.11}$$

where n_{failed} is the number of samples failed. If the item tested is composed of several devices—as is a laser module, which includes a laser diode, a monitor photodiode, and other passive components—the total failure rate is given by the sum of the failure rate for each device and component as indicated in Eq. (7.10).

7.1.4 Reliability Tests and Conditions

There are several kinds of reliability tests for the devices: long-term endurance tests, accelerated tests, high-stress tests, stepwise stress tests, marginal tests, environmental tests, and so forth. The most important point of these tests is to assess the reliability under conditions corresponding to those in the actual equipment or systems in which the devices will be used. The test conditions therefore need to be selected carefully because the degradation modes—and thus the device lives and failure rates—are different under different operating conditions. The reliability tests are usually performed under conditions in which aging is accelerated but the degradation mode and failure mechanism are the same observed in actual equipment and systems. Under these conditions, the next equation holds:

$$\Delta D = \Delta St, \tag{7.12}$$

where ΔD indicates the quantity of degradation, ΔS is the magnitude of the stress, and t is the aging time. Equation (7.12) indicates that the influence of the magnitude of the stress, such as current and temperature, and the aging time on the quantity of degradation is the same. Under this relation, the testing time can thus be shortened by increasing the magnitude of stress without changing the failure mechanisms of the devices. Acceleration beyond the range within which the failure mechanism differs from that in the actual equipment and systems is never indicative of the actual reliability.

The most important and widely used test for estimating device life and failure rate of optoelectronic pn-junction devices is the accelerated life test. A constant current aging (parameters: current and temperature) or a constant output power aging (parameters: output power and temperature) is used for testing forward-bias devices, such as laser diodes and LEDs, whereas a constant voltage aging (parameters: voltage and temperature) is frequently used for testing reverse-bias devices, such as photodiodes and modulators. The cumulative failure distribution under these conditions is obtained from the TTF data, and then the MTTF, median life (or MTF), and variance are calculated. These calculations are required for estimating the reliability characteristics under actual operating conditions. When the temperature is a parameter (temperature acceleration), the obtained MTTF and MTF are empirically related to the temperature by using an Arrhenius relation:

$$\text{MTTF}(T)\text{or MTF}(T) = A_\text{T} \exp(E_\text{a}/k_\text{B}T), \tag{7.13}$$

where E_a is the activation energy and A_T is a constant if the other accelerating parameter is constant. In the same way that the reliability characteristics are related to the temperature, they have to be related to the magnitude of other aging stresses. The relation is also empirically given by

$$\text{MTTF}(T)\text{or MTF}(T) = A_\text{stress} \, \Delta S^{-m}, \tag{7.14}$$

where ΔS is the magnitude of the stress (such as current or output power) and a constant m is greater than zero. The term A_stress is a constant if the other accelerating parameter is constant. MTTF and MTF of devices in equipment and systems can be estimated from the accelerated aging results by using Eqs. (7.13) and (7.14).

Another important meaning of the accelerated aging tests is to clarify the degradation modes and failure mechanisms. The clarification of the mechanisms provides a physical base for the statistical estimation and allows reliability to be increased by changing device designs appropriately or by improving fabrication processes and crystal growth techniques in ways targeted to prevent the identified degradation modes and failure mechanisms.

7.2 BASICS OF DEGRADATION AND RELIABILITY FOR OPTOELECTRONIC *pn*-JUNCTION DEVICES

The reliability of a device is determined by the weakest part of the device and depends on factors ranging from material properties to processing technologies. Failure modes can be roughly divided into three categories: sudden, rapid, and gradual in corresponding to the degrading part as shown in Fig. 7.3. The failure modes and their causes for laser diodes and LEDs (forward-bias devices) are summarized in Fig. 7.7, and those for photodiodes and modulators (reverse-bias devices) are summarized in Fig. 7.8. The main causes of failure and the factors enhancing the failure are summarized in Table 7.1 and 7.2.

Most optoelectronic devices have a *pn*-junction to inject carriers, to separate photo-induced carriers, or to generate a strong electric field as described in the

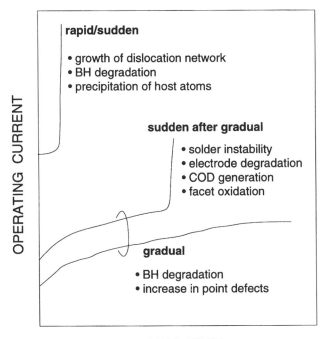

AGING TIME

FIGURE 7.7   Failure modes and causes for laser diodes and LEDs (forward bias devices).

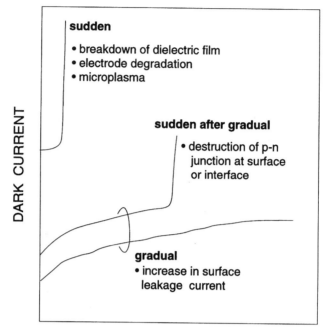

AGING TIME

FIGURE 7.8   Failure modes and causes for photodiodes and optical modulators (reverse bias devices).

TABLE 7.1   Main Causes of and Factors Enhancing the Failure of Laser Diodes and LEDs (Forward Bias Devices)

Part	Cause	Enhancement Factor
Inner region	Dislocation, precipitation	Current, ambient temperature (heat and light)
Facet (surface)	Oxidation	Light, moisture
Electrode	Metal diffusion, alloy reaction	Current, ambient temperature (heat)
Bonding part	Solder instability (reaction and migration)	Current, ambient temperature (heat)
Heat sink	Separation of metal	Ambient temperature (heat), current
BH interface	Defect at BH interface	Current, ambient temperature (heat)

TABLE 7.2   Main Causes of and Factors Enhancing the Failure of Photodiodes and Optical Modulators (Reverse-Bias Diodes)

Part	Cause	Enhancement Factor
pn-junction		
Inner region	Microplasma	Electrical field (current), ambient temperature (heat)
Interface (surface)	Interface (surface) state density, depletion layer narrowing	Electrical field, leakage current, ambient temperature (heat)
Dielectric film between electrode and semiconductor	Low quality of dielectric film	Electrical field, ambient temperature (heat)
Electrode	Metal diffusion, alloy reaction	Current, ambient temperature (heat)

previous chapters. Light-emitting diodes (LEDs) and laser diodes use radiative recombination of the injected carriers, pin-photodiodes and APDs detect photocurrent by dividing electrons and holes at the *pn*-junction, and modulators require a strong electric field at the *pn*-junction for normal operation. The degradation mechanisms of such devices can be roughly divided into two categories according to the direction of bias at the *pn*-junction. For devices used under a forward bias, the main factors in degradation are the current flow through the *pn*-junction and the emitted light, whereas for devices used under a reverse bias the main factor is the high electric field at the *pn*-junction. This strong field causes inhomogeneous current flow to increase at the weak part of the *pn*-junction and finally results in the destruction of the *pn*-junction. The following sections discuss the degradation modes and failure mechanisms.

7.2.1   Degradation of Devices Used Under Forward Bias: Laser Diodes and LEDs

As summarized in Table 7.1, the degradation of optoelectronic *pn*-junction devices used under forward bias occurs at electrodes, crystals, and bonding parts. Facet degradation is also a factor in laser diodes, and buried heterointerface (BH) degradation occurs in devices with a buried heterostructure. These degradations are observed in bulk, quantum well (QW), and strained QW devices and they determine device reliability.

7.2.1.1   Crystal: Dislocation and Precipitation   The main causes of rapid degradation are generation and growth of dislocations, and sometimes precipitation of host atoms. The growth of the dislocation is divided into two directions, $\langle 100 \rangle$ and $\langle 110 \rangle$, when the dislocation is observed from the direction perpendicular to the (001) substrate (normally the *pn*-junction plane). The $\langle 100 \rangle$

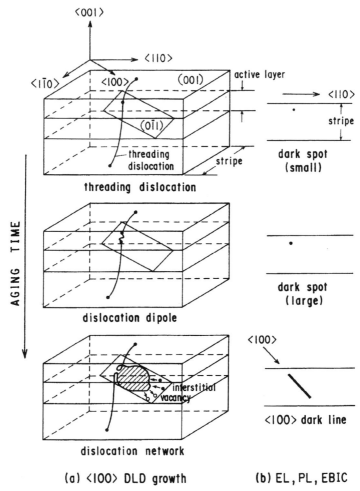

FIGURE 7.9   Illustration of (a) an elongation model of ⟨100⟩ dislocation network and (b) an EL topograph, an EBIC image, and a PL topograph.[3] Reprinted with permission from *Reliability and Degradation of Semiconductor Lasers and LEDs* by Artech House, Inc., Norwood, MA, USA, http://www.artech-house.com.

dislocation network grows by so-called nonradiative-recombination-enhanced defect motion.[4] The energy emitted by the nonradiative recombination at the dislocation and other defects, such as point defects, is transformed into lattice vibrations through multiple phonon emission and gives rise to the low-temperature defect motion. Here the defects are mainly interstitial atoms or vacancies. By absorbing an interstitial atom or emitting a vacancy, the dislocation network elongates and results in a very complicated structure even though the growth direction is the ⟨100⟩ direction.[3,5] One example of a mechanism by which a ⟨100⟩ dislocation network elongates is shown in Fig. 7.9(a).

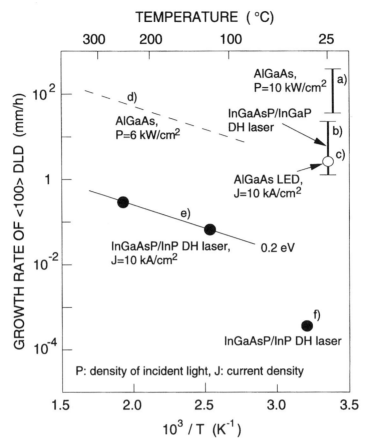

FIGURE 7.10 Growth rate of ⟨100⟩ dislocation networks in various materials. (a) R. Ito, et al., *IEEE, J. Quantum. Electron.*, **QE-11,** 551, 1975; (b) O. Ueda, et al., *J. Appl. Phys.*, **57,** 1523, 1985; (c) S. Yamakoshi, et al., *Tech. Digest Int. Electron. Device Mtg.*, Washington D.C., 1978, p. 642; (d) H. Imai, et al., *Japan. J. Appl. Phys.*, **18,** 589, 1979; (e) M. Fukuda, et al., *J. Appl. Phys.*, **54, ** 1246, 1983; (f) K. Ishida, et al., *Appl. Phys. Lett.*, **40,** 16, 1982.

The rate of dislocation network growth depends on the properties of the material used as shown in Fig. 7.10. AlGaAs/GaAs systems show high rates of dislocation growth but InGaAs(P)/InP systems show no or very slow growth. The main origins of the dislocations are threading dislocation originating from defects in the substrate and stacking faults introduced during crystal growth. Consequently, the generation of ⟨100⟩ dislocations can be reduced by improving crystal growth techniques and by using a high-quality substrate with a low etch pit density (defect density). These approaches have been used to prevent degradation resulting from dislocations. The ⟨100⟩ dislocation network is often observed in III-V compound semiconductors with relatively large band-gaps,

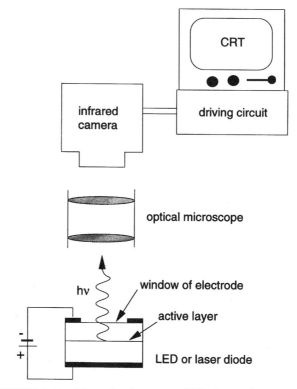

FIGURE 7.11   Electroluminescence (EL) observation system.

such as AlGaAs/GaAs or materials emitting visible light, and also appears in II-VI compound semiconductors. Laser diodes and LEDs with CdZnSe quantum well structures exhibit the ⟨100⟩ dislocation networks (dislocation loops) confined to the quantum well region.[6] The dislocation networks in materials showing nonradiative-recombination-enhanced defect motion act as a sink for injected carriers and as an absorber of emitted light. Consequently, as shown in Fig. 7.9 (b), such a network can be observed as a ⟨100⟩ dark line in an electroluminescence (EL) topograph, an EBIC (electron beam induced current) image, a photoluminescence (PL) topograph, and so forth. The dark line is called a dark line defect. An EL observation system is shown for reference in Fig. 7.11.

The ⟨110⟩ dislocation network is caused by mechanical stress, and its growth rate thus depends on the magnitude of the stress. The growth rate also depends on the bonding strength (energy) of host atoms in the crystal. In III-V compound semiconductors, the slip plane under the mechanical stress is the (111) plane and the projection of this plane to the (001) plane is the ⟨110⟩ dislocation (see Fig. 7.12). If the mechanical sress is more than about 10^8 dyn/cm^2, the slip dislocation grows from surface or interface, such as semiconductor-electrode interface, under current injection.[7] When the slip dislocation reaches the active layer,

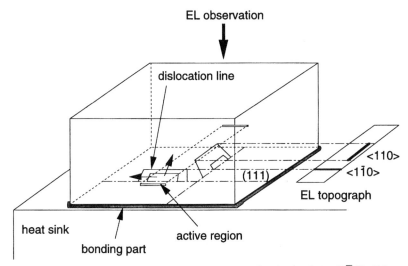

FIGURE 7.12   Illustration of (a) an elongation model for the $\langle 110 \rangle$ (or $\langle 1\bar{1}0 \rangle$) dislocation network, and (b) an EL topograph.[3] Reprinted with permission from *Reliability and Degradation of Semiconductor Lasers and LEDs* by Artech House, Inc., Norwood, MA, USA, http://www.artech-house.com.

$\langle 110 \rangle$ dark line defects appear in the EL topograph. This kind of dislocation is scarcely related to the defect motion but is a result of mechanical stress and therefore is found in most semiconductor materials. Ternary and quaternary materials, such as InGaAs and InGaAsP, tend to be resistant to dislocation growth. The generation of $\langle 110 \rangle$ dislocation networks can be suppressed by decreasing the amount of mechanical stress from the outside of the chips—for example, by improving the mounting process—but these dislocations are observed often even in recent devices.

The devices with degradation due to dislocation networks are ordinarily eliminated by screening tests before application. The long-term stability of the semiconductor crystals in devices which pass the screening is determined by point defects if the material shows nonradiative-recombination-enhanced defect motion.

Precipitation is mainly found in quaternary materials and is observed as dark spots in electroluminescence topographs. The host atoms, such as In and Ga, locally precipitate and act as absorbers of emitted light.[8,9] This type of precipitation, however, is scarcely found in recent devices fabricated from high-quality wafers.

7.2.1.2 *Electrode: Diffusion and Reaction*   The electrodes in optoelectronic *pn*-junction devices used practically are usually ohmic contacts. Just after metal is deposited on a semiconductor, an energy barrier within the semiconductor is induced at the interface between the metal and the semiconductor. The influ-

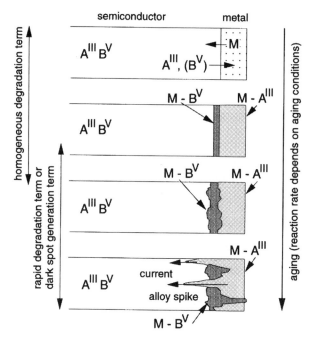

FIGURE 7.13   Conceptual illustration of the reaction of the interface between alloy-type electrodes and III-V compound semiconductors.

ence of this barrier needs to be eliminated in order to obtain ohmic contact. Two kinds of electrode, alloy-type and Schottky-type, are used for establishing ohmic contact (see Section 6.2.1).[10,11] A metal used in the alloy-type electrodes is required to easily react with semiconductors by sintering or alloying to reduce the height of the energy barrier. Devices with the alloy-type electrodes, such as AuZnNi, show electrode degradation under practical operating conditions. During operation the alloy reaction proceeds at the interface between the semiconductor and the electrode. The metals diffuse toward the inner region of the device, and alloy layers of the metal-III element and metal-V element are formed separately during operation. The electrical resistance of a metal-V element layer tends to be high,[3] and thus current concentrates in the thinner part of this layer, and then alloy spikes are formed under the current flow as shown in Fig. 7.13. Spikes on a 1300 nm InGaAsP/InP LED are shown in Fig. 7.14. When alloy spikes reach the active region, dark spot defects are evident in electroluminescence (EL) topographs and the *pn*-junction is partially damaged. The elemental metal atoms diffused from the electrode, such as Au, act as nonradiative recombination centers in the active layer and decrease device efficiency. During the penetration of the alloy spikes, a slip dislocation often grows as a result of the mechanical stress generated at the interface between the alloy region and semiconductor. In InGaAsP/InP sys-

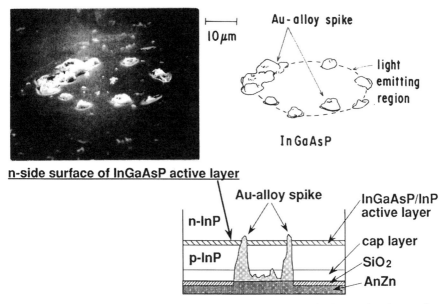

FIGURE 7.14   Alloy spikes on the *n*-side surface of an InGaAsP active layer after removal of an *n*-InP layer and substrate.[3] Reprinted with permission from *Reliability and Degradation of Semiconductor Lasers and LEDs* by Artech House, Inc., Norwood, MA, USA, http://www.artech-house.com.

tems, for example, when the alloy region extends through the InGaAsP cap layer and reaches the InP cladding layer, slip dislocations are generated and the device degrades rapidly.[12] This behavior for slip dislocations results from the difference between the threshold for mechanical stress on generation of slip dislocation in the InGaAsP cap layer and that in the InP cladding layer. The threshold is ordinarily higher in ternary and quaternary semiconductors than in binary semiconductors, and thus slip dislocations are generated in the InP cladding layer.

As discussed in Section 6.2.1.2, in the Schottky-type electrode the width of the barrier between the electrode and semiconductor is reduced by heavily doping the semiconductor and ohmic contact is established. A metal used in a Schottky-type electrode is required to form the inert interface between the metal and semiconductor. Under practical operating conditions, devices with Schottky-type electrodes, such as those made of Ti/Pt/Au, are quite stable for long times because of the inert interface. Under very severe operating conditions, however, for example, an ambient temperature of 200°C and a current density greater than 30 kA/cm^2, degradation similar to that seen in the case of alloy-type electrodes occurs as shown in Fig. 7.15. During operation under severe conditions, a depletion layer of the group-III element is formed at the interface because of out-diffusion of the group-III element. The diffusion here is the grain boundary diffusion in the Ti/Pt/Au layer. The electrical resistance

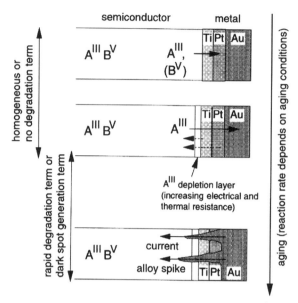

FIGURE 7.15   Conceptual illustration of the reaction at the interface between Schottky-type electrodes and III-V compound semiconductors.

of the depletion layer is high, and thus current tends to concentrate in the thinner part of the layer . Finally, alloy spikes are formed, dark spot defects become evident in electroluminescence (EL) topographs, and the *pn*-junction is partially damaged at the part corresponding to the dark spot defects.

A Schottky-type electrode, such as one made of Ti/Pt/Au, is commonly used in the most highly reliable devices. It is quite stable during long-term operation and shows no problems under practical conditions.

7.2.1.3   Facet (Surface): Oxidation and Catastrophic Optical Damage

Facet degradation is mainly a problem of laser diodes because the optical density at the facet is high (sometimes more than a few megawatts per square centimeter). As shown in Fig. 7.16, facet degradation is triggered by optical absorption of emitted light at the facet. Photo-induced electrons and holes nonradiatively recombine at the facet, and this recombination results in heating. The heating further decreases the band-gap energy at the facet, and this further increases light absorption at the facet. This positive feedback loop results in facet oxidation leading to catastrophic optical damage (COD). The generation of photoinduced electron-hole pairs is accompanied by bond-breaking at the facet. This bond-breaking increases the rate of semiconductor oxidation, and the light-emitting part of the facet is heavily oxidized as shown in Fig. 7.17. The depth profile of the laser facet can be obtained by using an Auger electron microprobe under Ar-ion sputtering. An oxide film about 50 nm thick is formed during 10^4 hours of aging in a 25°C atmosphere under a constant output power of 3 mW.

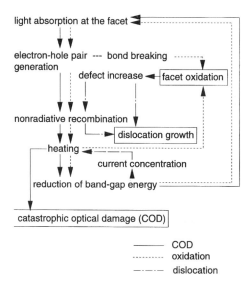

FIGURE 7.16   Positive feedback loop of facet oxidation and catastrophic optical damage.

(This thickness was calculated as the product of sputtering rate and time in the figure.) This enhancement of oxidation is called photo-enhanced oxidation, and the mechanisms of the resultant facet oxidation are basically similar to those of the thermal oxidation or oxidation at room temperature: the oxidation laws are the same as those for the thermal or low-temperature oxidation.[13] The stability of the facet against oxidation depends on the material used and roughly coincides with the material stability under thermal oxidation. From the viewpoint of reliability, facet oxidation can be understood as a process of defect injection into the active region.[3] Consequently, dislocation networks are often observed in the vicinity of the facet after severe facet oxidation of laser diodes composed of materials showing nonradiative-recombination-enhanced defect motion.

Catastrophic optical damage occurs suddenly when the facet temperature reaches the melting point of the semiconductor (see Fig. 7.18). The output power suddenly decreases after catastrophic optical damage. After the molten part cools, a dislocation loop extending from the facet in the direction perpendicular to the facet remains and is observed as a $\langle 110 \rangle$ dark line defect. This $\langle 110 \rangle$ dark line defect is, of course, different from that introduced by the slip dislocation caused by mechanical stress. As shown in Fig. 7.18, catastrophic optical damage is repeatedly generated in the part at which the output power density reaches the damage level.

Facet degradation depends very much on the properties of material used to make the laser diode. AlGaAs/GaAs systems are oxidized and heated by nonradiative recombination of photo-induced carriers more easily than are InGaAs(P)/InP systems. The rate of oxidation is usually high in 850 nm AlGaAs/GaAs laser

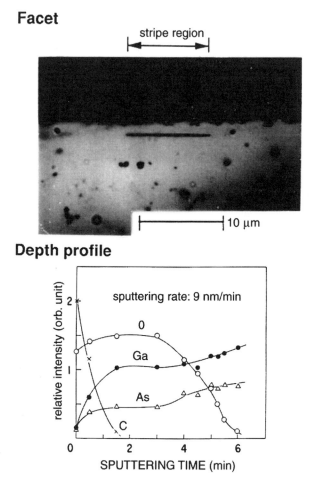

FIGURE 7.17   Facet oxidation and Auger depth profile for a planar-type (oxide stripe) AlGaAs/GaAs laser diode. The aging was carried out at 3 mW in a 25°C atmosphere for 10^4 hours.[3] Reprinted with permission from *Reliability and Degradation of Semiconductor Lasers and LEDs* by Artech House, Inc., Norwood, MA, USA, http://www.artech-house.com.

diodes and other laser diodes emitting visible light, whereas oxidation is very slow in 1300 nm-band and 1550 nm-band InGaAsP/InP laser diodes.[3] The output power level at which catastrophic optical damage occurs is generally lower and the facet oxidation rate higher in laser diodes composed of the materials showing the nonradiative-recombination-enhanced defect motion than in laser diodes composed of materials showing little or no nonradiative-recombination-enhanced defect motion. This is because a large amount of energy is emitted by nonradiative recombination and facet temperature therefore rises easily when laser diodes are made of these materials showing the enhanced defect

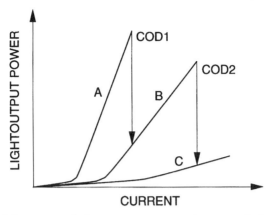

(a) current-light output power characteristics

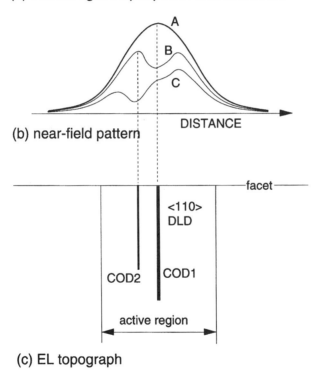

(b) near-field pattern

(c) EL topograph

FIGURE 7.18 Change in lasing characteristics when catastrophic optical gamage occurs.[3] Reprinted with permission from *Reliability and Degradation of Semiconductor Lasers and LEDs* by Artech House, Inc., Norwood, MA, USA, http://www.artech-house.com.

motion. Henry et al. have calculated that the facet temperature in a typical AlGaAs/GaAs double heterostructure rises by about 1500 K within about 100 ns under a light output power around 5 MW/cm.[14] The level at which catastrophic optical damage occurs tends to be low when the photon energy emitted is large. In addition, it is inversely proportional to the pulse duration (input optical energy) if laser diodes operate under pulsed operation. In 850 nm-band AlGaAs/GaAs laser diodes, the damage level decreases with increasing operating current pulse width (from around 1 to 15 MW/cm^2) and approaches the level under dc (a few hundred kilowatts per square centimeter to 1 MW/cm^2). For 1300 nm-band and 1550 nm-band InGaAsP/InP laser diodes, the level is beyond 10 MW/cm^2 even under dc operation.[15]

In compressively strained quantum well lasers, facet degradation is enhanced by the band-gap shrinkage. The strain in the quantum wells is introduced by lattice mismatching, and the stress changes from biaxial in the inner region to uniaxial at the facet. This strain release at the facet induces band-gap shrinkage at the facet, and this band-gap shrinkage is added into the feedback loops in Fig. 7.16. This problem of strain-release can be clearly seen in the facet degradation of 980 nm-band InGaAs/GaAs strained quantum well lasers.[16]

The facet degradation can be reduced or eliminated by cutting the feedback loops shown in Fig. 7.16 at any point. The most effective way for cutting them is to use a nonabsorbing mirror structure or to reduce the density of defects participating in the nonradiative recombination, and to use some kind of passivation before applying the dielectric film coating.[17] The band-gap energy at the facet is higher than that in the inner region when a nonabsorbing mirror structure is made by regrowth of a material having a larger band-gap energy on the facet or by impurity diffusion in regions other than the facet region.[17,18] The band-gap can be reduced by a few tens of millielectron-volts, for example, when zinc is diffused as an impurity. Consequently, the light emitted in the inner region is scarcely absorbed at the facet and band-gap shrinkage at the facet is not induced. A dielectric film coating on the facets is often used to suppress oxidation by preventing contact with the atmosphere and to increase the level at which catastrophic optical damage occurs by decreasing the surface state density. Decreasing the output power density at the facet also suppresses the damage and can be performed by making a thin active layer (reducing optical confinement to the active layer) or by using a structure with a large optical cavity.[19]

7.2.1.4 *Buried Heterointerface*

Buried heterointerface degradation is, of course, peculiar to the buried heterostructure type of devices, such as laser diodes and some edge-emitting-type LEDs. The buried heterostructure laser diodes can be divided into two types as shown in Fig. 7.19. In type I the double heterostructure is physically and chemically etched down to the mesa structure and then buried by the semiconductor layers. In type II the double heterostructure including an active layer is grown in the groove built on the semiconductor epitaxial layer and substrate. The buried heterointerface degradation is

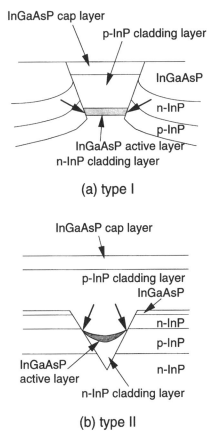

InGaAsP cap layer

p-InP cladding layer

InGaAsP

n-InP

p-InP

InGaAsP active layer

n-InP cladding layer

(a) type I

InGaAsP cap layer

p-InP cladding layer

InGaAsP

n-InP

p-InP

InGaAsP active layer

n-InP

n-InP cladding layer

(b) type II

FIGURE 7.19 Two types of buried heterostructure laser diodes. The arrows indicate the parts of the active region that are particularly susceptible to degradation.

induced by the increase in defects at the buried heterointerface during operation, although the kinds of defects responsible for this degradation have not been identified. The increased defect density leads to an increase in nonradiative recombination current (see Fig. 7.20). The rate of the buried heterointerface degradation depends on the device structure. The degradation of type-I laser diodes is sometimes faster than that of type-II laser diodes because the sides of the active layer in the type-I laser diodes are directly etched. It is difficult to apply the buried heterostructure of type I to laser diodes composed of materials showing nonradiative-recombination-enhanced defect motion. The most popular buried heterostructure can be seen in 1300 nm-band and 1550 nm-band InGaAsP/InP laser diodes, for which the buried heterointerface degradation governs the rate of degradation and thus reliability. The main cause of degradation in these devices is the increase in nonradiative recombination current described above. The injected carrier lifetime decreases because of the non-

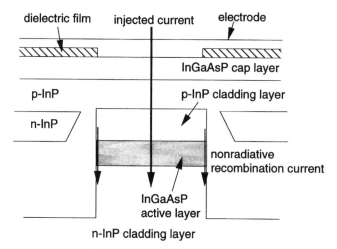

FIGURE 7.20   Illustration of buried heterointerface degradation.

radiative recombination, and the decrease in the lifetime results in the increase in threshold current [see Eq. (3.45)]. The threshold carrier density, however, is kept constant during the degradation,[20] and the slope efficiency (or external differential quantum efficiency) is also kept constant because the internal cavity (optical) loss scarcely changes. Consequently, lasing wavelength scarcely changes during buried heterointerface degradation, although the threshold current and operating current at a certain output power increase as shown in Fig. 7.21. This buried heterointerface degradation can be suppressed by improving

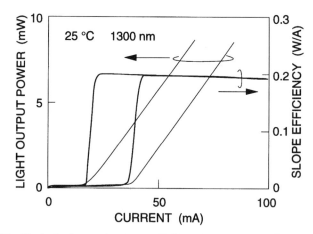

FIGURE 7.21   Typical change in current–light output power and current-slope efficiency characteristics during buried heterostructure degradation in InGaAsP/InP buried heterostructure laser diodes.

the quality of the buried heterointerface by cleaning or chemically etching the damaged layer on the side wall of the active layer just before regrowth of burying layers.[22]

7.2.2 Degradation of Devices Used Under Reverse Bias: Photodiodes and Modulators

The relation between the degraded parts, the main mechanisms, and the enhancement factors for devices used under reverse bias are summarized in Table 7.2 and Fig. 7.8. Most degradation modes are the same in the photodiodes and semiconductor modulators biased reversely. The main parts degraded are the *pn*-junctions, the electrodes, and the dielectric films between electrodes and semiconductors.

7.2.2.1 pn-Junction at the Surface or at the Interface Between the Semiconductor and the Dielectric Coating Film   The semiconductor surface and the interface between a semiconductor and a dielectric film are very defective because the crystal structure ends there and surface or interface states are introduced. Consequently, the surface or the interface tend to degrade under device operation more than does the inner region. If a *pn*-junction is located at the surface or at the interface between a semiconductor and a dielectric film, the increase in the defect density at the surface or at the interface induces leakage current. Under reverse bias conditions the electric field biased to the *pn*-junction leaks out to the dielectric film to some extent, and mobile ions (such as sodium ions) accumulate in the film and hot holes are injected from the semiconductor to the dielectric film at the *pn*-junction perimeter. These accumulations of electrical charge introduce an inversion layer as shown in Fig. 7.22, and then the depletion region narrows near the *pn*-junction at the surface (or interface). After the depletion layer narrows or the interfacial state density increases, the leakage

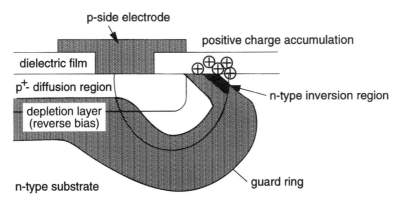

FIGURE 7.22 Schematic illustration of a model for the degradation of the *pn*-junction surface in devices used under reverse bias.

current increases. The leakage current due to the electrical charge accumulation decreases if the bias is stopped and the device is left for a certain term, because then the mobile charges spread again. The reverse (leakage) current is not large; it can, for example, be measured in microamperes, but most of it flows at the *pn*-junction perimeter. The leak path therefore melts suddenly when the current density reaches a critical value, and then the *pn*-junction is destroyed. These degradation modes have been observed mainly in *pin*-photodiodes and APDs, but similar modes are also observed in modulators because those devices are used under reverse bias.[23,24]

7.2.2.2 *pn-Junction in the Inner Region: Microplasma*

If there is a defect such as a threading dislocation through the *pn*-junction, the breakdown voltage of the device is decreased tremendously there, and current concentrates there. This current concentration induces so-called microplasma at the defective small area of the *pn*-junction.[25] As a result, an instability in current-voltage characteristics is induced. This instability introduces an electric noise in photodiodes, especially APDs. The total reverse current is small but, as in the case of the surface leakage, the density is quite high. When the current density reaches a critical value, the point generating the microplasma melts suddenly and the *pn*-junction is destroyed. Device failures due to this type of degradation are eliminated by decreasing the defect density during crystal growth or by screening tests.

7.2.2.3 *Electrode: Reaction*

As in laser diodes and LEDs, the alloy reaction between an electrode and the semiconductor occurs during operation when an alloy-type electrode is employed in devices used under reverse bias. The electrode, such as one made of AuZn, gradually alloys with the semiconductor cap layer during operation, and the leakage current increases when the alloy region reaches the *pn*-junction as shown in Fig. 7.23. The mechanisms of the electrode degradation have already been discussed in detail in Section 7.2.1.2. This degradation can be also eliminated by using a Schottky-type electrode.

7.2.2.4 *Dielectric Film Between the Electrode and the Semiconductor: Electric Breakdown*

During operation under reverse bias, a strong electric field is generated in the vicinity of the *pn*-junction as well as at the *pn*-junction. The electrodes and semiconductors are usually insulated from each other by a dielectric film except where they are intentionally in ohmic contact. The dielectric film is normally less than 1 μm thick and a potential of more than a few volts is applied between the electrode and the semiconductor. A strong electric field between the electrode and the semiconductor is thus generated, especially at the edge of the electrode. If the quality of the dielectric film is low or there are pin-holes in it, electric breakdown occurs as shown in Fig. 7.24. The breakdown point is usually at the edge of the electrode because the electric field is strongest there. The quality of the dielectric film tends to be low when the time and temperature of the processing after the dielectric film deposition are long

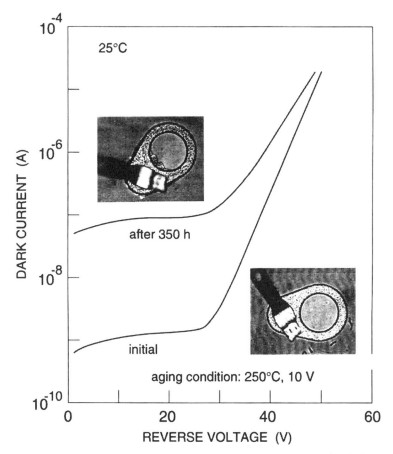

FIGURE 7.23   Change in dark current during electrode degradation of an InGaAs APD with an AuZn *p*-side electrode. The aging was carried out at 250°C for 1000 hours. The electrode metal penetrates into the window region after degradation in the photograph. (Courtesy of Mr. H. Sudo).

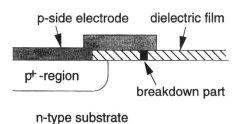

FIGURE 7.24   Schematic cross-sectional view of dielectric film breakdown (edge breakdown).

and high. This breakdown, which results in a sudden increase in the leakage (dark) current, can be suppressed by improving the process technology.

7.3   RELIABILITY

As described in the previous sections, there are many causes of failure (and degradation). This section qualitatively reviews the ways that those causes influence device reliability and the ways in which failure criteria are determined. Some examples of reliability estimation are also shown for each kind of device.

7.3.1   Light-Emitting Diodes (LEDs)

LEDs are generally used under constant current conditions, so their degradation is usually monitored by measuring the reduction of optical output power. This output power reduction is mainly caused by the dislocation network growth and the electrode degradation (for devices having alloy-type electrodes), both of which are degradations of the inner active region. In addition to those inner degradations, host atoms such as In and Ga sometimes precipitate in long-wavelength InGaAsP/InP LEDs made from low-quality wafers. AlGaAs/GaAs and InGaAsP/InP LEDs with Schottky-type electrodes, however, show no degradation when devices having precipitation or the origins of dislocation networks are eliminated in screening tests.

As LEDs are devices using spontaneous emission, they can operate over a temperature range wider than that in which laser diodes can be used. The aging conditions for LED reliability tests are therefore set at relatively high temperature (see Table 7.3). Here a case in which the reliability of 1300 nm InGaAsP/InP LEDs with AuZn p-side electrodes was estimated is described as an example. A log-normal plot of their wearout failure is shown in Fig. 7.25. The TTF here is the time at which the output power decreased by 1 dB and it is obtained from a plot like one shown in Fig. 7.3. From this plot the MTF and its standard deviation at an ambient temperature of 100°C and an injected current density of 12 kA/cm^2 are estimated to be about 7×10^4 hours and 0.5. From an Arrhenius plot of the MTF data, like that shown in Fig. 7.26 (where T_j and T_a are the junction temperature and the ambient temperature), the activation energy in Eq. (7.13) is estimated to be 0.8 eV. The linearity of the Arrhenius plot is very important here because it indicates that the degradation (failure) mechanism

TABLE 7.3   Examples of Accelerated Aging Conditions for LEDs

Current density	5–30 kA/cm^2
Ambient temperature	50–200°C

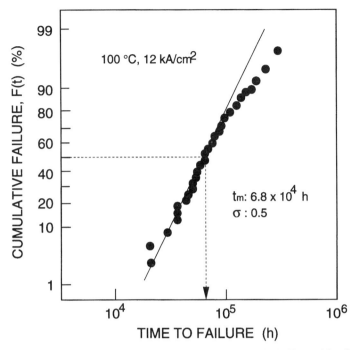

FIGURE 7.25 Log-normal failure distribution of LEDs with AuZn p-side electrodes.

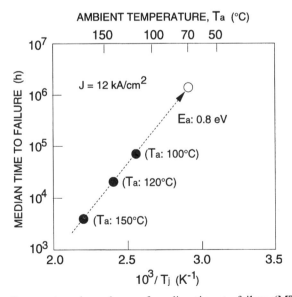

FIGURE 7.26 Temperature dependence of median time to failure (MTF) (Arrhenius plot) for LEDs with AuZn p-side electrodes and a device life estimation at $70°$C.

TABLE 7.4   Examples of Reliability of Surface-
Emitting InGaAsP/InP LEDs

p-Side Electrode	AuZn(Ni)	Ti/Pt/Au
Device life (h)[a]	$>10^6$	$\gg10^6$
Failure rate (FIT)	Wearout	<0.1
	Random	<100

[a]1 dB down from the initial light output power at 70°C.

is the same throughout the range of aging conditions tested. Extrapolating the linear trend to the actual operating temperature (here, $T_a = 70°C$) yields a device life (MTF) of more than 10^6 hours. From this MTF and the standard deviation the failure rate at a 10-year service term (at an ambient temperature of 70°C and an injected current density of 12 kA/cm^2) is calculated to be less than 0.1 FIT (see Fig. 7.6). This value is negligibly small. The rate of degradation depends on the square of the injected current density.[3] When the injected current changes, the MTF can also be estimated by using Eq. (7.14). The random failure rate can be estimated by using Eq. (7.9) and assuming an activation energy. If the activation energy is not obtained during reliability tests, a value of 0.35 eV is empirically used. The reliability test results for these LEDs are summarized in Table 7.4. The LEDs with Schottky-type p-side electrodes are generally more reliable than those with alloy-type p-side electrodes. For AlGaAs/GaAs LEDs and other LEDs emitting visible light, high reliability has also been estimated under various operating conditions.

Characteristics other than the current–light output power characteristics, such as modulation characteristics, have to be evaluated in order to confirm LED reliability in actual applications. The changes in the other characteristics, however, are small, and the change in the current–light output power characteristics can be used to monitor the failure of LEDs.

7.3.2   Laser Diodes

Lasers in equipment and systems are used under constant current or constant output power conditions. The degradation during constant current operation is usually monitored as a reduction of optical output power, whereas the degradation during constant output power operation is usually monitored as an increase in operating current. These changes are introduced by the increase in threshold current and the decrease in slope efficiency (external differential quantum efficiency). Roughly speaking, the main changes in device parameters during the degradation are the decrease in injected carrier lifetime for long-wavelength InGaAs(P)/InP devices including InGaAs/GaAs strained QW lasers and the increase in internal cavity loss (absorption coefficient) for AlGaAs/GaAs devices and other material systems emitting visible light. The failure criteria, and thus the device lives, are mainly determined by the change in current–light output characteristics.

The reliability of laser diodes differs from that of LEDs in that it is often determined by characteristics other than the current–light output power characteristics. Although the reliability of laser diodes used in most consumer electronics equipment (such as displays, printers, and bar-code readers) can be assured by monitoring current–light output power characteristics, various characteristics of the laser diodes used in optical fiber communication systems and some consumer electronics equipment need to be evaluated to assure reliability.

The range of conditions for reliability tests, however, is often narrower than those in reliability tests for LEDs because the lasing temperature is often below 100°C. This makes it difficult to estimate the reliability of laser diodes within a short period.

7.3.2.1 Changes in Lasing Characteristics

Transverse Mode: Far (Near)-Field Pattern and Optical Coupling   For transmitting information, the output power from laser diodes is coupled to fibers or waveguides. If this coupling is unstable, the transmitted power is also unstable and the transmission quality is poor. An unstable far-field pattern of the laser diode also degrades the performance of optical disks and printers. The stability of the transverse mode (far-field pattern) is therefore important for laser diodes. The transverse mode often changes in gain-guiding laser diodes, such as the planar types, because the lasing area is determined by the distribution of injected current and is influenced by device degradation. In refractive index-guiding laser diodes, such as the buried heterostructure types, the transverse mode is quite stable as long as there is no severe degradation, such as catastrophic optical damage.[3] This stability is due to the clearly defined lasing area.

Longitudinal Modes and the Pulse Output Response   Stability of the longitudinal modes and of the pulse output response of laser diodes strongly influences transmission quality. The stability of AlGaAs/GaAs laser diodes is governed by the behaviors of defects because nonradiative recombination and optical absorption occur easily at the defects. The defects in AlGaAs/GaAs material systems act as saturable absorbers and tend to generate self-sustained pulsation (see Section 3.5.2.1).[26–28] This pulsation degrades transmission quality tremendously. Normal refractive index-guiding InGaAsP/InP laser diodes scarcely show this pulsation because the InGaAs(P)/InP material systems are insensitive to defects. [The pulsation is intentionally produced in some laser diodes, such as 780 nm AlGaAs/GaAs laser diodes for optical disk systems, in order to suppress the noise caused by mode-hopping and optical feedback at a distant mirror (see Sections 3.5.4 and 8.2.1).]

The longitudinal modes are sensitive to device degradation, and the lasing wavelength always changes during degradation (see the next section). The spectral width corresponding to the FWHM (full width at half maximum) of the envelope of each mode peak increases because degradation reduces the optical cavity performance (Q-value). This increase in spectral width is a severe problem in high bit-rate or long-span systems. The transmission speed in a fiber is slightly different for different wavelengths. If the envelope of multiple longitudinal modes is broad, the shape of the modulated output pulse is deformed during transmission in the fiber. Under modulation this spectral broadening is itensified because of the chirping (see Section 3.5.2.1). These phenomena also reduce the transmission quality. Laser diodes with single modes, such as DFB laser diodes, are therefore usually used in systems when the bit-rate is on the order of gigabits per second or more. Longitudinal mode deterioration occurs in the single mode laser diodes during degradation. Severe degradation decreases the side-mode suppression ratio and often results in multimode operation (see in Fig. 7.27). In DFB laser diodes, the longitudinal mode deterioration tends to be intensified as the grating height or coupling constant κL (see Section 3.4.1) increases because of the decrease in crystal quality near the grating and because of the increasingly inhomogeneous optical field distribution within the cavity.[26]

Lasing Wavelength   As shown in Fig. 3.12, lasing wavelength is governed by several factors. Wavelength change for the envelope of each mode peak can be roughly expressed with the following equations:

$$\delta\lambda_{env} = -\delta\lambda_{bf} + \delta\lambda_{Joule}, \qquad (7.15)$$

where $\delta\lambda_{bf}$ is the envelope shift due to the change in the transition energy (or the band-gap energy) caused by band-filling and $\delta\lambda_{Joule}$ is the envelope shift due to Joule's heating, and for each mode,

$$\delta\lambda_m = -\delta\lambda_{n_r\text{-plasma}} + \delta\lambda_{n_r\text{-Joule}}, \qquad (7.16)$$

where $\delta\lambda_{n_r\text{-plasma}}$ is mode shift due to the change in the refractive index caused by the plasma effect and $\delta\lambda_{n_r\text{-Joule}}$ is the mode shift due to Joule's heating. During aging under a constant output power, the wavelength usually increases because the operating current increases to keep output power constant and thus the Joule's heating is dominant. During aging under a constant current, however, the wavelength tends to decrease. This decrease is due to the plasma effect and the band-filling effect, which are intensified as the threshold carrier density increases during degradation. Both wavelength and mode shifts occur in Fabry–Perot laser diodes, whereas in DFB laser diodes the longitudinal mode varies according to Eq. (7.16). Typical changes in lasing wavelength are shown in Figs. 7.28 and 7.29 for 1550 nm buried heterostructure QW DFB laser diodes with a 900 μm-long cavity. There is very little change during the initial stage, although the threshold current clearly increases. The main cause of the degradation in InGaAsP/InP laser diodes is the decrease in injected carrier lifetime due to the degradation of the buried heterointerface. During the degradation, the injected threshold carrier density is kept constant.[29] After severe degradation, the wavelength begins to change because of the increase in internal optical loss

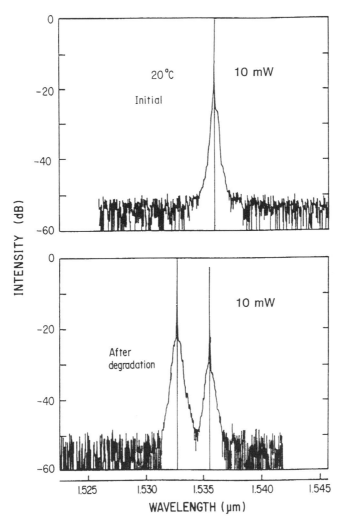

FIGURE 7.27   Change in the lasing spectrum of a DFB laser diode degraded under a constant power of 20 mW at 25°C.[3] The κL value is about 2. Reprinted with permission from *Reliability and Degradation of Semiconductor Lasers and LEDs* by Artech House, Inc., Norwood, MA, USA, http://www.artech-house.com.

(increase in the injected threshold carrier density). The wavelength shift during degradation of DFB lasers is determined by the change in effective grating pitch, which change is introduced by the refractive index change in the active region [see Eq. (3.32)]. The shift, $\delta\lambda_{n_r\text{-plasma}}$, is a function of the injected carrier density and may be expressed, using Eqs. (3.27) and (3.32), for 1550 nm-band InGaAs(P) systems as follows:

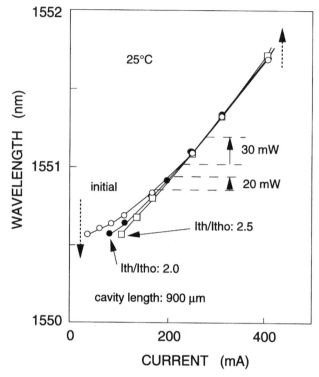

FIGURE 7.28 Typical change in lasing wavelength of DFB laser diodes. The arrows indicate the wavelength change at 20 mW and 30 mW.[29]

$$-\delta\lambda_{n_r\text{-plasma}} = -2\Lambda(dn_r/dn)\Delta n = -2\Lambda(6 \times 10^{-21})\Delta n, \qquad (7.17)$$

where Δn is the change in carrier density (which after lasing is on the order of 10^{18} cm^{-3}) and Λ is the grating pitch (on the order of 10^{-5} cm). The wavelength change due to the plasma effect would therefore be more than 1 nm after a 50% increase in threshold current—if the increase in the threshold current were due solely to the increase in threshold carrier density. Figures 7.28 and 7.29, however, show wavelength shifts of less than 0.1 nm. This small amount of wavelength shift is due to the constant threshold carrier density in the initial stage of the buried heterointerface degradation as discussed above. The shift, $\delta\lambda_{n_r\text{-Joule}}$, is about 0.1 nm/°C. Therefore, the wavelength lengthening becomes dominant under constant output power operation as shown in Fig. 7.28, whereas the influence of the Joule's heating is quite small under constant current operation as shown in Fig. 7.29.

The time course of the wavelength change due to heating depends on mounting configurations. The wavelength of devices with a junction-up configuration

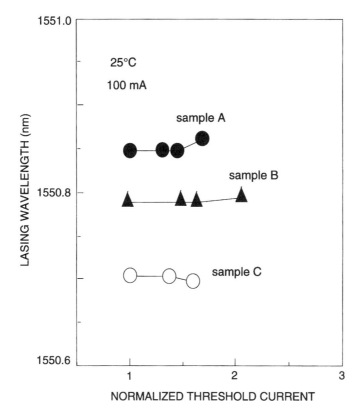

FIGURE 7.29   Behaviors of wavelength at a constant current of 100 mA for DFB laser diodes during degradation.

tends to increase more rapidly during degradation than does that of devices with a junction-down configuration. In any case, these wavelength shifts depend on the device degradation: devices with no degradation show scarcely any wavelength change. If the optical cavity loss increases initially (and thus the threshold carrier density increases initially) as observed often in AlGaAs/GaAs lasers, the decrease in wavelength will be greatly larger than those shown in Figs. 7.28 and 7.29. These behaviors of lasing wavelength during degradation often need to be considered when determining failure criteria.

Intensity Noise (AM Noise)   The increase in the intensity noise in systems employing an intensity modulation technique is a more severe problem in analog transmission systems than in digital transmission. The signals in digital systems (1 and 0) correspond to the ON and OFF states of laser diodes, and the optical power difference between the two levels in large, while in analog systems many signals are superimposed within a small range of optical power. The carrier-to-noise ratio (CNR) in analog systems is inversely proportional to the

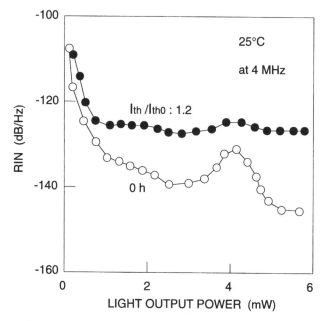

FIGURE 7.30   Change in the relative intensity noise at 4 MHz during degradation for a 1300 nm-band buried heterostructure InGaAsP/InP Fabry–Perot laser diode. The threshold current increased by 20% after degradation.

relative intensity noise. As shown in Fig. 7.30, the intensity noise increases as laser diodes degrade because the longitudinal mode becomes unstable and other noises (such as $1/f$ noise) increase. These behaviors of intensity noises should also be considered when determining failure criteria for laser diodes used in transmission systems (especially analog systems), optical disks, and so forth.

Spectral Linewidth and Phase Noise (FM Noise)   The increase in the intensity noise is accompanied by an increase in the phase noise. These two noises are related by the linewidth enhancement factor discussed in Section 3.5.3.3. For DFB and DBR laser diodes, the behavior of spectral linewidth is a measure of the change in phase noise and differs during degradation as shown in Fig. 7.31 for a 1550 nm buried heterostructure QW DFB laser with a 900 μm-long cavity. Spectral linewidth at various stages of degradation is shown in the figure as a function of inverse output power, $1/P_{\text{out-st}}$. In the initial stage of degradation, a curve showing the spectral linewidth as a function of the inverse output power increases smoothly and monotonically. With aging, there is a parallel shift of the curve toward larger linewidths. In the next stage, the slope increases gradually because of the internal optical loss. Finally, the nonlasing side mode becomes unstable and rebroadening of the spectral linewidth occurs in the relatively low output power range. These behaviors correspond to the degradation modes of laser diodes, and if the internal optical loss increases initially (as observed in

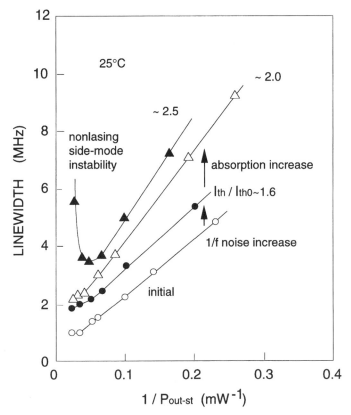

FIGURE 7.31 Typical changes in spectral linewidth during the aging of a DFB laser diode with a 900 μm-long cavity. The aging was carried out at 50°C and under a constant current of 450 mA.[29]

AlGaAs/GaAs laser diodes, for example), the increase in the slope is generated first.

The residual linewidth, which is obtained by the linear extrapolation of the relation between linewidth and $1/P_{out-st}$ to $1/P_{out-st} = 0$, also increases in any stage, especially the initial stage.[30] This residual linewidth increase is mainly due to the increase in $1/f$ noise, and this $1/f$ noise mainly increases in the initial stage of degradation, during which the nonradiative-recombination current increases and the threshold carrier density is kept nearly constant. The FM noise spectrum determining spectral linewidth is expressed by the following equation:[30]

$$S_{FM}(f) = C_{wn} + K(c_0/4\pi n_g)^2 \alpha^2 \alpha_H q l_{BH} S_{int}/4wf, \qquad (7.18)$$

where C_{wn} is a term related to white (shot) noise, K is a constant determined by the active volume, gain factor, and so on, c_0 is the velocity of light in vacuum,

n_g is the group index, α is the linewidth enhancement factor, α_H is Hooge's parameter, q is the electron charge, S_{int} is the interfacial (or surface) recombination velocity, w is the thickness of the carrier confinement (depletion) layer at the buried heterointerface, and I_{BH} is the nonradiative-recombination current per unit length under high bias at the buried heterointerface (or surface). This current is given by

$$I_{BH} = q s_0 L_{BH} n_i \exp(q V/2 k_B T), \tag{7.19}$$

where s_0 is the intrinsic recombination velocity at the buried heterointerface (or surface), L_{BH} is the diffusion length along the interface (or surface), and n_i is the intrinsic carrier density. The initial stage of the degradation of buried heterostructure InGaAsP/InP laser diodes is mainly due to the buried heterointerface degradation, and the increase in $1/f$ noise is governed by the quality of the buried heterointerface.

These degradation behaviors of spectral linewidth (phase noise) determine the operating range of DFB laser diodes in coherent systems. Consequently, the phase noise also needs to be assured when the failure criteria for such systems are defined.

7.3.2.2 Lasing Characteristics and Failure Criteria   The failure criteria are basically determined according to the requirements of equipment and systems in which the devices are to be used. Each device characteristic, however, cannot actually be monitored continuously during its operation. Even in reliability tests or screening tests, the monitoring of each characteristic is difficult because it would take too long. Changes in critical characteristics therefore need to be inferred from changes in other characteristics which can be easily monitored in the equipment and systems. What is usually monitored in equipment and systems is the operating current or output power. In 1300 nm-band and 1550 nm-band buried heterostructure InGaAsP/InP laser diodes, for example, the transverse mode, longitudinal mode, and pulse output respose (modulation characteristics) are stable within the operating range in which the operating current increases, under constant output power operation, to 150% of the initial current. Spectral linewidth, however, slowly increases within the operating range in which the operating current increases up to 120% of the initial current under constant output power operation. A 150% increase in operating current can thus be set as the failure criterion for digital transmission systems using intensity modulation techniques, while for coherent systems the criterion should be set at a 120% increase in operating current. The TTF for each device is estimated to be the time at which the operating current increases to 150% and 120% of the initial values.

7.3.2.3 Wavelength (Material) and Reliability   The reliability of laser diodes, like that of the other optoelectronic *pn*-junction devices, is rather complicated and essentially depends on the properties of material from which they are made.

Long-Wavelength InGaAsP/InP Laser Diodes   The reliability of a long-wavelength buried heterostructure InGaAsP(P)/InP laser diode with a Schottky-type electrode is mainly determined by the degradation of the buried heterointerface degradation and thus depends on the quality of the interface. For bulk, quantum well and strained quantum well laser diodes with a high quality interface—that is, one for which the interfacial state density is low—device lives are estimated to be more than 10^5 hours, even at temperatures between $50°C$ and $70°C$, when the laser diodes are used in digital and analog transmission systems employing intensity modulation techniques or in coherent systems. The buried heterointerface degradation has been suppressed even in laser diodes in which the heterointerface is buried by Fe-doped semi-insulating InP.[31] For high-temperature operation, 1300 nm-band InAsP/InP and InGaAlAs/InP strained quantum well laser diodes have been developed.[32,33] These laser diodes, fabricated by using the technology developed for InGaAs(P)/InP laser diodes, show the feasibility of device lives greater than 10^5 hours even at temperatures around $100°C$. These long-wavelength devices are the main sources used in fiber transmission systems, and laser diodes with high-temperature characteristics are especially promising sources for access networks and so forth.

Strained QW InGaAs/GaAs Laser Diodes   Strained quantum well InGaAs/GaAs lasers lasing at 980 nm are used as pumping sources in Er-doped fiber amplifiers. The output power required for this application is normally to be more than 100 mW, and thus sudden failure due to catastrophic optical damage often occurs during long-term operation if the facet is not passivated before coating or light absorption is not suppressed. This failure is caused by the band-gap shrinkage resulting from the release of strain at the facet. As in the long-wavelength InGaAsP/InP laser diodes, degradation in the inner region scarcely occurs, and this indicates that the rate of nonradiative recombination is not so high as that in AlGaAs/GaAs material systems. These strained InGaAs/GaAs laser diodes are very stable if the strain release at the facet is eliminated. The laser diodes released from facet degradation show device lives longer than 10^5 hours and less than 1000 FIT.[34] The failure criterion for these devices is set at an increase in operating current of around 120% of the initial value. This criterion is determined by the change in lasing wavelength influencing the pumping efficiency (see Section 8.1.1.5).

AlGaAs/GaAs Laser Diodes and Laser Diodes Emitting Visible Light   The materials used for short-wavelength laser diodes tend to show the nonradiative-recombination-enhanced defect motion. Consequently, these laser diodes show some reliability problems in the inner region and at the facets (described in Section 7.2.1). Those problems can be eliminated by reducing the number of crystal defects in the inner region and by passivating the facets. The 780 nm-band AlGaAs/GaAs laser diodes usually used in compact disks show stable operating characteristics, and the possibility of using them in economical communication systems (such as a local area network, LAN) has often been examined. Their device lives at 50°C are estimated to be more than 10^5 hours, but sudden failures can occur during long-term operation if the devices having defects in the active region are not screened out. Laser diodes with lasing wavelengths less than 700 nm are important for consumer electronics, such as bar-code readers and optical disks. Strained quantum well InGa(Al)P/GaAs laser diodes emitting visible (630 nm-band) light have shown stable operating characteristics and have become practical sources. These lasers also show long-term stability as optical sources in plastic fiber communication systems for automotive use and so forth. They are also applied to consumer electronics, such as laser disks. The reliability of blue-green laser diodes composed of InGaN material systems will be assessed in the near future.[35]

7.3.3 Photodiodes

Optoelectronic *pn*-junction devices used under reverse bias are ordinarily more reliable than those used under forward bias. This mainly results from the little energy for accelerating degradation because of the reverse bias. Their simple operating mechanism and simple device structures also contribute to their reliability. The input energy is the electrical energy given by the product of the applied reverse voltage and the photocurrent (or dark current when there is no light illumination). The dark current is given by the sum of diffusion current, generation-recombination current, tunneling current, and surface leakage current. The deterioration in the performance of photodiodes is mainly due to the increase in leakage (dark) current because the increase in leakage current decreases the ratio of photocurrent (signal) to dark current (noise). The dark current usually used as a measure of degradation, and an increase in dark current is caused by *pn*-junction degradation. As shown in Table 7.2 and Fig. 7.8, various kinds of degradation modes, from rapid to gradual, are observed.

In accelerated aging tests, the aging parameters are temperature and reverse bias voltage (or reverse current). The ambient temperature is set at a very high temperature, usually between 100°C and 300°C, because the temperature limit for device operation is not as strict as the limit for laser diodes and LEDs and the rate of degradation is quite low. The statistical treatment can, like that for laser diodes and LEDs, be performed by using log-normal failure distribution functions for wearout modes and using exponential failure distribution functions for random modes. The failure criteria are of course determined by the performance required of the application equipment and systems. They are, for example, set at the point at which dark current increases by two orders of magnitude from the initial value at a certain reverse voltage for pin-photodiodes or at the dark current of 1 μA at a certain reverse voltage for APDs. The raw data from aging tests of InGaAs/InP APDs and the log-normal plots of TTF are shown in Figs. 7.32 and 7.33.[23] The failure criterion was set at the dark current of 1 μA at 0.9 V_B at 25°C (V_B = breakdown voltage).

The reliability of photodiodes has already been established. The causes of degradation discussed above have been eliminated or suppressed by improving

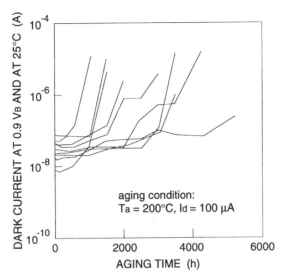

FIGURE 7.32 Change in dark current at 0.9 V_B (V_B = breakdown voltage). The devices are InGaAs APDs with Ti/Pt/Au electrode. The aging was carried out at 200°C and at a constant dark current of 100 μA. From H. Sudo et al.[23] ©1988 IEEE. Reprinted by permission of the authors and IEEE.

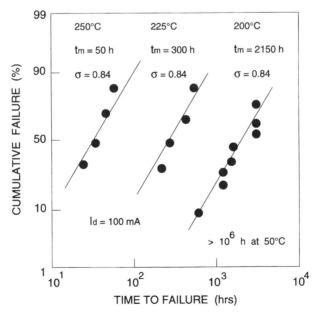

FIGURE 7.33 Log-normal plots of TTF for InGaAs/InP APDs with Ti/Pt/Au electrodes. The aging was carried out at 200°C and under a constant dark current of 100 μA. From H. Sudo et al.[23] ©1988 IEEE. Reprinted by permission of the authors and IEEE.

TABLE 7.5 Examples of Reliability of Various Kinds
of Photodiodes under Practical Uses

Si pin-photodiodes and APDs	
Mean time to failure	$\gg 10^6$ h
Failure rate	$\ll 100$ FITs
Ge pin-photodiodes and APDs	
Mean time to failure	$> 10^6$ h
Failure rate	< 100 FITs
InGaAs pin-photodiodes and APDs	
Mean time to failure	$> 10^6$ h
Failure rate	$\gg 100$ FITs

crystal growth and fabrication technologies. Some examples of device lives and failure rates for pin-photodiodes and APDs are summarized in Table 7.5. The main site of the degradation determining reliability is in the *pn*-junction at the surface or at the interface between the semiconductor and a dielectric film. Solar cells are more reliable than photodiodes and APDs. InGaAs(P)/InP and Ge pin-photodiodes and APDs are mainly used in optical communication systems with silica fibers, and Si photodiodes operating at short wavelengths are mainly used in consumer electronics and in optical communication systems with plastic fibers.

7.3.4 Modulators

Both kinds of optical semiconductor modulators, the electroabsorption types and the phase modulation types, are used in equipment and systems. They are used under reverse bias, and the main failure modes are thus similar to those of photodiodes. The energy accelerating degradation is therefore small, and the modulators are basically highly reliable. As with photodiodes, the input electrical energy is given by the product of applied reverse voltage and photocurrent (or dark current when there is no light incidence). During the operation of the electroabsorption types, more than a few milliamperes of photocurrent is often generated in the bias circuit. The photocurrent flows through *pn*-junction inhomogeneously and is usually larger on the optical input side than on the other side. Most of the degradation of such waveguide-type devices occurs on the input side. Photocurrent (dark) current can be used as a measure of degradation because it reflects the *pn*-junction situation and because the extinction ratio in electroabsorption-type modulators is inversely related to the photocurrent.

One example of device life estimation for a Mach–Zender-type semiconductor modulator is shown in Fig. 7.34.[24] The device life decreases as the product of applied bias voltage and photocurrent increases. If the optical input power is +10 dBm, which corresponds to a photocurrent of 0.25 mA at bias voltage of −4 V, the product is about 1 mW. The device life under this condition is more than 10^5 hours (see Fig. 7.34). Device lives greater than 10^5 hours under

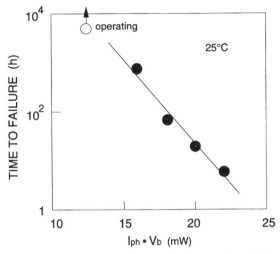

FIGURE 7.34   Relationship between TTF and the product of applied voltage (V_b) and photocurrent (I_{ph}) for a Mach–Zehender type of modulator.[24]

actual application conditions have also been estimated for electroabsorption-type modulators.[36] These modulators are key components in high bit-rate and long-span optical fiber communication systems because of the very small chirping.

7.4 PACKAGING PROBLEMS

Optoelectronic *pn*-junction devices are mounted on heat sinks or stems and packaged or mounted in modules before they are actually used in application equipment and systems. This mounting and packaging often leads to reliability problems, and those problems are mainly related to the bonding part. In addition, as described in Section 6.3.3, other devices, such as electrical coolers, are mounted to control the operating temperature of devices. Those additional devices also introduce reliability problems, and the total failure rate for an optoelectronic device module is calculated as the sum of the failure rates for the device, bonding part, optical coupling, and so forth.

7.4.1 Bonding Part: Solder Instability

The degradation of bonding parts, like that of electrodes, depends very much on the material. Soft solders (low-melting point solders) such as In, Sn, and Sn-rich Au-Sn, introduce reliability problems into devices even though they help relieve mechanical stress in the initial stage of aging (see Fig. 7.35). This absorption of mechanical stress at the bonding part induces thermal fatigue, and

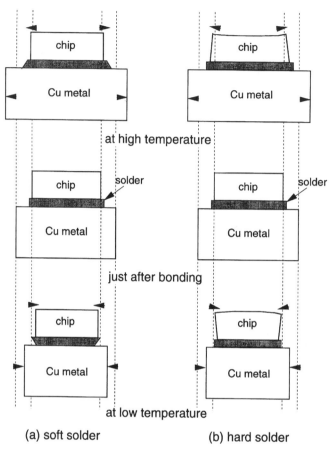

FIGURE 7.35 Illustration of the difference in thermal distortion between (a) soft and (b) hard solder.

the bonding part is eventually broken because cracks are generated during long-term operation. Other problems related to the solders are whisker growth, void formation at the bonding part, and reaction between the semiconductor and the solder metal if they touch directly.[3] These kinds of solder-related degradation result in sudden failure during operation that is otherwise stable (because of no degradation in the optoelectronic *pn*-junction devices themselves). Whisker growth and reactions between the semiconductor and solder can cause sudden short circuits between the *p*- and *n*-sides of devices. The thermal fatigue and void formation at the bonding part cause rapid degradation due to increases in electrical and thermal resistance. This instability at the bonding part is increased tremendously by electrical bias or current flow. The images of the instability are summarized in Fig. 7.36 for a case of inhomogeneous current density. This inhomogeneous current density often appears because the current flow path is

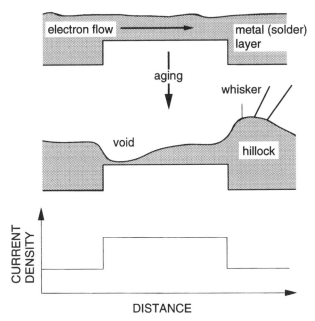

FIGURE 7.36 Schematic illustration for images of void formation and hillock and whisker growth due to inhomogeneous current density.

not homogeneous. Soft solder metal is carried from where the current density is high to where the current density is low. Consequently, voids, hillocks, and whiskers tend to form at the bonding part. As shown in Fig. 7.37, another type of void formation also occurs at the bonding part. Bulk diffusion and grain-boundary diffusion occur at the interface between the electrode metal (such as Au) and the layer of soft solder, and the rate of grain-boundary diffusion is higher than that of bulk diffusion. As a result, voids are formed at the interface.[37,38] A void of this type is called the Kirkendall void.

A hard solder (high-melting point solder), such as an Au-rich AuSn solder, is usually used in most optoelectronic *pn*-junction devices in order to reduce or eliminate the instability induced by the solder metal. Devices with hard solder show very stable aging characteristics. Soft solder, however, is still used in some equipment because the bonding process with soft solder is easier than that with hard solder.

7.4.2 Electric (Peltier) Coolers and Thermistors

Optoelectronic *pn*-junction devices, especially laser diodes and modules, are often used under temperatures that are controlled by using Peltier coolers and thermistors as described in Section 6.3.3. The stability of the coolers and thermistors is important and often influences the reliability of modules.

The instability of a Peltier cooler is mainly caused by the soft solder used

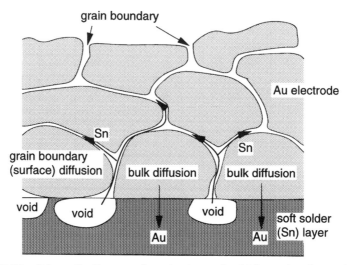

FIGURE 7.37   Schematic diagram of mechanisms of void formation due to the difference between the diffusion rates in the bulk material and at the grain boundary.

for bonding to thermoelectric material to the electrode. Degradation similar to that in the case of the bonding of laser diodes and LEDs is induced by the soft solder instability under current flow. The solder instability increases the electrical resistance of the cooler and reduces its ability to control temperature. The thermistor itself is quite stable but reliability problems are introduced by the instability of the soft solder used for bonding to the package. These problems are eliminated by using hard solder. Degradation of the Peltier cooler and thermistor changes the operating conditions of optoelectronic devices and often leads to module failure even when the optoelectronic devices themselves are not degraded.[39,40]

7.4.3   Optical Coupling

Optoelectronic *pn*-junction devices are optically coupled to fibers and waveguides by fixing each optical element (lens, fiber, and the device). This problem on optical coupling is mainly for modules used in optical fiber communication systems including data links. Either laser welding or fixing with some kind of solder or plastic is generally used to fabricate modules. Fixing with plastic is quite easy, but the plastic is easily degraded by heat and humidity, the elements are displaced, and the optical coupling is decreased.[41] A package sealed with plastic is also easily degraded by heat and humidity. The stability of fixing with solder is better than that of fixing with plastic, but creep due to thermal and mechanical stress gradually occurs at the bonding part during operation. An example of such bonding parts is shown in Fig. 7.38. Creep here means gradual deformation of the bonding part under mechanical stress, and it results in dis-

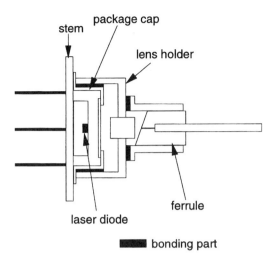

FIGURE 7.38   Cross-sectional view of a coaxial type laser module.

placement of the optical elements. If a mechanical stress is generated between the stem and fiber code, creep occurs at the bonding parts between the cap and the lens holder and between the lens holder and ferrule holder in Fig. 7.38. The creep tends to be more severe for a solder with a lower melting point, and the optical coupling is stable when hard solder is used. The failure due to creep is described by a log-normal failure distribution function, and the failure rate at room temperature can be decreased very much by using hard solders.[42] The fixing problem is eliminated by using laser welding. Parts bonded or fixed with laser welding are very stable, and displacement is scarcely observed. Most modules of optoelectronic *pn*-junction devices for optical fiber communication systems are fabricated with laser welding.

A problem on the optical fiber cables themselves is generated in the optoelectronic device modules. So-called pistoning occurs when the fiber is not sufficiently fixed to the frame of packages or modules. The fiber is pushed into the module under a mechanical force generated by the difference between the thermal expansion coefficients of the glass fiber and the coated or covered materials (such as nylon coating).[40,41] The fiber can be broken in the modules or the optoelectronic device is destroyed by the pushed fiber. This fiber pistoning can be eliminated by reducing the difference between the thermal expansion coefficients of the fiber and its covering materials or improving the fixing method of fiber.

7.4.4 Packaging-Induced Failure

Short-wavelength laser diodes hermetically sealed under dry nitrogen or some other inactive gas, sometimes fail suddenly after gradual degradation when they are operated at a relatively high output power. This problem is called packaging-induced failure (PIF).[43] If the residual gas in a package or module contains a volatile hydrocarbon, the hydrocarbon can build up over the emitting region at the facet of the laser diodes during operation because of a photochemical reaction. Heat generated when the resultant hydrocarbon layer absorbs the light emitted from the laser diode eventually results in catastrophic facet degradation (melting of the facet).[44,45] PIF can be eliminated by sealing the module or the package in an atmosphere containing oxygen.[43–46] Oxygen gas either prevents the deposition of the hydrocarbon or else oxidizes the deposited hydrocarbon to CO_2 and H_2O. The 980 nm strained QW laser diodes and laser diodes emitting visible light are sometimes sealed in an atmosphere containing oxygen. PIF can be also suppressed by using clean packaging parts baked at high temperature and not treating the devices and the parts of the package or module with alcohol.

7.4.5 Plastic Encapsulation

As described in Section 6.3.4.2, optoelectronic *pn*-junction devices used in consumer electronics are usually encapsulated with plastics, such as epoxy resin and silicone resin. The reliability of devices encapsulated with plastic depends very much on the properties of the plastic, and these devices are always less reliable than devices sealed hermetically. The devices encapsulated with plastic show instability when subjected to humidity because, as shown in Fig. 7.39, plastic cannot provide perfect protection from humidity. The data shown

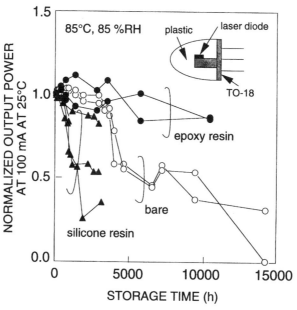

FIGURE 7.39 Stability of laser diodes encapsulated with silicone resin or epoxy resin. The 1300 nm-band buried heterostructure InGaAsP/InP Fabry–Perot laser diodes without facet coating films were tested at 85°C under 85% RH. The results of testing bare laser diodes (without plastic covers) are also shown for reference.

in Fig. 7.39 is for laser diodes mounted on TO-18 stems. They were 1300 nm InGaAsP/InP Fabry–Perot types without any facet coating films and were encapsulated with plastic transparent to the emitted light. Humidity easily penetrates into silicone and accumulates as water at the interface between the devices and the silicone, and then facet oxidation occurs easily.[47] Consequently, the laser diodes encapsulated with silicone degrade more rapidly than bare laser diodes. The laser diodes covered with certain kinds of epoxy-resin are relatively stable. With regard to the test results shown in Fig. 7.39, the failure of the samples encapsulated with an epoxy resin was caused by erosion of the TO-18 stem. The degradation under such a high humidity is to some extent suppressed by facet coating, and surface passivation is indispensable for devices in which facets or surfaces degrade easily such as 850 nm-band AlGaAs/GaAs laser diodes. Mechanical stress caused by the different thermal expansion coefficients of the plastic and the devices is also a key factor because the plastic is in direct contact with the devices. As shown in Fig. 7.40, devices encapsulated with certain kinds of plastic operate stably over a few thousand cycles in tests during which temperature is cycled between $-40°C$ and $+70°C$. This stability is also sufficient for actual applications. The stability of the other devices, such as photodiodes, in plastic packaging is similar to that of the laser diodes.

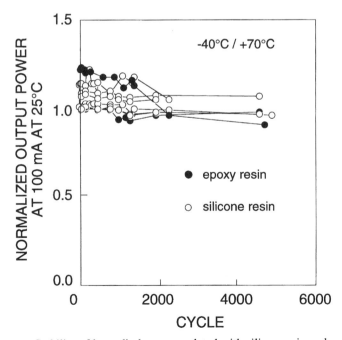

FIGURE 7.40 Stability of laser diodes encapsulated with silicon resin and epoxy resin. The samples were similar to those whose test results are shown in Fig. 7.39, and they were tested by cycling the temperature between $-40°C$ and $70°C$.

7.5 ELECTRICAL SURGE AND ELECTROSTATIC DISCHARGE

Most devices which fail during handling or measuring are influenced by surge or electrostatic discharge (ESD). The endurance level and failure mode are different with the direction of the surge because in *pn*-junction devices the surge in the forward direction is a current surge, while that in the reverse direction is a voltage surge. The weakest part in a device is destroyed under the electrical surge or ESD, and the part degraded during long-term operation is usually the one destroyed and the surge endurance gradually decreases during operation. Forward surge generally causes catastrophic optical damage in short-wavelength laser diodes and often results in melting of the interface between the electrode and the semiconductor in long-wavelength laser diodes, photodiodes, and modulators. A reverse surge commonly destroys the *pn*-junction. The voltage surge is biased first at the *pn*-junction, and then a large amount of current is concentrated within the narrow area where the *pn*-junction is weakest (often at a dislocation or defect). Finally, melting occurs at the part. The endurance level gradually decreases as the input energy per unit time increases as shown in Fig. 7.41. The figure shows the results of tests in which, at each pulse width, amplitude of square current pulses was increased in 0.2-A step and 50 pulses at each step level were applied until the laser diodes were destroyed. The endurance

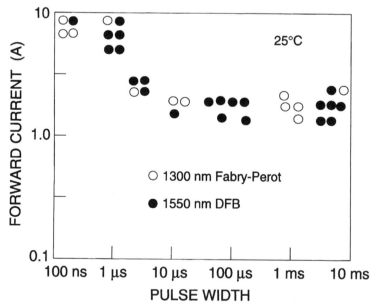

FIGURE 7.41 Forward surge endurance level as a function of pulse width for 1300 nm-band buried heterostructure InGaAsP/InP Fabry–Perot laser diodes and for 1550 nm-band buried heterostructure InGaAsP/InP DFB laser diodes.

level decreases as the pulse width increases and saturates at around 2 A when the pulse width is longer than 10 μs. The surge endurance level can be increased by improving the stability of each part degraded during long-term operation.

REFERENCES

1. W. G. Irenson, *Reliability Handbook*, McGraw-Hill, New York, 1966.

2. A. S. Jordan, A comprehensive review of the lognormal failure distribution with application to LED reliability, *Microelectron. Reliability*, **18**, 267, 1978; A. S. Jordan, Confidence limits on the failure rate and a rapid projection nomogram for the lognormal distribution, *Microelectron. Reliability*, **24**, 101, 1984.

3. M. Fukuda, *Reliability and Degradation of Semiconductor Lasers and LEDs*, Artech House, Norwood, MA, 1991.

4. C. H. Henry and D. V. Lang, Nonradiative capture and recombination by multiphonon emission in GaAs and GaP, *Phys. Rev., B.*, **15**, 989, 1977.

5. P. W. Hutchinson and P. S. Dobson, Defect structure of degraded GaAlAs-GaAs double heterojunction lasers, *Phil. Mag.*, **32**, 745, 1975.

6. For example, G. D. U'Ren, G. M. Haugen, P. F. Baude, M. A. Haase, K. K. Law, T. J. Miller, and B. J. Wu, Transmission electron microscopy of $\langle 100 \rangle$ dark line defects in CdZnSe quantum well structures, *Appl. Phys. Lett.*, **67**, 3862, 1995.

7. T. Kamejima, K. Ishida, and J. Matsui, Injection-enhanced dislocation glide under uniaxial stress in GaAs-(GaAs)As double heterostructure laser, *Japan. J. Appl. Phys.*, **16**, 233, 1977.

8. S. Yamakoshi, M. Abe, O. Wada, S. Komiya, and T. Sakurai, Reliability of high radiance InGaAsP/InP LEDs operating in the 1.2–1.3 μm wavelength, *IEEE J. Quantum. Electron.*, **QE-17**, 167, 1981.

9. M. Fukuda, K Wakita, and G. Iwane, Dark defects in InGaAsP/InP double heterostructure lasers under accelerated aging, *J. Appl. Phys.*, **54**, 1246, 1983.

10. See for example, B. L. Sharma, *Semiconductor and Semimetals*, Academic Press, New York, 1981, vol. 15, ch. 1; A. Piotorowska, A. Guivarch, and G. Pelous, Ohmic contacts to III-V compound semiconductors: A review of fabrication techniques, *Solid-State Electron.*, **26**, 179, 1983; V. G. Keramidas, H. Temkin, and S. Mahajan, Ohmic contacts to InP and InGaAsP, *Inst. Phys. Conf. Ser.*, 56, ch. 5, p. 293, 1981.

11. See for example, M. Ettenberg and I. Ladany, Metallization for diode lasers, *J. Vacuum Sci., Technol.*, **19**, 799, 1981; A. K. Shinha, T. E. Smith, M. H. Read, and J. M. Poate, n-GaAs Schottky diodes metallized with Ti and Pt/Ti, *Solid-State Electron.*, **19**, 489, 1976.

12. K. Endo, S. Matsumoto, H. Kawano, I. Sakuma, and T. Kamejima, Rapid degradation of InGaAsP/InP double heterostructure lasers due to $\langle 110 \rangle$ dark line defect formation, *Appl. Phys. Lett.*, **40**, 921, 1982.

13. M. Fukuda and K. Takahei, Optically enhanced oxidation of III-V compound semiconductors, *J. Appl. Phys.*, **57**, 129, 1985.

14. C. H. Henry, P. M. Petroff, R. A. Logan, and F. R. Meritt, Catastrophic damage of $Al_xGa_{1-x}As$ double-heterostructure laser material, *J. Appl. Phys.*, **50**, 3721, 1979.

15. Y. Nakano, K. Takahei, Y. Noguchi, Y. Suzuki, H. Nagai, and K. Nawata, High power output InGaAsP/InP buried heterostructure lasers, *Electron. Lett.*, **17**, 782, 1981.

16. For example, M. Fukuda, M. Okayasu, J. Temmyo, and J. Nakano, Degradation behavior of 0.98 μm strained quantum well InGaAs/AlGaAs lasers under high-power operation, *IEEE J. Quantum Electron.*, **30**, 471, 1994.

17. H. Yonezu, M. Ueno, T. Kamejima, and I. Hayashi, An AlGaAs window structure laser, *IEEE J. Quantum Electron.*, **QE-15,** 775, 1979.

18. K. Sasaki, M. Matsumoto, M. Kondo, T. Ishizumi, T. Takeoka, S. Yamamoto, and T. Hijikata, Highly reliable 150 mW cw operation of single-stripe AlGaAs lasers with window grown on facets, *Japan. J. Appl. Phys.*, **30,** part 2 (5B), L904, 1991.

19. G. D. Henshall, Gallium aluminium arsenide heterostructure lasers: Factors affecting catastrophic degradation, *Solid-State Electron.*, **20**, 595, 1977.

20. M. Fukuda and G. Iwane, Degradation of active region in InGaAsP/InP buried heterostructure laser, *J. Appl. Phys.*, **58**, 2932, 1985.

21. M. Fukuda, Y. Noguchi, G. Motosugi, Y. Nakano, N. Tsuzuki, and O. Fujita, Suppression of interface degradation in InGaAsP/InP buried heterostructure lasers, *J. Lightwave Technol.*, **5**, 1778, 1987.

22. H. Mawatari, M. Fukuda, S. Matsumoto, K. Kishi, and Y. Itaya, Reliability and degradation behaviors of semi-insulating Fe-doped InP buried heterostructure lasers fabricated by RIE and MOVPE, *J. Lightwave Technol.*, **15**, 534, 1997.

23. H. Sudo and M. Suzuki, Surface degradation mechanism of InP/InGaAs APDs, *J. Lightwave Technol.*, **6**, 1496, 1988.

24. M. Yuda, M. Fukuda, and H. Miyazawa, Degradation mode in semiconductor optical modulators, *Electron. Lett.*, **31**, 1778, 1995.

25. F. Capasso, P. M. Petroff, W. B. Bonner, and S. Sumski, Investigation of microplasmas in InP avalanche photodiodes, *IEEE Electron. Dev. Lett.*, **EDL-1,** 27, 1980.

26. K. D. Chik, J. C. Dyment, and B. A. Richardson, Self-sustained pulsations in semiconductor lasers: Experimental results and theoretical confirmation, *J. Appl. Phys.*, **51,** 4029, 1980.

27. T. L. Paoli, Changes in the optical properties of cw (AlGa)As junction lasers during accelerated aging, *IEEE J. Quantum Electron.*, **QE-13,** 351, 1977.

28. R. L. Hartman, R. A. Logan, L. A. Koszi, and W. T. Tsang, Pulsations and absorbing defects in (AlGa)As injection lasers, *J. Appl. Phys.*, **50,** 4616, 1979.

29. M. Fukuda, F. Kano, T. Kurosaki, and J. Yoshida, Reliability and degradation behavior of highly coherent 1.55-μm long-cavity multiple quantum well (MQW) DFB lasers, *J. Lightwave Technol.*, **10**, 1097, 1992.

30. M. Fukuda, T. Hirono, T. Kurosali, and F. Kano, $1/f$ noise behavior in semiconductor laser degradation, *Photon. Technol. Lett.*, **5**, 1165, 1993.

31. S. Matsumoto, M. Fukuda, K. Sato, Y. Itaya, and M. Yamamoto, Highly reliable 1.55 μm GaInAsP laser diodes buried with semi-insulating iron-doped InP, *Electron. Lett.*, **30**, 1305, 1994.

32. M. Yamamoto, N. Yamamoto, and J. Nakano, MOVPE growth of strained InAsP/InGaAsp quantum-well structures for low-threshold 1.3-μm lasers, *IEEE J. Quantum Electron.*, **30**, 554, 1994.

33. C. E. Zah, M. C. Wang, R. Bhat, T. P. Lee, S. L. Chuang, Z. Wang, D. Darby, D. Flanders, and J. J. Hsieh, High temperature modulation dynamics of 1.3 μm $Al_xGa_yIn_{1-x-y}$As-InP compressive-strained multiple-quantum-well lasers, *Proc. 14th IEEE Int. Semiconduct. Laser Conf.*, Maui, HI, 1994, p. 215.

34. G. Grasso, F. Magistrali, G. Salmini, A. Oosenbrug, A. Jakubowicz, D. Darby, and P. Whitney, Qualification testing and field reliability results of 980 nm pump lasers, *Tech. Dig. Conf. Optical Fiber Communication*, San Diego, Feb.-March, 1995, Paper ThC2, OSA Tech. Dig. series, vol. 8, Optical Society of America, Washington, D.C., 1995, p. 232.

35. S. Nakamura, M. Senoh, S. Nagahama, N. Iwasa, T. Yamada, T. Matsushita, H. Kiyoku, and Y. Sugimoto, Characteristics of InGaN multi-quantum-well-structure laser diodes, *Appl. Phys. Lett.*, **68**, 3269, 1996; Nakamura, M. Senoh, S. Nagahama, N. Iwasa, T. Yamada, T. Matsushita, Y. Sugimoto, H. Kiyoku, Room-temperature continuous-wave operation of InGaN multi-quantum-well-structure laser diodes, *Appl. Phys. Lett.*, **69**, p. 4056, 1996.

36. T. Watanabe, Y. Saito, K. Sato, and H. Soda, High reliability modulator-integrated distributed-feedback lasers, *Tech. Dig. Conf. Optical Fiber Comm.* (OFC), WH9, 1993, p. 119.

37. S. Nakahara and R. J. McCoy, Interfacial void structure of Au/Sn/Al metallizations on Ga-Al-As light-emitting diodes, *Thin Solid Films*, **72**, p. 457, 1980.

38. S. Nakahara and R. J. McCoy, Kirkendall void formation in thin-film diffusion couples, *Appl. Phys. Lett.*, **37**, 42, 1980.

39. S. P. Sim, A review of the reliability of III-V opto-electronic components, in A. Christou and B. A. Unger, eds., *Semiconductor Device Reliability*, NATO ASI Series E: Applied Science, vol. 175, Kluwer Academic Pub., Dordrecht 1990, p. 301.

40. P. Su and B. A. Unger, Temperature cycling tests of laser modules, in A. Christou and B. A. Unger, eds., *Semiconductor Device Reliability*, NATO ASI Series E: Applied Science, vol. 175, Kluwer Academic Pub., 1990, p. 363.

41. J. L. Spencer, Assuring the reliability of lasers intended for the uncontrolled environment, in A. Christou and B. A. Unger, eds., *Semiconductor Device Reliability*, NATO ASI Series E: Applied Science, vol. 175, Kluwer Academic Pub., Dordrecht 1990, p. 75.

42. O. Mitomi, T. Nozawa, and K. Kawano, Effects of solder creep on optical component reliability, *IEEE Trans. Comp. Hybrids, and Manufacturing Technol.*, **CHMT-9**, 265, 1986.

43. P. A. Jakobson, J. A. Sharps, and D. H. Hall, Requirements to avert packaging induced failure (PIF) of high power 980 nm laser diodes, *Conf. Proc., IEEE Lasers and Electro-Optics Society 1993 Annual Meetings*, San Jose, 1993, PD2.1, p. 21.

44. J. A. Sharps, P. A. Jakobson, and D. W. Hall, Effects of packaging atmosphere and organic contamination on 980 nm laser diode reliability, *Conf. Optical Amplifier and their Applications, OSA Technical Digest Series*, vol. 14, Optical Society of America, Washington D.C., 1994, WD5-1, p. 46.

45. J. A. Sharps, Reliability of hermetically packaged 980 nm diode lasers, *Conf. Proc. IEEE Lasers and Electro-optics Society 1994 Annual Meeting*, Boston, Oct.–Nov., 1994, DL1.1, p. 35.

46. P. Whitney, D. Darby, L. Scheffel, D. Flanders, and R. Pusch, Influence of package atmosphere on 980 nm high power chip reliability, *Proc. 5th European Symp. on Reliability of Electron Devices, Failure Physics and Analysis, ESREF'94*, Glasgow, U.K., Oct. 1994, p. 257.

47. M. Fukuda, F. Ichikawa, H. Sato, S. Tohno, and T. Sugie, Pigtail type laser modules entirely molded in plastic, *Electron. Lett.*, **31,** 1745, 1995; M. Fukuda, F. Ichikawa, H. Sato, Y. Hibino, K. Moriwaki, S. Tohno, T. Sugie, and J. Yoshida, Plastic packaging of semiconductor laser diodes, *Proc. IEEE 1996 Electron. Comp. Technol. Conf.*, Florida, May 1996, p. 1101.

CHAPTER EIGHT

Application of Optoelectronic *pn*-Junction Devices

Uses of semiconductor lasers, LEDs, photodiodes, and optical semiconductor modulators have spread to fields ranging from consumer electronics to optical fiber communication systems. Some examples of their application fields are shown in Fig. 8.1. Optical fiber communication systems, from trunk systems to subscriber systems, and data links form the infrastructure of the recent information networks. Typical examples of consumer electronics equipment using optoelectronic *pn*-junction devices are printers, autofocus cameras, and optical disk systems. The characteristics required of optoelectronic *pn*-junction devices are specified to the equipment or system, and many kinds of optoelectronic *pn*-junction devices have been developed for the various applications. This chapter describes the application equipment and systems from the device point of view.

8.1   OPTOELECTRONIC *pn*-JUNCTION DEVICES IN FIBER OPTICS

The main applications for optoelectronic *pn*-junction devices in fiber optics are in optical fiber communication systems. These systems are roughly divided into two categories: public communication systems and data links. The public optical fiber communication systems are widely used throughout the world, and the optical fiber data links are applied to local area networks, intermachine, and so forth. In both the public systems and data links, information is transmitted through the optical fibers, and there are optoelectronic *pn*-junction devices at both ends of the fibers. There are various methods used in the transmission of information in such optical fiber systems, and some typical methods are described in the next section.

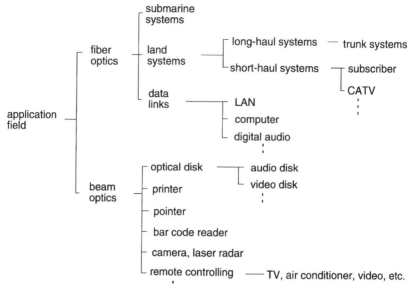

FIGURE 8.1 Application fields for optoelectronic *pn*-junction devices.

8.1.1 Public Optical Fiber Communication Systems

Optical fiber communication systems, such as public communication networks and data links, are the basic infrastructure of the information society and are indispensable in our daily lives. Some of the advantages of the optical fiber systems over metallic transmission systems are the following: (1) High-frequency signal (high-bit-rate data) can be easily transmitted long distances without large losses. (2) Electromagnetic induction noise is never induced during transmission through optical fiber cables. (3) Optical fiber cable is light, flexible, and economical (and thus the compact packaging is possible). As shown in Fig. 8.1, the public optical fiber communication systems are divided into submarine systems and land systems. Submarine systems have already been used to connect countries all over the world as well as to interconnect the parts of individual countries. The submarine systems let us talk with people in foreign countries without time delay. In land systems, long-haul trunk systems have been connected between large cities. The land systems also include systems such as subscriber systems and CATV [community (or common) antenna television, cable and telecommunication television system, or cable television system].

In such optical fiber communication systems, LEDs and laser diodes emit light modulated with a signal, and the resultant optical signal is received with photodiodes, such as pin-photodiodes and APDs, as shown in Fig. 8.2. The optical modulators modulate the light emitted from laser diodes and then the modulated light is transmitted through the optical fiber [Fig. 8.2(b)]. In long-haul systems, repeaters which include photodiodes and laser diodes and elec-

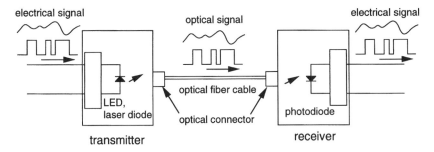

(a) transmitter and receiver (direct modulation)

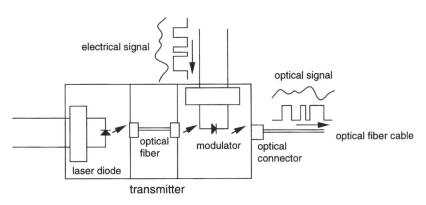

(b) semiconductor optical modulator (external modulation)

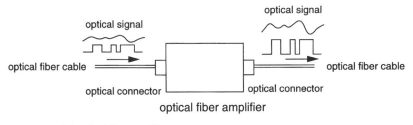

(c) optical fiber amplifier

FIGURE 8.2 Basic constitution of optical fiber communication systems.

tronic circuits are inserted. In the repeater, the weak optical signal transmitted is detected with a photodiode and then electrically reformed and amplified. Then the electrical signal is converted with a laser diode into an optical signal and transmitted again through the optical fiber cable. For long-haul transmission without repeaters, the optical signals are often amplified directly by optical fiber amplifiers. The optical fiber communication systems can be divided into digital systems and analog systems from the modulation point of view. Most long-haul

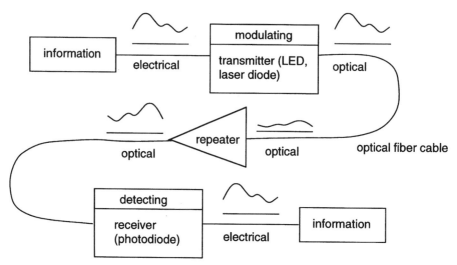

FIGURE 8.3 Basic constitution of an analog system using intensity modulation.

and large-capacity optical communication systems are digital systems, and the analog systems are sometimes used for transmitting information over a short distance.

8.1.1.1 Analog Systems Analog optical fiber communication systems are more basic and simpler than digital systems (see Section 8.1.1.2). The analog systems are sometimes used for short-haul transmission of video signals. In such systems, laser diodes or LEDs are modulated with analog signals. The analog systems can be divided into two kinds: those using direct intensity modulation and those using pulse modulation.

Analog systems employing intensity modulation are the simplest (see Fig. 8.3). The laser diodes or LEDs are directly modulated with an electrical analog signal, and an optical analog signal having the same shape as the electrical signal is transmitted through the optical fiber cable. Repeaters are inserted for transmission of long distance. The transmitted optical analog signal is detected with a pin-photodiode or an APD and is translated back into an electrical signal having the same shape as the optical signal. High linearity of current–light output power characteristics is required so that the laser diodes and LEDs can trace the input electrical analog signal exactly as illustrated in Fig. 8.4. The buried heterostructure laser diodes are the ones usually used because their linearity is much better than that of the laser diodes with gain guiding (see Section 3.2.4). The linearity of laser diodes and LEDs is mainly degraded by Joule's heating (see Sections 2.2.5.1 and 3.5.1) and reduction of injected carrier lifetime with increasing injected current.[1] The distortion in LEDs, for example, is electrically compensated by using an electronic circuit. The linearity (or distortion) is exactly expressed with the linearity of the amplitude and phase of output optical

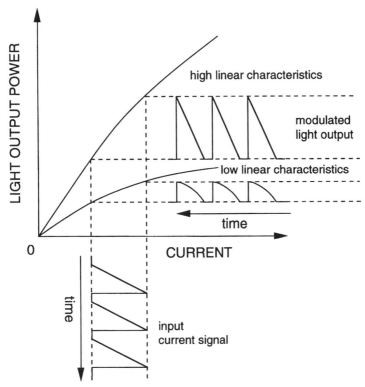

FIGURE 8.4 Linearity of current–light output power characteristics and distortion of output power for LEDs.

signal against the amplitude of input electrical signal (differential gain DG and differential phase DP). The signal-to-noise ratio for the analog transmission, basically given by Eq. (4.44) for pin-photodiodes and by Eq. (4.72) for APDs, is usually required to be more than 40 dB in analog systems. The transmission quality in the analog systems is determined mainly by the DG, DP, and then signal-to-noise ratio.

An important factor influencing the signal-to-noise ratio in systems using laser diodes is the intensity fluctuation (noise) of optical output power (see Section 3.5.4). If the intensity noise is large, the electrical analog signal is not accurately translated to an optical signal by a laser diode and the signal-to-noise ratio decreases. The intensity noise is often expressed with the RIN given by Eq. (3.61). The RIN of the Fabry–Perot type of laser diodes having multiple longitudinal modes is around -130 dB/Hz and that of DFB laser diodes is less than -140 dB/Hz. DFB laser diodes are therefore frequently used for analog transmission.

If the laser diodes are combined with multimode optical fibers, especially graded-index multimode fibers, another optical intensity fluctuation is intro-

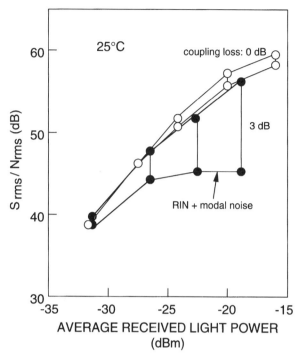

FIGURE 8.5 Signal-to-noise ratio and modal noise in an analog system. The laser diode is a 1300 nm-band buried heterostructure InGaAsP/InP Fabry-Perot laser diode.

duced during transmission. This fluctuation is called the modal noise and is induced by speckle (pattern) and optical coupling loss.[2,3] The speckle is the intensity distribution in a cross section of multimode fiber, and the distribution is induced by the interference between the modes transmitted. The speckle changes very quickly with time. Consequently, if there is optical coupling loss at the connecting parts of the fiber, some of the optical power leaks at the connecting parts and results in the optical intensity fluctuation of transmitted power. The modal noise reduces the signal-to-noise ratio as shown in Fig. 8.5, and the signal-to-noise ratio is determined by the sum of the RIN and the modal noise. When a coupling loss of 3 dB is intentionally introduced, the signal-to-noise ratio fluctuates tremendously between the minimum and the maximum levels.[4] The modal noise has been reduced by using a modulation technique employing a superimposed high-frequency pulse or sinusoidal wave. The high frequency is usually of more than 600 MHz (VHF) and the bottom of the modulation current amplitude is set below the threshold current of laser diode in order to reduce the coherence of the laser diode and thus the interference in the optical fiber (see Section 3.5.3.1). The laser diodes used for this purpose are therefore required to have a large modulation bandwidth, about 1 GHz, as well as a linear current–light output relation.

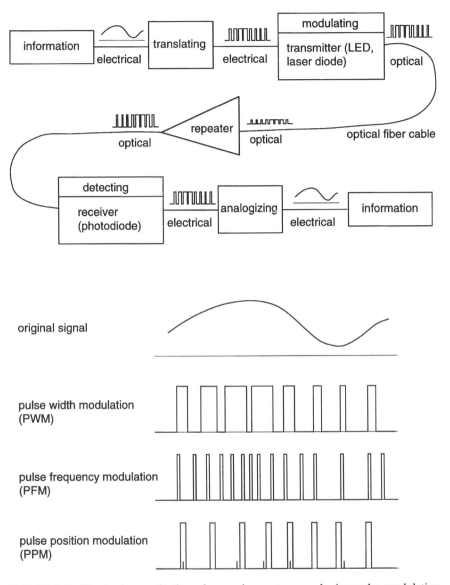

FIGURE 8.6   The basic constitution of an analog system employing pulse modulation, and some pulse signals translated from analog signals.

The analog systems employing pulse modulations can suppress the influence of distortions and noises. The amplitude of the input electrical analog signal is translated to pulse width (PWM), pulse frequency (PFM), pulse position on time scale (PPM), and so forth (see Fig. 8.6), and then laser diodes and LEDs are modulated with the pulsed signals. The optical pulse signals are transmitted

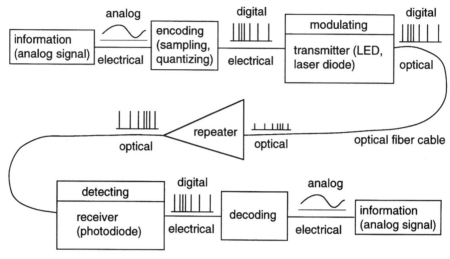

FIGURE 8.7 Basic constitution of a digital system.

through optical fiber cable. The signal detected by an APD or pin-photodiode, and then the electrical pulse signal from the photodiode, is translated again into an electrical analog signal.

8.1.1.2 Digital Systems Digital optical fiber communication systems are the most popular ones and are widely used in submarine and land systems and in data links. Laser diodes and LEDs are directly modulated with electrical digital signals as optical sources in transmitters or output powers of laser diodes are modulated by optical modulators biased with the electrical digital signals (see Fig. 8.2). The electrical digital signals are rectangular pulsed signals and thus the modulated optical signals are also rectangularly shaped pulses. Either pin-photodiodes or APDs receive the transmitted signals and convert them into electrical signals. When the electrical input is a digital signal, the signal can directly be transmitted to the receiver. The electrical input, however, is ordinarily an analog signal proportional to the intensity of the information such as voice. As shown in Fig. 8.7, the analog signal is translated into a digital signal (quantizing and coding) and then bias to laser diodes, LEDs, or optical modulators. The received optical digital signal is converted into an electrical digital signal and then translated into an electrical analog signal (decoding). The coding signals are generally multiplexed and then the laser diodes, LEDs, or optical modulator is modulated by the multiplexed signals. The multiplexed signals transmitted are demultiplexed before decoding. The signal multiplication results in the increase in bit-rate of the signal transmitted, and thus the laser diodes, LEDs, and optical modulator are required to operate at high frequencies (for example, frequencies of more than a few gigabits per second).

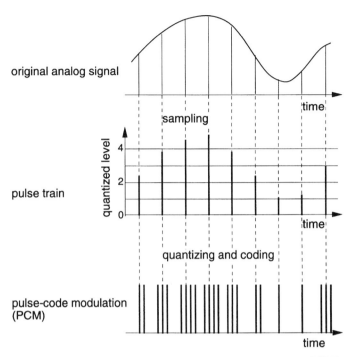

FIGURE 8.8 Simplified image of pulse-code modulation (PCM).

There are several modulation methods for translating analog signals to digital signals. For digital transmission in optical fiber communication systems, pulse code modulation (PCM) is usually employed (see Fig. 8.8). The analog signal is translated into a train of pulses whose heights correspond to the amplitude of the analog signal within certain time intervals (sampling). Each pulse is quantized and coded with a pulse number and position on the time scale. Some types of digital signals can be used for coding. Those are mainly RZ (return-to-zero), NRZ (nonreturn-to-zero), and CMI (code-mark-inversion) as shown in Fig. 8.9. When coded signal is "1", the pulse continues through the whole frame in the NRZ case, whereas it returns to 0 within the frame in the RZ case. The coded signal "0" corresponds to vacancy (no electrical pulse) within the frame for both cases. In the CMI case, the frame is fully occupied with an electrical pulse if the coded signal is "1" and half occupied with an electrical pulse if the signal is "0". The digital code of NRZ is mainly used in various digital communication systems, and the CMI is sometimes used in relatively low-bit-rate systems (less than a few tens of megabits per second).

The wavelength of the light used is chosen to be within the low loss range of the optical fiber in order to reduce transmission loss. Most optical fibers used in public communication systems consist of silica, for which the low loss range corresponds to the 1300 nm band and 1550 nm band. Consequently, the

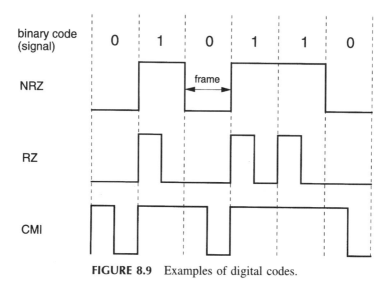

FIGURE 8.9   Examples of digital codes.

optical sources and detectors are usually devices made of InGaAs(P)/InP material systems. For economy, some data links use plastic optical fibers, and then optoelectronic devices emitting and responding to red light are used.[5]

The laser diodes, LEDs, and optical modulators in these systems are modulated with electrical digital pulse signals. This is the intensity modulation and the light intensity is modulated according to the signals (see Figs. 2.13 and 3.43). The devices used are required to have large modulation bandwidths and stable pulse responses (see Sections 2.2.5.3, 3.5.2, 5.2.3.3). The modulation bandwidths of LEDs are usually small because the bandwidth is determined by a spontaneous emission rate of the order of a nanosecond (see Section 2.2.5.3). The LEDs are mainly used in low-bit-rate systems, systems in which information is often sent at rates that are not more than a few tens of megabits per second. LEDs are mainly used in economical short-haul systems because their available optical power is low (see Section 2.2.3). The optical sources in long-haul and high-bit-rate systems are usually laser diodes with a single transverse modes because of their high output powers and high frequency responses. The modulation bandwidth is determined by the relaxation frequency (see Section 3.5.2.2) and parasitic factors such as the CR-time constant. The factors limiting transmission distance in the systems using laser diodes are chirping and loss. The transmitted pulse is deformed by the chirping and the pulse width gradually increases during transmission in an optical fiber (chromatic dispersion). In addition, the optical pulse height gradually decreases because of the fiber loss. The chirping also limits the bit-rate because the pulse interval gradually narrows as the bit-rate increases. Single longitudinal mode laser diodes, such as DFB laser diodes, are therefore used in the high-bit-rate systems (more than about 1 Gb/s in order to reduce the influence of the chirping (see Section 3.5.2).

To further reduce chirping, external optical modulators are used (see Section 5.2.3).

Laser diodes showing pulsations cannot be used in digital systems because of large intensity noises, although the influence of optical noises from laser diodes, such as quantum noise, in digital systems is much smaller than that in analog systems. In addition, optical feedback noise (see Section 3.5.4) is suppressed by cutting the optical feedback to the laser diode with an optical isolator set in the module.

Either pin-photodiodes or APDs are used as optical detectors, and the signal detection with the photodiode is performed by discriminating as to whether there is an optical pulse or not. This results in high receiver sensitivity (see Sections 4.3.5.1, 4.3.5.2, 4.5, and 4.6), and the receiver sensitivity determines bit error rate (BER), which is the ratio of the number of bits received incorrectly to the total number of bits received. The signal-to-noise ratio is given by Eq. (4.47) for pin-photodiodes and by Eq. (4.74) for APDs. The BER in trunk systems, for example, is required to be less than 10^{-11} and that in subscriber systems is required to be less than 10^{-9}. The BER increases as the transmission distance increases and as chirping increases.

8.1.1.3 Wavelength (Frequency)-Division-Multiplexing Systems   In the analog systems and the digital systems described in the previous sections, intensity modulation of laser diodes and LEDs, which is not amplitude modulation, is combined with direct detection by photodiodes (intensity modulation-direct detection, IM-DD). In systems using wavelength(or frequency)-division-multiplexing techniques (WDM or FDM systems), intensity modulation-direct detection and several coherent detections (see Section 8.1.1.4) are applicable. The intensity modulation-direct detection is mainly used for WDM systems in which the wavelength interval (allocation) is large because of the simple constitution. The basic constitution of such a WDM (or FDM) system is shown in Fig. 8.10. Each electrical signal is converted into an optical signal having a wavelength different from that into which any of the other signals are converted, and then those optical signals are multiplexed (at the multiplexer) and transmitted through only one optical fiber cable. The optical signals multiplexed are divided again into each wavelength (at the demultiplexer) and detected with photodiodes. In the WDM systems, the digital and analog systems and their combinations are used, and thus the modulation method for each wavelength (channel) is the same as described in Sections 8.1.1.1 and 8.1.1.2. Laser diodes and LEDs are used as optical sources at each wavelength (from λ_1 to λ_N in Fig. 8.10) and pin-photodiodes and APDs are used as optical detectors. The characteristics required of laser diodes, LEDs, and photodiodes are basically similar to those required when these devices are used in analog and digital systems.

The optical multiplexer and the optical demultiplexer have the same constitution. The multiplexer (or demultiplexer) usually consists of either the interference filters, diffraction grating, interferometers, optical fiber grating, or array waveguides. The input ports and the output port of the multiplexer are installed

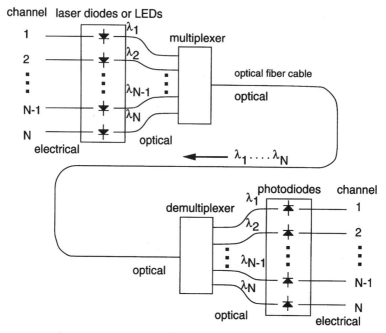

FIGURE 8.10 Basic constitution WDM (and FDM) systems.

as the output ports and the input port in the demultiplexer, respectively, as recognized in Fig. 8.10. Each signal (wavelength) is multiplexed by the optical multiplexer and demultiplexed exactly by the optical demultiplexer. If the emission spectrum of each LED or laser diode is broad, each wavelength ($\lambda_1, \lambda_2 \ldots, \lambda_N$ in Fig. 8.10) is mixed with each other at the demultiplexer and the mixed signals are introduced to each channel (1, 2, $\ldots$, N in Fig. 8.10). Consequently, the spectral width of laser diodes and LEDs and the wavelength-filtering characteristics of the multiplexer and demultiplexer are taken into consideration when the channel allocation is set. The width of the emission spectrum is about 40 nm for an 850 nm-band AlGaAs/GaAs LED and is about 120 nm for a long-wavelength InGaAsP/InP LED (see Section 2.2.5.2). For Fabry–Perot laser diodes, that width is at least a few nanometers (see Sections 3.2.3 and 3.5.3.1). These LEDs and laser diodes are therefore used in WDM systems where the wavelength interval between the channels is set sufficiently wide. LEDs and Fabry–Perot laser diodes, for example, can be applied to a WDM system in which the channel wavelengths are set at 850, 1300, and 1550 nm.[6] Those systems having wide channel intervals are convenient but cannot be used in long-haul and high-frequency systems because the transmission characteristics of the optical fiber, such as transmission loss and chromatic dispersion, are different at different wavelengths (see Section 8.1.1.2). To increase the transmission rate and reduce the deterioration of transmission quality due to the chromatic dispersion, laser diodes with single longitudinal modes, such as DFB laser diodes, are used. The spectral linewidth of such single-mode laser diodes is sufficiently small and is much less than 1 GHz; here the value of 1 GHz corresponds to about 8×10^{-3} nm in the 1550 nm band (see Section 3.5.3.2). In systems employing DFB laser diodes, the channels can be allocated with a separation of less than 0.1 nm. Many channels can therefore be allocated within a narrow wavelength range and the transmission quality difference with different wavelengths can be avoided. Chirping (see Sections 3.5.2.1 and 3.5.3.3), however, is generated if the laser diodes are directly modulated, and the channel spacing is required to be sufficiently wider than the spread due to chirping. The temperature of laser diodes used in such WDM(FDM) systems also needs to be controlled because the lasing wavelength of DFB laser diodes changes by about 0.1 nm/°C (see Section 3.4.1.3). The temperature is usually controlled by using a Peltier cooler installed in the module (see Section 6.3.3).

8.1.1.4 Coherent Systems   In coherent optical fiber communication systems, the emitted light is treated as a lightwave and either the frequency (in frequency-shift keying, FSK), phase (in phase-shift keying, PSK), or amplitude (in amplitude shift keying, ASK) is modulated according to the input electrical signal.[7] The detection sensitivity can be improved by about 10 dB over that obtained with IM-DD, and this means that the transmission distance can be lengthened. There are two kinds of detection methods used in coherent systems: optical heterodyne detection and optical homodyne detection. As shown in Fig. 8.11 for the heterodyne system, the system constitutions are quite different from those of systems using IM-DD. The frequency, phase, or amplitude of the lightwave (optical output power) from a highly coherent laser diode is modulated by an optical modulator according to electrical signal (see Sections 5.2.1 and 5.2.2). For FSK the laser diode is often modulated directly by the electrical signal because the chirping corresponds to the frequency shift. In such a case no optical modulator is used. At the front of the optical detector, a beat signal is generated by coupling the transmitted optical signal with the light from a local oscillator. The weak transmitted signal is therefore amplified by the light from a local oscillator because the amplitude of the beat signal is proportional to the product of the amplitude of the transmitted light and light from the local oscillator. The signal detection is called homodyne detection if the lasing wavelength (frequency) of the local oscillator coincides with the wavelength of the transmitted light and is called heterodyne detection if there is a constant frequency difference between the transmitted signal and the light from the local oscillator. Various coherent systems can be constructed by combining the detection and modulation methods.

The great advantages of the coherent systems are based on the coherence of the light from laser diodes. The coherence is usually expressed with the spectral linewidth given by Eq. (3.58). The spectral linewidth needs to be very small for coherent systems use and is ordinarily less than 1 MHz (although the exact linewidth required depends on the system). A long-cavity DFB laser diode—one that has a (strained) quantum well structure about 1 mm long for reducing the

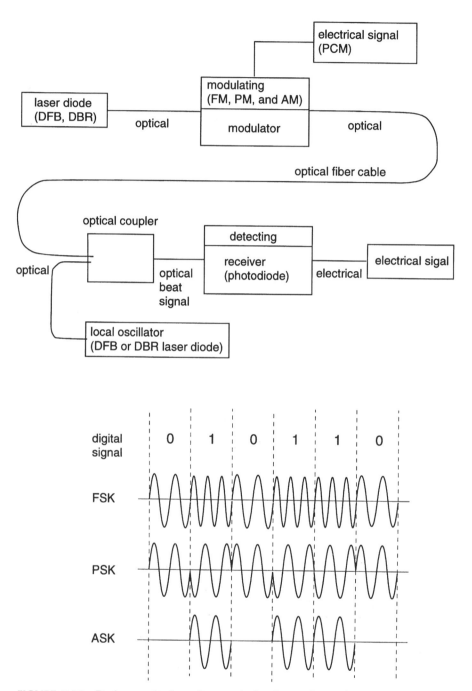

FIGURE 8.11 Basic constitution of an optical coherent heterodyne system and the modulated signal.

linewidth enhancement factor (see Section 3.5.3.3.)—is mainly used in a coherent system. When the spectral linewidth needs to be further reduced, an external mirror is also used.[8] Multisection DFB laser diodes (see Section 3.4.1.5) and DBR laser diodes are sometimes applied to the local oscillator because their wavelengths can be tuned electrically. The tuning is also carried out by temperature controlling because the lasing wavelength changes with temperature. Temperature control is essential for stabilizing the wavelength (frequency) of the laser diodes in the transmitter and local oscillator. It is carried out by using Peltier coolers (see Section 6.3.3).

When laser diodes are modulated directly, the optical intensity, wavelength (frequency), and phase of their output all change with the modulated carrier density. Consequently, the pure FSK and PSK ("pure" here means that the optical intensity is constant under signal current modulation) is not obtained. The FSK and PSK without optical intensity variation can be obtained by using an optical modulator where a suitable wavelength is chosen for the modulated light wave (see Section 5.2.3.2). In a highly coherent DFB laser diode the chirping (frequency shift) is small because the spectral linewidth enhancement factor is also small (see Sections 3.5.3.3). Such a laser diode, however, is often used for FSK under direct modulation, although the modulation efficiency (FM-response) is not large. The FM-response is expressed as the ratio of frequency deviation to injected current (in units of Hz/mA). This FM-response varies with modulation frequency as shown in Fig. 8.12. The thermal effect due to Joule's heating (red-shift) is dominant at low frequencies (a few hundred of kilohertz or less), whereas the plasma effect (blue-shift) is dominant at higher frequencies (see Section 3.4.1 and Fig. 3.12). The phase of the emitted lightwave also changes according to the change in FM-response. At the frequency at which the quantity of blue-shift is equal to that of red-shift, the frequency shift is cancelled out and a dip appears in the FM-response (see Fig. 8.12). The frequency at the dip needs to be lowered because the electrical input signal corresponding to the frequency disappears in the optical output (FM) signal. The thermal effect is related to the bonding configuration, especially the bonding of the laser diode chip to the heat sink. As shown in Fig. 6.18, the thermal effect influenced by the junction temperature of the laser diode gradually increases when the frequency decreases. When the modulation frequency is more than a few hundred kilohertz, the thermal effect is determined by the thermal conduction within the laser diode chip. In the lower range of modulation frequency, the thermal conductivities of heat sink and stem and bonding configuration influence the conduction of heat generated in the laser diode. The flatness of FM-response therefore increases by improving bonding configuration and materials.

8.1.1.5   *Optical Fiber Amplifiers*

There are several types of optical amplifiers, and popular ones are the rare-earth-doped optical fiber amplifiers.[9] As shown in Fig. 8.13, optical amplification is generated with the stimulated emission between two energy levels. Those levels are introduced by ions doped into the optical fiber, and those ions are rare-earth ions (such as Er^{3+}, Nd^{3+}, and

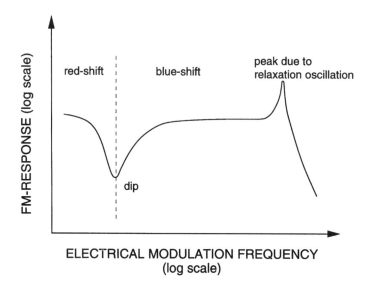

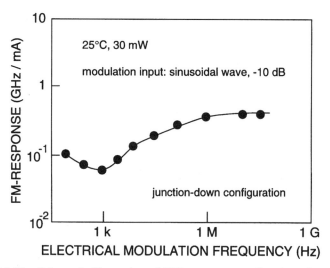

FIGURE 8.12 Schematic illustration of FM-response as a function of modulation frequency and actual data for a 1550 nm-band buried-heterostructure-type multiple-quantum well InGaAsP/InP DFB laser diode with a 900 μm-long cavity.

Pr^{3+}). The energy levels introduced depend on the ions doped, and an Er-doped optical fiber amplifier (EDFA) is used in the 1550 nm band and a Pr-doped fiber amplifier (PDFA) is used in the 1300 nm band. Pumping is required for stimulated emission (see Section 1.1.3.1) and thus for amplifying the input optical signal. The basic configuration of an EDFA is shown in Fig. 8.14. The optical

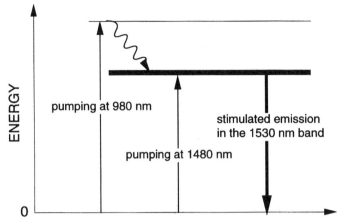

FIGURE 8.13   Simplified pumping schemes for Er-doped fiber amplifiers.

input signal and the light from the pumping source are coupled at the optical multiplexer and are then forwarded to the Er-doped fiber. There the input optical signal is amplified without being converted to an electrical signal. In an EDFA two wavelength bands, the 980 nm and 1480 nm bands (see Fig. 8.13), are applicable for pumping. Consequently, 980 nm-band strained quantum well laser diodes, such as InGaAs/GaAs laser diodes, and 1480 nm-band (strained) quantum well InGaAsP/InP laser diodes are used for the pumping. In a PDFA the 1060 nm-band is the wavelength band used for pumping, and 1060 nm-band strained InGaAs/GaAs laser diodes are the main pumping sources. An output power of more than 100 mW is ordinarily required from pumping sources, but other characteristics (such as high frequency response, high coherence, and so forth) are not important. The pumping sources are of course high-power laser diodes with long cavities, thin active layers, highly reflective films on the rear facets, and so forth. The wavelength stability, however, is important for

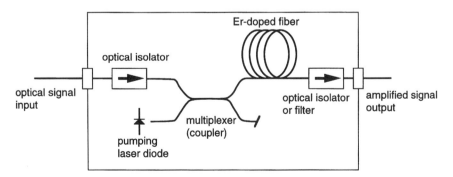

FIGURE 8.14   Basic configuration of Er-doped fiber amplifier.

pumping an EDFA efficiently. A Peltier cooler is therefore usually used to prevent wavelength fluctuation and temperature increases that would reduce output power.

The optical fiber amplifiers have some excellent characteristics—large bandwidth, high gain, and low noise—and are applicable in all communication systems, including analog systems and the systems using WDM techniques.[9] The optical fiber amplifiers, for example, enable the transmission distance in high-bit-rate long-haul systems to be lengthened without using repeaters and enable the number of branches in access networks to be increased. Optical fiber amplifiers have thus been one of the key components in optical fiber communication systems.

8.1.2   Optical Fiber Data Links

The use of optical fiber data links has spread rapidly from local area networks (LANs) to the computer, audio, and mobile fields. Many kinds of laser diodes and LEDs emitting light at wavelengths ranging from the visible to the infrared are used as the optical sources. Some examples of transmission data rates in some popular data links are shown in Fig. 8.15 as a function of transmission

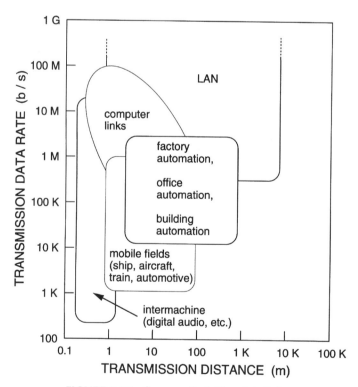

FIGURE 8.15   Some optical fiber data links.

distance. The following sections discuss some typical links from the viewpoint of the application of optoelectronic devices.

8.1.2.1 Optical Fiber Local Area Networks

The optical fiber LANs are similar to the public communication systems. Some of their advantages over the systems using metallic cables are their higher transmission capacity and bit rate and their longer transmission distance. The scale of the LANs, however, is locally restricted, and they are used within factories, offices, buildings, and so forth.

Computers, terminals, and other office equipment (such as printers) are connected with each other by optical fiber cables as shown in Fig. 8.16(a). At the

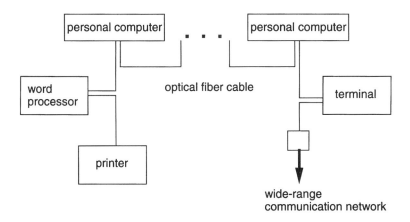

(a) example of an optical fiber LAN

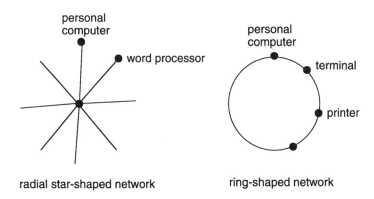

(b) network structures

FIGURE 8.16   Simplified examples of (a) an optical fiber local area network in an office and (b) network structures.

end of the fiber cables, laser diodes and LEDs are used for transmitting information and photodiodes are used for receiving information. The laser diode or LED transforms an electrical signal into an optical signal, and the optical signal is transmitted to the photodiode which transforms the optical signal back into an electrical signal. The output data from personal computers are transmitted very quickly in the LAN, and the optical data addressed to a printer, for example, are extracted by the printer and printed out. An optical ethernet having a radial-shape network [see Fig. 8.16(b)] and a fiber-distributed data interface (FDDI) having a ring-shape network [see Fig. 8.16(b)] are typical LANs. The FDDI is a relatively large-scale and high-speed LAN (100 Mb/s), and the total fiber length is often over a few tens of kilometers. The fiber optic LAN is very suitable for use in factories because no noise is introduced by electromagnetic induction.

8.1.2.2 Digital Audio Field   Data links are used frequently in our daily lives. One example of their use in the audio field can be seen in Fig. 8.17. Digital audio equipment is connected by optical fibers. The scale of these links is smaller than that of the optical fiber LANs: the fiber length is no more than a few meters. The rate at which digital signals such as NRZ and RZ signals are transferred is ordinarily less than 10 Mb/s. LEDs are the optical sources usually used in these data links because of the low bit rate and for reasons of economy. When a plastic optical fiber is used, LEDs emitting visible light with a wavelength near 660 nm (red) are used. The connection between the amplifier and everything except the digital audio tape is unidirectional, and the connection between the amplifier and the audio tape is bidirectional. The audio digital signals from each of the players, the tape, and the tuner are converted into optical signals by LEDs (or laser diodes) at one end of the optical fiber and are transmitted to the opposite end. There the optical digital signal is converted back into an electrical signal by a photodiode (usually a Si pin-photodiode), and the digital electrical signal is transformed into an analog signal and sent to the speaker. As in the optical fiber LAN, the LEDs (or laser diodes) and pho-

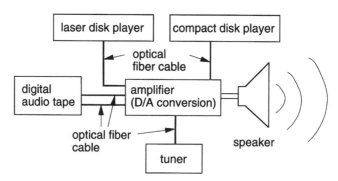

FIGURE 8.17   An example of a data link in digital audio.

todiodes used in digital audio equipment are required to be low-noise devices. For recording, an optical signal is sent to the digital audio tape and is recorded after being translated into an electrical signal.

8.1.2.3 Mobile Fields Optical fiber data links are lighter and more compact than metallic data links, in addition to the advantage that they are not subjected to noise induced by electromagnetic induction. The optical fiber data links are therefore used in mobile equipment, such as that on automobiles, train, and aircraft. The data rate in these mobile applications is low and is often less than 1 Mb/s.

8.2 OPTOELECTRONIC *pn*-JUNCTION DEVICES IN BEAM OPTICS

Various kinds of optoelectronic *pn*-junction devices have been widely used in audiovisual fields, office-automation fields, and home-application fields in addition to the telecommunication fields described in the previous sections. In addition, we also use optoelectronic *pn*-junction devices in our daily lives without being aware of them, for example, when we use automatic vending machines or television sets. We can buy something to drink from a vending machine. When paper money is inserted into a vending machine, a sensor for paper money discrimination starts to operate. It contains LEDs or an LED array emitting visible and infrared light and photodiodes. The display for the charge is also carried out with LEDs emitting visible light. For detecting a cup, an LED emitting infrared light and a photodetector are used. The other indicators of the machine are also formed with various kinds of LEDs. In TV sets, a combination consisting of an LED emitting infrared light and a photodiode is used for remote control, and LEDs emitting visible light are used in the channel display and in the other indicators. These are only two examples, and there are many kinds of equipment and systems using optoelectronic *pn*-junction devices.

8.2.1 Optical Disks

The major application field for laser diodes is optical disks. Optical disk equipment is used for reading or recording information and can be roughly divided into two groups: reading-only and recording-and-reading types, as shown in Fig. 8.18. The optical disk head in this equipment consists of a laser diode, lenses, photodiodes, and so forth. During recording, it changes electrical information into optical information and then records the optical information on the disk. During reading (or playback), the head optically reads the recorded information and changes the optical information into electrical information. Disks 90, 120, 130, and 300 mm in diameter are commonly used. A minidisk, 64 mm in diameter, is also used for digital audio. Some of the advantages that optical disks have over other kinds of memories are their larger capacity, shorter access time, and smaller size. They are therefore used in the terminal equipment of com-

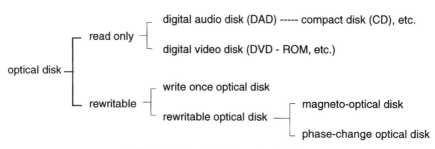

FIGURE 8.18 Typical optical disks.

puters as well as in audiovisual equipment. Some optical disks are described in the following sections.

8.2.1.1 Read-Only Optical Disks   As an example of read-only optical disks, a schematic illustration of the optical systems for digital audio (compact) disks is shown in Fig. 8.19. Compact disks (CDs), which are 120 mm in diameter, are typical digital audio disks. Compact disk usually means compact disk–digital audio (CD-DA), but it also includes the compact disk–read-only memory (CD-ROM) for data memory and the compact disk–interactive (CD-I) for multimedia use. The 780 nm light emitted from AlGaAs/GaAs laser diodes is collimated by a lens and focused to a diameter of about 1 μm on the disk. If there is no pit where the light is incident, it is reflected at the aluminum mirror of the disk and returns to the lens. The depth of a pit is set at a value such that the difference between the path of the light reflected at a pit and the path of the light reflected at the mirror is an integer multiple of the half-wavelength. Consequently, if there is a pit where the light is incident, the amount of reflected light decreases tremendously because the reflected lights are almost cancelled by interference. The incident and reflected beams pass through the quarter-wave plate and the whole reflected light is introduced to the photodiode by the (polarization) beam splitter because of the polarization rotation due to the quarter-wave plate. By the photodiode, the reflected light which has a signal whether a pit is on the disk or not is changed into an electrical digital signal (pulse code modulation). Here, the accuracy of the optical beam tracking on the disk is less than 0.1 μm.

The beam splitter in the equipment prevents the laser diode from an incidence of the light reflected at the disk. This is required in order to suppress the optical feedback noise (see Section 3.5.4.3) and then the resultant noise in an electrical output of the photodiode. Laser diodes with self-sustained pulsation are also used as optical sources in order to suppress further the optical feedback noise. The diameter of the focused optical beam on the disk needs to be about 1 μm and this diameter cannot be obtained when optical sources with low spatial coherence, such as LEDs (see Section 3.5.3.1), are used. The astigmatism in the laser diode (see Section 3.2.4.2) is also important for focusing, and only

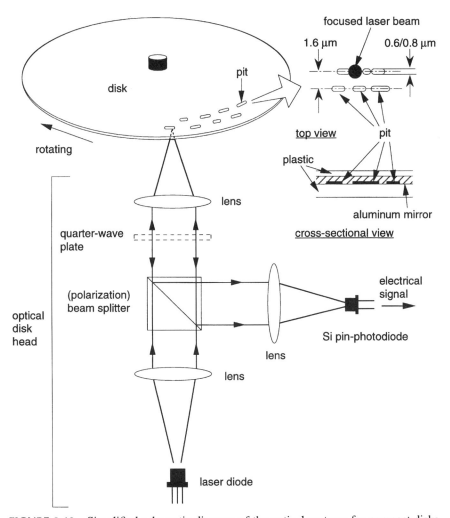

FIGURE 8.19 Simplified schematic diagram of the optical systems for compact disks.

laser diodes with a low astigmatism difference can be used. The lasers used for the digital audio disks, of course, have stable single-transverse modes (see Section 3.2.4). Low noise is an important factor for the photodiodes and Si pin-photodiodes are usually used (see Section 4.3.5.1).

8.2.1.2 Digital Video Disks   There are three broad categories of digital video (or versatile) disks: digital video disk–read only memory (DVD-ROM), digital video disk–write once (DVD-R), and digital video disk–random access memory (DVD-RAM). The lasing wavelength used with digital video disks is set at 650 nm or at 635 (red light) and laser diodes composed of AllnGaP/InGaP are the ones usually used as optical sources. The storage capacity of these disks is about seven times that of digital audio disks because of the shorter wavelength of the laser diode when the diameter of the disks is 120 mm. The characteristics required of the laser diode and photodiode are the same as in the case of digital audio disks. The optical systems of digital video disks are similar to those of digital audio disks for DVD-ROMs and are described in the next section for DVD-RAMs.

8.2.1.3   Rewritable Optical Disks   There are two types of rewritable optical disks: phase-change optical disks and magneto-optical disks. The phase-change optical disks are frequently used as rewritable disks because their optical systems can be simply constructed and are basically similar to those in Fig. 8.19. The disks composed of chalcogenide material such as GeSbTe have a low-melting point. The information to be recorded is biased to the laser diodes as an electrical digital signal. The modulated beam from the laser diode is focused to a diameter of about 1 μm of the disk and heats the disk by an amount corresponding to the modulated signal. If the temperature at the irradiated part exceeds the melting point and then quickly decreases, the phase of the irradiated part is changed from crystal into amorphous. The reflectivity of the amorphous is lower than that of the crystal, and thus the information can be recorded on the disk.

The optical system for reading is similar to that for compact disks shown in Fig. 8.19. When the disk is irradiated with light the power of which is lower than that used for writing, an intensity of the reflected light is modulated according to the difference in the refractivity. The signal recorded on the disk can therefore be monitored as an optical digital signal which is converted into an electrical signal by a photodiode. The information recorded on the disk is erased by irradiating a beam the power of which is the same as or slightly less than that used for writing. The disk partially melts under the irradiation and the phase of the molten part changes from amorphous into crystal during slow cooling.

A schematic illustration of the optical systems of magneto-optical disks is shown in Fig. 8.20. Magneto-optical recording materials such as TbFbCo and NdFeCo are used to make the disks. The beam from the laser diode modulated with an electrical digital signal is focused to a diameter of about 1 μm on the disk and heats the disk by an amount corresponding to the modulated signal. If the temperature at the irradiated part exceeds the Curie temperature, the direction of magnetization is changed by an external magnetic field (see insert in Fig. 8.20). The information can thus be recorded by changing the direction of magnetization in the disk.

When magnetized material is irradiated with linearly polarized light the power of which is lower than that used for writing, the polarization of the reflected light is rotated and the direction of its rotation is different with the direction of the magnetization of the material. Depending on the direction of polarization, the reflected light is separated by using the polarization beam splitter. If the intensity of a light separated by the polarization beam splitter is monitored, it is modulated as a result according to the direction of polarization. The signal recorded on the disk can therefore be monitored as an optical digital signal which is converted into an electrical signal by a photodiode.

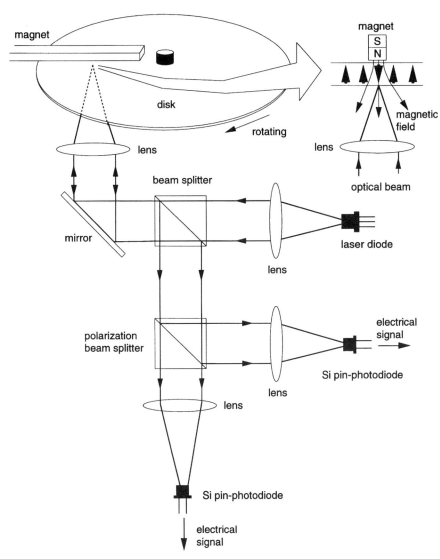

FIGURE 8.20   Simplified schematic diagram of the optical systems for magneto-optical disks.

For these rewritable optical disks, low-noise laser diodes and photodiodes must be used, and a small astigmatism difference is important for the laser diodes. More optical power is needed for heating disk and recording, and laser diodes with an output power of about 30 mW or more are used to record information on the disks. The laser diodes with self-sustained pulsation that tend to be used for reading cannot be used for writing because it is difficult to get high output power from those laser diodes. Superimposition of a high-frequency signal is often used for the high-power laser diodes in order to suppress the optical feedback noise by lowering their coherence (see Section 3.5.4.3).

8.2.2 Printers

There are two types of optical sources usually used in the printers: laser diodes and LED arrays. The printers using laser diodes are called laser beam printers and are one of the most attractive types of equipment used in office automation. Words and figures can be printed rapidly and clearly more easily by a laser beam printer than by the other types of printers. Laser diodes and gas lasers (such as an He-Ne laser) are used as optical sources for laser beam printers, but laser diodes enable the printer to nonetheless be compact and to consume little electricity.

A schematic diagram of a laser beam printer is shown in Fig. 8.21(a). The

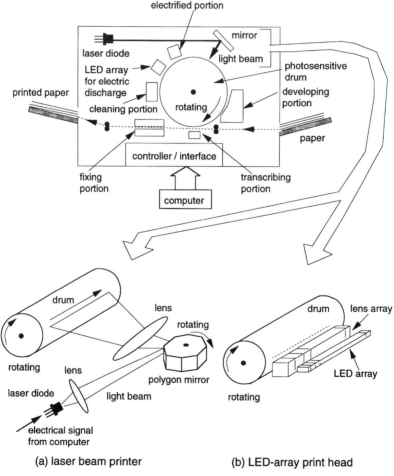

(a) laser beam printer   (b) LED-array print head

FIGURE 8.21 Simplified schematic diagram of a laser beam printer.

laser diode is driven by modulated signals from a computer. The modulated optical beam collimated by the lens is reflected by the rotating polygon-mirror and scanned on the photosensitive drum. The drum is homogeneously charged when it passes through the charging portion consisting of an LED array. The homogeneous electrification is partially erased in accordance with the scanned optical beam because the electrical resistance at the light-irradiated part decreases and the electric charge is released. The signals from the computer are therefore written onto the drum. At the developing portion, an electrically charged colored plastic powder (toner) is electrostatically attached to the written parts. At the transcribing portion, the colored powder is transferred to paper, and the transferred pattern is fixed by heating and pressing at the fixing portion. The information from the computer is thus printed on the paper. The drum is then cleaned at the cleaning portion.

The 780 nm-band laser diode used for printers must have a stable transverse mode and a small astigmatism difference (see Section 3.2.4). Those characteristics are required to focus the beam from the laser diode to a small area and to write information clearly on the drum. The temperature dependence of output power is also an important factor in high-quality printing. The output power of a laser diode gradually decreases as the width of the input current pulse increases because of the Joule's heating (see Sections 3.5.1 and 6.3.1.3). If there is a large change in the laser beam power used for erasing the electrification, shading appears in the printed pattern. The rate of change in output power is expressed by a droop as shown in Fig. 8.22. The droop for laser diodes used in printers is normally between 3% and 5%. LED arrays emitting visible light with the peak wavelength in the red range (see Section 2.2.5.4) are also used for erasing the electric charge on the drum. LED arrays are usually used in other equipment, such as facsimile and copier, for erasing the electric charge.

An LED array can also be used in the print head by combining it with a lens

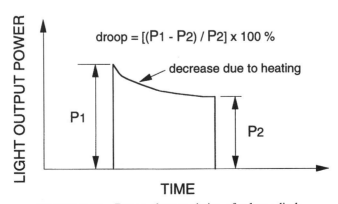

FIGURE 8.22 Droop characteristics of a laser diode.

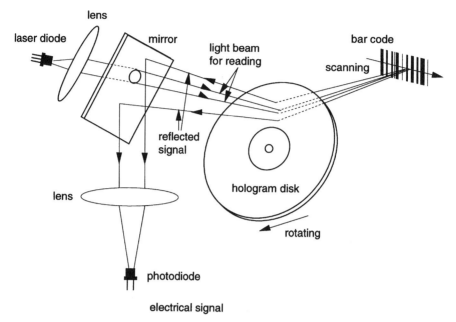

FIGURE 8.23 Simplified schematic diagram of a hologram scanner.

array [see Fig. 8.21(b)]. The LED print head is simple but shading appears in the printed pattern if the output power level of each LED is scattered. The emission peak of the LED array used in printers is in the 700 nm wavelength band.

8.2.3 Hologram Scanners

The hologram scanner is widely used in various equipment and is ordinarily used in bar-code readers in POS (point-of-sale) systems. A hologram scanner is also often used in laser beam printers for scanning the laser beam on the drum precisely. A simplified illustration of the hologram scanner is shown in Fig. 8.23. The optical beam for reading the bar code is focused by a lens through the hologram disk and scanned on the bar code by rotating the hologram disk. Gratings with coaxial circles are formed on the hologram disk, and the incident laser beam is bent at the grating by an amount determined by the grating pitch. The reflected light modulated according to the bar code is reflected by the mirror and monitored by the photodiode. The monitored optical signal is then translated into an electrical signal. The optical sources for bar-code readers are mainly laser diodes but are often LEDs emitting red light in the 600 nm band. A small astigmatism difference is also required for the laser diodes. The AlInGaP/InGaP material systems are usually used in those devices.

In bar-code readers, a polygon-mirror is sometimes used instead of the holo-

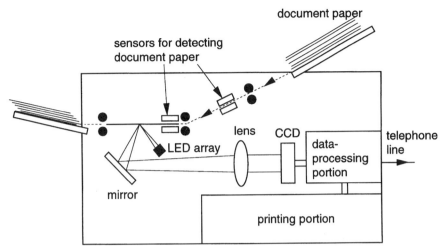

FIGURE 8.24   Simplified schematic diagram of a facsimile machine.

gram disk. The light reflected by the polygon-mirror scans the bar code, and the light modulated according to the bar code is detected with a photodiode. The following processes are the same as those in the case of a hologram disk.

8.2.4   Facsimile Machines

The facsimile machine is described here as an example of LED (array) application. A simplified illustration of a facsimile machine is shown in Fig. 8.24. The light from the LED array is focused on the document paper. The light reflected at the paper is focused on a charge-coupled device (CCD) by a lens and the optical information is converted into electrical information. The electrical information is sent through the data-processing portion to a telephone line.

The LED array consists of LEDs emitting visible light at wavelengths ranging from green to red. Some AlGaAs/GaAs LEDs emitting infrared light in the 850 nm band and Si photodiodes are often used like a photointerrupter for the sensor of document paper detection.

8.2.5   Other Applications

LEDs emitting visible light are widely used in displays and indicators, often in automobiles, trains, and airplanes. For example, LED (array) lamps are used in cars as stop lamps, brake lamps, and signal lamps. A photodiode and an LED emitting infrared light is widely used for the remote control of air conditioners, audio equipment, cameras, and so forth. Various kinds of optoelectronic devices are also used as sensors. A combination consisting of an LED and a photodiode is often used as a paper sensor in facsimiles, as a tape-end senser in videotape records (VTRs), as a dust sensor in cleaners (see Fig. 8.25), as a dirt

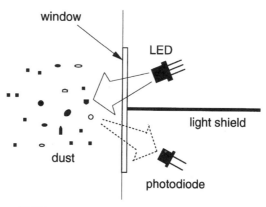

FIGURE 8.25 An example of a dust sensor.

detector for rinsing in washers, and so forth. Laser diodes emitting visible light are used as pointers and those emitting visible and infrared light are used to measure distance (see Fig. 8.26) because of their good directivity and high output power. The laser diode is modulated with high pulse current and the pulsed high-power beam emitted is reflected at a certain substance to be measured. The reflected pulse beam is detected with the photodetector. The reflected light is ordinarily weak, and thus the APD is often used as the photodiode to widen the detection power limit. The distance is calculated as the difference between the time the light was emitted from the laser diode and the time it was detected by the photodiode. In addition to the edge-emitting type of laser diodes, the surface-emitting types have begun to be widely used in the parallel processing of information and in parallel interconnections between computers.

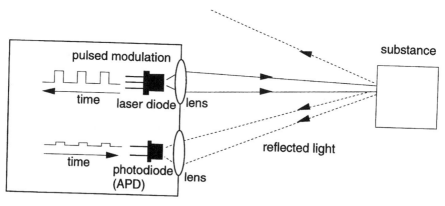

FIGURE 8.26 An example of a system measuring distance by using a laser diode and a photodiode.

REFERENCES

1. K. Asatani and T. Kimura, Analyses of LED nonlinear distortions, *IEEE Trans. Electron Dev.*, **ED-25,** 199, 1978.

2. J. Vanderwall and J. Blackburn, Suppression of some artifacts of modal noise in fiber-optic systems, *Opt. Lett.*, **4,** 295, 1979.

3. K. Sato and K. Asatani, Speckle noise reduction in fiber optic analog video transmission using semiconductor laser diodes, *IEEE Trans.*, **COM-29,** 1017, 1981.

4. M. Fukuda and G. Iwane, Modal and relative intensity noise under VHF modulation in InGaAsP/InP lasers before and after degradation, *Electron. Lett.*, **20,** 964, 1984.

5. See, for example, Y. Koike, T. Ishigure, and E. Nihei, High-bandwidth graded-index polymer optical fiber, *J. Lightwave Technol.*, **13,** 1475, 1995.

6. See, for example, S. Shimada and N. Uchida, Field trial of medium/small capacity optical fiber transmission systems, *Review of the ECL, NTT*, Japan, **29,** 1087, 1981; Special Feature (2): Technologies for economical optical access system components, *NTT Review*, **9**(6), 1997.

7. See, for example, T. Okoshi and K. Kikuchi, *Coherent Optical Fiber Communication*, Kluwer Academic Pub., Dordrecht, 1988.

8. See, for example, R. W. Tkach and A. R. Chraplyvy, Regimes of feedback effects in 1.5-μm distributed feedback lasers, *J. Lightwave Technol.*, **4,** 1655, 1986.

9. See, for example, T. P. Lee, ed., *Current Trends in Optical Amplifiers and Their Applications*, World Scientific Pub., Singapore, 1996.

Index